Lecture Notes in Mathematics

T0092539

Editors:
J.-M. Morel, Cachan
B. Teissier, Paris

For further volumes:
http://www.springer.com/series/304

Gani T. Stamov

Almost Periodic Solutions of Impulsive Differential Equations

 Springer

Gani T. Stamov
Technical University of Sofia
Department of Mathematics
Sliven
Bulgaria

ISBN 978-3-642-27545-6   e-ISBN 978-3-642-27546-3
DOI 10.1007/978-3-642-27546-3
Springer Heidelberg New York Dordrecht London

Lecture Notes in Mathematics ISSN print edition: 0075-8434
                              ISSN electronic edition: 1617-9692

Library of Congress Control Number: 2012933111

Mathematics Subject Classification (2010): 34A37; 34C27; 34K14; 34K45

Printed on acid-free paper

Springer is part of Springer Science+Business Media (www.springer.com)

*To my wife, Ivanka, and our sons,
Trayan and Alex, for their support and
encouragement*

Preface

Impulsive differential equations are suitable for the mathematical simulation of evolutionary processes in which the parameters undergo relatively long periods of smooth variation followed by a short-term rapid changes (i.e., jumps) in their values. Processes of this type are often investigated in various fields of science and technology.

The question of the existence and uniqueness of almost periodic solutions of differential equations is an age-old problem of great importance. The concept of almost periodicity was introduced by the Danish mathematician Harald Bohr. In his papers during the period 1923–1925, the fundamentals of the theory of almost periodic functions can be found. Nevertheless, almost periodic functions are very much a topic of research in the theory of differential equations. The interplay between the two theories has enriched both. On one hand, it is now well known that certain problems in celestial mechanics have their natural setting in questions about almost periodic solutions. On the other hand, certain problems in differential equations have led to new definitions and results in almost periodic functions theory. Bohr's theory quickly attracted the attention of very famous researchers, among them V.V. Stepanov, S. Bochner, H. Weyl, N. Wiener, A.S. Besicovitch, A. Markoff, J. von. Neumann, etc. Indeed, a bibliography of papers on almost periodic solutions of ordinary differential equations contains over 400 items. It is still a very active area of research.

At the present time, the qualitative theory of impulsive differential equations has developed rapidly in relation to the investigation of various processes which are subject to impacts during their evolution. Many results on the existence and uniqueness of almost periodic solutions of these equations are obtained.

In this book, a systematic exposition of the results related to almost periodic solutions of impulsive differential equations is given and the potential for their application is illustrated.

Some important features of the monograph are as follows:

1. It is the first book that is dedicated to a systematic development of *almost periodic theory* for *impulsive differential equations.*
2. It fills a void by making available a book which describes existing literature and authors results on the *relations* between the *almost periodicity* and *stability* of the solutions.
3. It shows the manifestations of *direct constructive methods*, where one constructs a uniformly convergent series of almost periodic functions for the solution, as well as of *indirect methods* of showing that certain bounded solutions are almost periodic, by demonstrating how these effective techniques can be applied to investigate almost periodic solutions of impulsive differential equations and provides interesting *applications* of many practical problems of diverse interest.

The book consists of four chapters.

Chapter 1 has an introductory character. In this chapter a description of the systems of impulsive differential equations is made and the main results on the fundamental theory are given: conditions for absence of the phenomenon "beating," theorems for existence, uniqueness, continuability of the solutions. The class of piecewise continuous Lyapunov functions, which are an apparatus in the almost periodic theory, is introduced. Some comparison lemmas and auxiliary assertions, which are used in the remaining three chapters, are exposed. The main definitions and properties of almost periodic sequences and almost periodic piecewise continuous functions are considered.

In Chap. 2, some basic existence and uniqueness results for almost periodic solutions of different classes of impulsive differential equations are given. The hyperbolic impulsive differential equations, impulsive integro-differential equations, forced perturbed impulsive differential equations, impulsive differential equations with perturbations in the linear part, dichotomous impulsive differential systems, impulsive differential equations with variable impulsive perturbations, and impulsive abstract differential equations in Banach space are investigated. The relations between the strong stability and almost periodicity of solutions of impulsive differential equations with fixed moments of impulse effect are considered. Many examples are considered to illustrate the feasibility of the results.

Chapter 3 is dedicated to the existence and uniqueness of almost periodic solutions of impulsive differential equations by Lyapunov method. Almost periodic Lyapunov functions are offered. The existence theorems of almost periodic solutions for impulsive ordinary differential equations, impulsive integro-differential equations, impulsive differential equations with time-varying delays, and nonlinear impulsive functional differential equations are stated. By using the concepts of uniformly positive definite matrix functions and Hamilton–Jacobi–Riccati inequalities, the existence theorems for almost periodic solutions of uncertain impulsive dynamical systems are proved.

Finally, in Chap. 4, the applications of the theory of almost periodicity to impulsive biological models, Lotka–Volterra models, and neural networks are presented. The impulses are considered either as means of perturbations or as control.

The book is addressed to a wide audience of professionals such as mathematicians, applied researches, and practitioners.

The author has the pleasure to express his sincere gratitude to Prof. Ivanka Stamova for her valuable comments and suggestions during the preparation of the manuscript. He is also thankful to all his coauthors, the work with whom expanded his experience.

Sliven, Bulgaria   *G. T. Stamov*

Contents

Introduction

I. Impulsive Differential Equations

The necessity to study impulsive differential equations is due to the fact that these equations are useful mathematical tools in modeling of many real processes and phenomena studied in optimal control, biology, mechanics, bio-technologies, medicine, electronics, economics, etc.

For instance, impulsive interruptions are observed in mechanics [13, 25, 85], in radio engineering [13], in communication security [79, 80], in Lotka–Volterra models [2–4, 76, 99, 103, 106, 188, 191, 192], in control theory [75, 104, 118, 146], in neural networks [5, 6, 36, 162, 169, 175–177]. Indeed, the states of many evolutionary processes are often subject to instantaneous perturbations and experience abrupt changes at certain moments of time. The duration of these changes is very short and negligible in comparison with the duration of the process considered, and can be thought of as "momentary" changes or as impulses. Systems with short-term perturbations are often naturally described by impulsive differential equations [15, 20, 62, 66, 94, 138].

Owing to its theoretical and practical significance, the theory of impulsive differential equations has undergone a rapid development in the last couple of decades.

The following examples give a more concrete notion of processes that can be described by impulsive differential equations.

Example 1. One of the first mathematical models which incorporate interaction between two species (predator–prey, or herbivore-plant, or parasitoid-host) was proposed by Alfred Lotka [109] and Vito Volterra [184]. The classical "predator–prey" model is based on the following system of two differential equations

$$\begin{cases} \dot{H}(t) = H(t)[r_1 - bP(t)], \\ \dot{P}(t) = P(t)[-r_2 + cH(t)], \end{cases} \tag{A}$$

where $H(t)$ and $P(t)$ represent the population densities of prey and predator at time t, respectively, $t \geq 0$, $r_1 > 0$ is the intrinsic growth rate of the prey, $r_2 > 0$ is the death rate of the predator or consumer, b and c are the interaction constants. More specifically, the constant b is the per-capita rate of the predator predation and the constant c is the product of the predation per-capita rate and the rate of converting the prey into the predator.

The product $p = p(H) = bH$ of b and H is the predator's functional response (response function) of type I, or rate of prey capture as a function of prey abundance.

There have been many studies in literatures that investigate the population dynamics of the type (A) models. However, in the study of the dynamic relationship between species, the effect of some impulsive factors, which exists widely in the real world, has been ignored. For example, the birth of many species is an annual birth pulse or harvesting. Moreover, the human beings have been harvesting or stocking species at some time, then the species is affected by another impulsive type. Also, impulsive reduction of the population density of a given species is possible after its partial destruction by catching or poisoning with chemicals used at some transitory slots in fishing or agriculture. Such factors have a great impact on the population growth. If we incorporate these impulsive factors into the model of population interaction, the model must be governed by an impulsive differential system.

For example, if at the moment $t = t_k$ the population density of the predator is changed, then we can assume that

$$\Delta P(t_k) = P(t_k^+) - P(t_k^-) = g_k P(t_k), \tag{B}$$

where $P(t_k^-) = P(t_k)$ and $P(t_k^+)$ are the population densities of the predator before and after impulsive perturbation, respectively, and $g_k \in \mathbb{R}$ are constants which characterize the magnitude of the impulsive effect at the moment t_k. If $g_k > 0$, then the population density increases and if $g_k < 0$, then the population density decreases at the moment t_k.

Relations (A) and (B) determine the following system of impulsive differential equations

$$
\begin{cases}
\dot{H}(t) = H(t)[r_1 - bP(t)], \quad t \neq t_k, \\[2mm]
\dot{P}(t) = P(t)[-r_2 + cH(t)], \quad t \neq t_k, \\[2mm]
H(t_k^+) = H(t_k), \quad P(t_k^+) = P(t_k) + g_k P(t_k),
\end{cases}
\tag{C}
$$

where t_k are fixed moments of time, $0 < t_1 < t_2 < \dots$, $\lim_{k \to \infty} t_k = \infty$.

In mathematical ecology the system (C) denotes a model of the dynamics of a predator–prey system, which is subject to impulsive effects at certain moments of time. By means of such models, it is possible to take into account

the possible environmental changes or other exterior effects due to which the
population density of the predator is changed momentary.

Example 2. The most important and useful functional response is the Holling
type II function of the form

$$p(H) = \frac{CH}{m+H},$$

where $C > 0$ is the maximal growth rate of the predator, and $m > 0$ is
the half-saturation constant. Since the function $p(H)$ depends solely on prey
density, it is usually called a *prey-dependent* response function. Predator–prey
systems with prey-dependent response have been studied extensively and the
dynamics of such systems are now very well understood [77, 88, 125, 135, 181,
192].

   Recently, the traditional prey-dependent predator–prey models have been
challenged, based on the fact that functional and numerical responses over
typical ecological timescales ought to depend on the densities of both prey and
predators, especially when predators have to search for food (and therefore
have to share or compete for food). Such a functional response is called a
ratio-dependent response function. Based on the Holling type II function,
several biologists proposed a ratio-dependent function of the form

$$p\left(\frac{H}{P}\right) = \frac{C\frac{H}{P}}{m+\frac{H}{P}} = \frac{CH}{mP+H}$$

and the following ratio-dependent Lotka–Volterra model

$$\begin{cases} \dot{H}(t) = H(t)\left[r_1 - aH(t) - \dfrac{CP(t)}{mP(t)+H(t)}\right], \\ \dot{P}(t) = P(t)\left[-r_2 + \dfrac{KH(t)}{mP(t)+H(t)}\right], \end{cases} \tag{D}$$

where K is the conversion rate.

   If we introduce time delays in model (D), we will obtain a more
realistic approach to the understanding of predator–prey dynamics, and it
is interesting and important to study the following delayed modified ratio-
dependent Lotka–Volterra system

$$\begin{cases} \dot{H}(t) = H(t)\left[r_1 - a\displaystyle\int_{-\infty}^{t} k(t-u)H(u)du - \dfrac{CP(t-\tau(t))}{mP(t)+H(t)}\right], \\ \dot{P}(t) = P(t)\left[-r_2 + \dfrac{KH(t-\tau(t))}{mP(t)+H(t-\tau(t))}\right], \end{cases} \tag{E}$$

where $k : \mathbb{R}^+ \to \mathbb{R}^+$ is a measurable function, corresponding to a delay kernel or a weighting factor, which says how much emphasis should be given to the size of the prey population at earlier times to determine the present effect on resource availability, $\tau \in C[\mathbb{R}, \mathbb{R}^+]$.

However, the ecological system is often affected by environmental changes and other human activities. In many practical situations, it is often the case that predator or parasites are released at some transitory time slots and harvest or stock of the species is seasonal or occurs in regular pulses. By means of exterior effects we can control population densities of the prey and predator.

If at certain moments of time biotic and anthropogeneous factors act on the two populations "momentary", then the population numbers vary by jumps. In this case we will study Lotka–Volterra models with impulsive perturbations of the type

$$
\begin{cases}
\dot{H}(t) = H(t)\Big[r_1 - a \int_{-\infty}^{t} k(t-u)H(u)du \\[2mm]
\qquad\qquad - \dfrac{CP(t-\tau(t))}{mP(t)+H(t)}\Big], \ t \neq t_k, \\[4mm]
\dot{P}(t) = P(t)\Big[-r_2 + \dfrac{KH(t-\tau(t))}{mP(t)+H(t-\tau(t))}\Big], \ t \neq t_k, \\[4mm]
H(t_k^+) = (1+h_k)H(t_k), \quad k = 1, 2, \ldots, \\[2mm]
P(t_k^+) = (1+g_k)P(t_k), \quad k = 1, 2, \ldots,
\end{cases}
\tag{F}
$$

where h_k, $g_k \in \mathbb{R}$ and t_k, $k = 1, 2, \ldots$ are fixed moments of impulse effects, $0 < t_1 < t_2 < \ldots$, $\lim\limits_{k\to\infty} t_k = \infty$.

By means of the type (F) models it is possible to investigate one of the most important problems of the mathematical ecology - the problem of stability of the ecosystems, and respectively the problem of the optimal control of such systems.

Example 3. Chua and Yang [42, 43] proposed a novel class of information-processing system called Cellular Neural Networks (CNN) in 1988. Like neural networks, it is a large-scale nonlinear analog circuit which processes signals in real time.

The key features of neural networks are asynchronous parallel processing and global interaction of network elements. For the circuit diagram and connection pattern, implementation the CNN can be referred to [44]. Impressive applications of neural networks have been proposed for various fields such as optimization, linear and nonlinear programming, associative memory, pattern recognition and computer vision [30, 31, 35, 40–44, 72].

The mathematical model of a Hopfield type CNN is described by the following state equations

$$\dot{x}_i(t) = -c_i x_i(t) + \sum_{j=1}^{n} a_{ij} f_j\left(x_j(t)\right) + I_i, \tag{G}$$

or by delay differential equations

$$\dot{x}_i(t) = -c_i x_i(t) + \sum_{j=1}^{n} a_{ij} f_j\left(x_j(t)\right) + \sum_{j=1}^{n} b_{ij} f_j\left(x_j(t - \tau_j(t))\right) + I_i, \tag{H}$$

where $i = 1, 2, \ldots, n$, n corresponds to the numbers of units in the neural network, $x_i(t)$ corresponds to the state of the ith unit at time t, $f_j(x_j(t))$ denotes the output of the jth unit at time t, a_{ij} denotes the strength of the jth unit on the ith unit at time t, b_{ij} denotes the strength of the jth unit on the ith unit at time $t - \tau_j(t)$, I_i denotes the external bias on the ith unit, $\tau_j(t)$ corresponds to the transmission delay along the axon of the jth unit and satisfies $0 \leq \tau_j(t) \leq \tau$ ($\tau = const$), c_i represents the rate with which the ith unit will reset its potential to the resting state in isolation when disconnected from the network and external inputs.

On the other hand, the state of CNN is often subject to instantaneous perturbations and experiences abrupt changes at certain instants which may be caused by switching phenomenon, frequency change or other sudden noise, that is, do exhibit impulsive effects.

Therefore, neural network models with impulsive effects should be more accurate in describing the evolutionary process of the systems.

Let at fixed moments t_k the system (G) or (H) be subject to shock effects due to which the state of the ith unit gets momentary changes. The adequate mathematical models in such situation are the following impulsive CNNs

$$\begin{cases} \dot{x}_i(t) = -c_i x_i(t) + \sum_{j=1}^{n} a_{ij} f_j\left(x_j(t)\right) + I_i, \quad t \neq t_k, \ t \geq 0, \\[4mm] \Delta x_i(t_k) = x_i(t_k^+) - x_i(t_k) = P_{ik}(x_i(t_k)), \quad k = 1, 2, \ldots, \end{cases} \tag{I}$$

or the impulsive system with delays

$$
\begin{cases}
\dot{x}_i(t) = -c_i x_i(t) + \sum_{j=1}^{n} a_{ij} f_j\left(x_j(t)\right) \\
\qquad\qquad + \sum_{j=1}^{n} b_{ij} f_j\left(x_j(t - \tau_j(t))\right) + I_i, \quad t \neq t_k,\ t \geq 0, \qquad \text{(J)} \\
\Delta x_i(t_k) = x_i(t_k^+) - x_i(t_k) = P_{ik}(x_i(t_k)), \quad k = 1, 2, \ldots,
\end{cases}
$$

where t_k, $k = 1, 2, \ldots$ are the moments of impulsive perturbations and satisfy $0 < t_1 < t_2 < ldots$, $\lim\limits_{k \to \infty} t_k = \infty$ and $P_{ik}(x_i(t_k))$ represents the abrupt change of the state $x_i(t)$ at the impulsive moment t_k.

Such a generalization of the CNN notion should enable us to study different types of classical problems as well as to "control" the solvability of the differential equations (without impulses).

In the examples considered the systems of impulsive differential equations are given by means of a system of differential equations and conditions of jumps. A brief description of impulsive systems is given in Chap. 1.

The mathematical investigations of the impulsive ordinary differential equations mark their beginning with the work of Mil'man and Myshkis [119], 1960. In it some general concepts are given about the systems with impulse effect and the first results on stability of such systems solutions are obtained. In recent years the fundamental and qualitative theory of such equations has been extensively studied. A number of results on existence, uniqueness, continuability, stability, boundedness, oscillations, asymptotic properties, etc. were published [14–18, 20, 61, 62, 87, 91–96, 124, 128, 129, 136, 138, 148–151, 153–161, 165, 166, 178, 199]. These results are obtained in the studying of many models which are using in natural and applied sciences [2, 10, 11, 59, 75, 103–105, 121, 130, 131, 152, 163, 164, 167, 168, 170–176, 178].

II. Almost Periodicity

The concept of almost periodicity is with deep historical roots. One of the oldest problems in astronomy was to explain some curious behavior of the moon, sun, and the planets as viewed against the background of the "fixed stars". For the Greek astronomers the problem was made more difficult by the added restriction that the models for the solar system were to use only uniform linear and uniform circular motions. One such solution, sometimes attributed to Hipparchus and appearing in the Almagest of Ptolemy, is the method of epicycles.

Let P be a planet or the moon. The model of motion of P can be written as

$$r_1 e^{i\lambda_1 t} + r_2 e^{i\lambda_2 t},$$

where r_1, λ_1 and r_2, λ_2 are real constants. When applied to the moon, for example, this not very good approximation.

Copernicus showed that by adding a third circle one could get a better approximation to the observed data. This suggests that if $\varphi(t)$ is the true motion of the moon, then there exist $r_1, r_2, \ldots, r_n$ and $\lambda_1, \lambda_2, \ldots, \lambda_n$ such that for all $t \in \mathbb{R}$,

$$\left| \varphi(t) - \sum_{j=1}^{n} r_j e^{i\lambda_j t} \right| \leq \varepsilon,$$

where $\varepsilon > 0$ is the observational error. If the numbers $\lambda_1, \lambda_2, \ldots, \lambda_n$ are not all rational multiplies of one real number, then the finite sum is not periodic. It would be almost periodic in the sense of the following Condition: If for every positive $\varepsilon > 0$, there exists a finite sum

$$\sum_{j=1}^{n} r_j e^{i\lambda_j t} \equiv p(t)$$

and for all $t \in \mathbb{R}$,

$$|f(t) - p(t)| < \varepsilon,$$

then the function $f(t)$ is *almost periodic*.

The idea of Ptolemy and Copernicus was to show that the motion of a planet is described by functions of this type. The main aspects of the historical development of this problem can be found in [122, 179].

The formal theory of almost periodic functions was developed by Harald Bohr [26]. In this paper Bohr was interested in series of the form

$$\sum_{n=1}^{\infty} e^{-\lambda_n s}$$

called Dirichlet series, one of which is the Riemann-zeta function. In his researches he noticed that along the lines $Re(s) = const.$, these functions had a regular behavior. He apparently hoped that a formal study of this behavior might give him some insight into the distribution of values of the Dirichlet series. The regular behavior he discovered we shall consider in the following way [55].

The continuous function f is *regular*, if for every $\varepsilon > 0$ and for every $t \in \mathbb{R}$, the set $T(f, \varepsilon)$,

$$T(f, \varepsilon) \equiv \left\{ \tau; \ \sup_{t \in \mathbb{R}} |f(t + \tau) - f(t)| < \varepsilon \right\}$$

is relatively dense in $\mathbb{R}$, i.e. if there is an $l > 0$ such that every interval of length l has a non-void intersection with $T(f, \varepsilon)$.

It is easy to see that the sum of two regular functions is also regular, and a uniform limit of functions from this class will converge to a regular function. Consequently, all regular functions are *almost periodic in the sense of Bohr*.

Later, Bohr considered the problem of when the integral of an almost periodic function is almost periodic. Many applications of this theory to various fields became known during the late 1920s. One of the results connected with the work of Bohr and Neugebauer [27] is that the bounded solutions of the system of differential equations in the form

$$\dot{x} = Ax + f$$

are almost periodic by necessity, when A is a scalar matrix, and f is almost periodic in the sense of Bohr.

The single most useful property of almost periodicity for studying differential equations is investigated from Bochner [22]. In this paper he introduced the following definition.

The continuous function f is *normal*, if from every sequence of real numbers $\{\alpha_n\}$ one can extract a subsequence $\{\alpha_{n_k}\}$ such that

$$\lim_{k \to \infty} f(t + \alpha_{n_k}) = g(t)$$

exists uniformly on $\mathbb{R}$. Bochner also proved the equivalence between the classes of *normal* and *regular* functions.

The utility of this definition for different classes of differential equations was exploited by Bochner in [23, 24].

Later on, in 1933 in Markoff's paper [114] on the study of almost periodic solutions of differential equations, it was recognized that almost periodicity and stability were closely related. Here for the fist time it was considered that strong stable bounded solutions are almost periodic.

After the first remarkable results in the area of almost periodicity in the middle of the twentieth century a number of impressive results were achieved. Some examples may be found are in the papers [12, 21, 47, 49, 53, 55–57, 69, 84, 97, 98, 143, 180, 183, 193–195, 201].

The beginning of the study of almost periodic piecewise continuous functions came in the 1960s. Developing the theory of impulsive differential equations further requires an introduction of definitions for these new objects. The properties of the classical (continuous) almost periodic functions can be greatly changed by impulses.

The first definitions and results in this new area were published by Halanay and Wexler [63], Perestyuk, Ahmetov and Samoilenko [8, 9, 127, 136–141], Hekimova and Bainov [67], Bainov, Myshkis, Dishliev and Stamov [17, 18].

Chapter 1
Impulsive Differential Equations and Almost Periodicity

The present chapter will deal with basic theory of the impulsive differential equations and almost periodicity.

Section 1.1 will offer the main classes of *impulsive differential equations*, investigated in the book. The problems of existence, uniqueness, and continuability of the solutions will be discussed. The piecewise continuous Lyapunov functions will be introduced and some main impulsive differential inequalities will be given.

Section 1.2 will deal with the *almost periodic sequences*. The main definitions and properties of these sequences will be considered.

Finally, in Sect. 1.3, we shall study the main properties of the *almost periodic piecewise continuous functions*.

1.1 Impulsive Differential Equations

Let $\mathbb{R}^n$ be the n-dimensional Euclidean space with norm $||.||$, and let $\mathbb{R}^+ = [0, \infty)$. For $J \subseteq \mathbb{R}$, we shall define the following class of functions: $PC[J, \mathbb{R}] = \{\sigma : J \to \mathbb{R}^n, \ \sigma(t) \text{ is a piecewise continuous function with points}$ of discontinuity of the first kind $\tilde{t} \in J$ at which $\sigma(\tilde{t}^+)$ and $\sigma(\tilde{t}^-)$ exist, and $\sigma(\tilde{t}^-) = \sigma(\tilde{t})\}$.

We shall make a brief description of the processes that are modeling with systems of impulsive differential equations.

Let $\Omega \subseteq \mathbb{R}^n$ be the phase space of some evolutionary process, i.e. the set of its states. Denote by P_t the point mapping the process at the moment t. Then the mapping point P_t can be interpreted as a point (t, x) of the $(n+1)$−dimensional space $\mathbb{R}^{n+1}$. The set $\mathbb{R} \times \Omega$ will be called an extended phase space of the evolutionary process considered. Assume that the evolution law of the process is described by:

G.T. Stamov, *Almost Periodic Solutions of Impulsive Differential Equations*,
Lecture Notes in Mathematics 2047, DOI 10.1007/978-3-642-27546-3_1,
© Springer-Verlag Berlin Heidelberg 2012

(a) A system of differential equations

$$\dot{x} = f(t, x), \tag{1.1}$$

where $t \in \mathbb{R}$; $x = col(x_1, x_2, \ldots, x_n) \in \Omega$; $f : \mathbb{R} \times \Omega \to \mathbb{R}^n$.

(b) Sets M_t, N_t of arbitrary topological structure contained in $\mathbb{R} \times \Omega$.

(c) An operator $A_t : M_t \to N_t$.

The motion of the point P_t in the extended phase space is performed in the following way: the point P_t begins its motion from the point $(t_0, x(t_0))$, $t_0 \in \mathbb{R}$, and moves along the curve $(t, x(t))$ described by the solution $x(t)$ of the (1.1) with initial condition $x(t_0) = x_0$, $x_0 \in \Omega$, till the moment $t_1 > t_0$ when P_t meets the set M_t. At the moment t_1 the operator A_{t_1} "instantly" transfers the point P_t from the position $P_{t_1} = (t_1, x(t_1))$ into the position $(t_1, x_1^+) \in N_{t_1}$, $x_1^+ = A_{t_1} x(t_1)$. Then the point P_t goes on moving along the curve $(t, y(t))$ described by the solution $y(t)$ of the system (1.1) with initial condition $y(t_1) = x_1^+$ till a new meeting with the set M_t, etc.

The union of relations (a), (b), (c) characterizing the evolutionary process will be called a *system of impulsive differential equations*, the curve described by the point P_t in the extended phase space—an *integral curve* and the function defining this curve—a *solution* of the system of impulsive differential equations. The moments $t_1, t_2, \ldots$ when the mapping point P_t meets the set M_t will be called *moments of impulse effect* and the operator $A_t : M_t \to N_t$— a *jump operator*.

We shall assume that the solution $x(t)$ of the impulsive differential equation is a left continuous function at the moments of impulse effect, i.e. that

$$x(t_k^-) = x(t_k), \ k = 1, 2, \ldots.$$

The freedom of choice of the sets M_t, N_t and the operator A_t leads to the great variety of the impulsive systems. The solution of the system of impulsive differential equation may be:

- A continuous function, if the integral curve does not intersect the set M_t or intersects it at the fixed points of the operator A_t.
- A piecewise continuous function with a finite number of points of discontinuity of first type, if the integral curve intersects M_t at a finite number of points which are not fixed points of the operator A_t.
- A piecewise continuous function with a countable set of points of discontinuity of first type, if the integral curve intersects M_t at a countable set of points which are not fixed points of the operator A_t.

In the present book systems of impulsive differential equations will be considered for which the moments of impulse effect come when some spatial-temporal relation $\Phi(t, x) = 0$, $(t, x) \in \mathbb{R} \times \Omega$ is satisfied, i.e. when the mapping point (t, x) meets the surface with the equation $\Phi(t, x) = 0$. Such systems can be written in the form

$$\dot{x} = f(t, x), \quad \Phi(t, x) \neq 0,$$
$$\Delta x(t) = I(t, x(t)), \quad \Phi(t, x(t)) = 0.$$

The sets M_t, N_t and the operator A_t are defined by the relations

$$M_t = \{(t, x) \in \mathbb{R} \times \Omega : \Phi(t, x) = 0\}, \quad N_t = \mathbb{R} \times \Omega,$$
$$A_t : M_t \to N_t, \quad (t, x) \to (t, x + I(t, x)),$$

where $I : \mathbb{R} \times \Omega \to \Omega$ and $t = t_k$ is a moment of impulse effect for the solution $x(t)$ if $\Phi(t_k, x(t_k)) = 0$.

We shall give a more detailed description of the following two classes of impulsive differential equations which have particular interest.

Class I. Systems with fixed moments of impulse effect. For these systems, the set M_t is represented by a sequence of hyperplanes $t = t_k$ where $\{t_k\}$ is a given sequence of impulse effect moments. The operator A_t is defined only for $t = t_k$ giving the sequence of operators $A_k : \Omega \to \Omega$, $x \to A_k x = x + I_k(x)$.

The systems of this class are written as follows:

$$\dot{x} = f(t, x), \quad t \neq t_k, \tag{1.2}$$
$$\Delta x(t_k) = I_k(x(t_k)), \quad k = \pm 1, \pm 2, \ldots, \tag{1.3}$$

where $\Delta x(t_k) = x(t_k^+) - x(t_k)$, $I_k : \Omega \to \mathbb{R}^n$, $k = \pm 1, \pm 2, \ldots$.

Let $\ldots < t_{-1} < t_0 < t_1 < t_2 < \ldots$, and $\lim\limits_{k \to \pm\infty} t_k = +\infty$. Denote by $x(t) = x(t; t_0, x_0)$ the solution of system (1.2), (1.3), satisfying the initial condition

$$x(t_0^+) = x(t_0). \tag{1.4}$$

The solution $x(t) = x(t; t_0, x_0)$ of the initial value problem (1.2), (1.3), (1.4), is characterized by the following:

(a) For $t = t_0$ the solution $x(t)$ satisfies the initial condition (1.4).
(b) For $t \in (t_k, t_{k+1}]$, the solution $x(t)$ satisfies the (1.2).
(c) For $t = t_k$ the solution $x(t)$ satisfies the relations (1.3).

Class II. Systems with variable impulsive perturbations. For these systems of impulsive differential equations, the set M_t is represented by a sequence of hypersurfaces $\sigma_k : t = \tau_k(x)$, $k = \pm 1, \pm 2, \ldots$.

Assume that $\tau_k(x) < \tau_{k+1}(x)$ for $x \in \Omega, k = \pm 1, \pm 2, \ldots$ and $\lim\limits_{k \to \infty} \tau_k(x) = \infty$ for $x \in \Omega$ ($\lim\limits_{k \to -\infty} \tau_k(x) = -\infty$).

We shall assume that the restriction of the operator A_t to the hypersurface σ_k is given by the operator $A_k x = x + I_k(x)$, where $I_k : \Omega \to \mathbb{R}^n$. The systems of impulsive differential equations of this class are written in the form

$$\begin{cases} \dot{x} = f(t, x), \quad t \neq \tau_k(x), \\ \Delta x(t) = I_k(x(t)), \quad t = \tau_k(x(t)), \quad k = \pm 1, \pm 2, \ldots. \end{cases} \tag{1.5}$$

The solution $x(t) = x(t; t_0, x_0)$ of the initial value problem (1.5), (1.4) is a piecewise continuous function but unlike the solution of (1.2), (1.3), (1.4) it has points of discontinuity depending on the solution, i.e. the different solutions have different points of discontinuity. This leads to a number of difficulties in the investigation of the systems of the form (1.5). One of the phenomena occurring with such systems is the so called "beating" of the solutions. This is the phenomenon when the mapping point $(t, x(t))$ meets one and the same hypersurface σ_k several or infinitely many times [94, 138]. Part of the difficulties are related to the possibilities of "merging" of different integral curves after a given moment, loss of the property of autonomy, etc.

It is clear that the systems of impulsive differential equations with fixed moments of impulse effect can be considered as a particular case of the systems with variable impulsive perturbations. Indeed, if $t = t_k$, $k = \pm 1, \pm 2, \ldots$ are fixed moments of time and we introduce the notation $\sigma_k = \{(t, x) \in \mathbb{R} \times \Omega : t = t_k\}$, then the systems of the second class are reduced to the systems of the first class.

Let $f : \mathbb{R} \times \Omega \to \mathbb{R}^n$, $\tau_k : \Omega \to \mathbb{R}$, $I_k : \Omega \to \mathbb{R}^n$, $\Omega \subset \mathbb{R}^n$, $\tau_k(x) < \tau_{k+1}(x)$, $\lim\limits_{k \to \pm\infty} \tau_k(x) = \pm\infty$ for $x \in \Omega$.

Now we shall give one such result.

Let $\mu > 0$, $(t_0, x_0) \in \mathbb{R} \times \Omega$ and

$$K(t_0, x_0, \mu) = \{(t, x) \in \mathbb{R} \times \Omega : \mu|x - x_0| < t - t_0\}.$$

Theorem 1.1 ([15]). *Let the following conditions hold.*

1. *The function $f(t, x)$ is continuous in $\mathbb{R} \times \Omega$ and is locally Lipschitz continuous with respect to $x \in \Omega$.*
2. *The integral curve $(t, x(t))$, $t \in \mathbb{R}$, of (1.5) is contained in $\mathbb{R} \times \Omega$ and*

$$(\tau_{k+1}, x_{k+1}) \in K(\tau_k, x_k^+, \mu), \quad \tau_k < \tau_{k+1}, \tag{1.6}$$

   where (τ_k, x_k) are the points at which the integral curve $(t, x(t))$, $t \in \mathbb{R}$, meets the hypersurfaces $\sigma_j \equiv t = \tau_j(x)$ and $x_i^+ = x_i + I_j(x_i)$ if $(\tau_i, x_i) \in \sigma_j$.
3. *The functions $\tau_k(x)$ are Lipschitz continuous with respect to $x \in \Omega$ with constants $L_k \leq \mu$ and*

$$\tau_k(x + I_k(x)) \leq \tau_k(x) \quad for \ x \in \Omega, \ k = \pm 1, \pm 2, \ldots.$$

Then the integral curve $(t, x(t))$ meets each hypersurface σ_k at most once.

Remark 1.1. Condition 1 of Theorem 1.1 is imposed in order to guarantee the existence and uniqueness of the solution of (1.5).

Corollary 1.1. *Theorem 1.1 still holds, if condition (1.6) is replaced by the condition*

$$||f(t, x)|| \leq M, \quad for \ (t, x) \in \mathbb{R} \times \Omega, \tag{1.7}$$

where $ML_k < 1, \ k = \pm 1, \pm 2, \ldots$.

Corollary 1.2. *Theorem 1.1 still holds, if condition (1.6) is replaced by the condition (1.7), and condition 3 is replaced by the following condition*
  3'. The functions $\tau_k(x)$ are differentiable in Ω and

$$||\frac{\partial \tau_k(x)}{\partial x}|| \leq L_k < \frac{1}{M}, \quad for \ x \in \Omega, \ k = \pm 1, \pm 2, \ldots, \tag{1.8}$$

$$\sup_{0 \leq s \leq 1, \ x \in \Omega} \left\langle \frac{\partial \tau_k}{\partial x}(x + sI_k(x)), I_k(x) \right\rangle \leq 0, \ k = \pm 1, \pm 2, \ldots,$$

where $\langle ., . \rangle$ is the scalar product in $\mathbb{R}^n$.

In fact, from (1.8) it follows that the functions $\tau_k(x), \ k = \pm 1, \pm 2, \ldots$ are Lipschitz continuous with constants L_k, and

$$\tau_k(x + I_k(x)) - \tau_k(x) = \int_0^1 \left\langle \frac{\partial \tau_k}{\partial x}(x + sI_k(x)), I_k(x) \right\rangle ds \leq 0,$$

i.e. condition 3 of Theorem 1.1 holds.

1.1.1 Existence, Uniqueness and Continuability

Let $\alpha < \beta$. Consider the system of impulsive differential equations (1.5).

Definition 1.1. The function $\varphi : (\alpha, \beta)$ is said to be a *solution* of (1.5) if:

1. $(t, \varphi(t)) \in \mathbb{R} \times \Omega$ for $t \in (\alpha, \beta)$.
2. For $t \in (\alpha, \beta), \ t \neq \tau_k(\varphi(t)), \ k = \pm 1, \pm 2, \ldots$ the function $\varphi(t)$ is differentiable and $\dfrac{d\varphi(t)}{dt} = f(t, \varphi(t))$.
3. The function $\varphi(t)$ is continuous from the left in (α, β) and if $t = \tau_k(\varphi(t))$, $t \neq \beta$, then $\varphi(t^+) = \varphi(t) + I_k(\varphi(t))$, and for each $k = \pm 1, \pm 2, \ldots$ and some $\delta > 0, \ s \neq \tau_k(\varphi(s))$ for $t < s < t + \delta$.

Definition 1.2. Each solution $\varphi(t)$ of (1.5) which is defined in an interval of the form (t_0, β) and satisfies the condition $\varphi(t_0^+) = x_0$ is said to be a *solution of the initial value problem* (1.5), (1.4).

We note that, instead of the initial condition $x(t_0) = x_0$, we have imposed the limiting condition $x(t_0^+) = x_0$ which, in general, is natural for (1.5) since (t_0, x_0) may be such that $t_0 = \tau_k(x_0)$ for some k. Whenever $t_0 \neq \tau_k(x_0)$, for all k, we shall understand the initial condition $x(t_0^+) = x_0$ in the usual sense, that is, $x(t_0) = x_0$.

It is clear that if $t_0 \neq \tau_k(x_0)$, $k = \pm 1, \pm 2, \ldots$, then the existence and uniqueness of the solution of the initial value problem (1.5), (1.4) depends only on the properties of the function $f(t, x)$. Thus, for instance, if the function $f(t, x)$ is continuous in a neighborhood of the point (t_0, x_0), then there exists a solution of the initial problem (1.5), (1.4) and this solution is unique if $f(t, x)$ is Lipschitz continuous in this neighborhood.

If, however, $t_0 = \tau_k(x_0)$ for some k, that is, (t_0, x_0) belongs to the hypersurface $\sigma_k \equiv t = \tau_k(x)$, then it is possible that the solution $x(t)$ of the initial value problem

$$\dot{x} = f(t, x), \quad x(t_0) = x_0 \tag{1.9}$$

lies entirely in σ_k.

Consequently, we need some additional conditions on $f(t, x)$ and $\tau_k(x)$ to guarantee the existence of solution $x(t)$ of the initial value problem (1.9) in some interval $[t_0, \beta)$, and the validity of the condition

$$t \neq \tau_k(x(t)) \text{ for } t \in (t_0, \beta) \text{ and } k = \pm 1, \pm 2, \ldots.$$

Conditions of this type are imposed in the next theorem.

Theorem 1.2 ([15]). *Let the following conditions hold:*

1. *The function $f : \mathbb{R} \times \Omega \to \mathbb{R}^n$ is continuous in $t \neq \tau_k(x)$, $k = \pm 1, \pm 2, \ldots$.*
2. *For any $(t, x) \in \mathbb{R} \times \Omega$ there exists a locally integrable function $l(t)$ such that in a small neighborhood of (t, x)*

$$\|f(s, y)\| \leq l(s).$$

3. *For each $k = \pm 1, \pm 2, \ldots$ the condition $t_1 = \tau_k(x_1)$ implies the existence of $\delta > 0$ such that*

$$t \neq \tau_k(x)$$

for all $0 < t - t_1 < \delta$ and $\|x - x_1\| < \delta$.

Then for each $(t_0, x_0) \in \mathbb{R} \times \Omega$ there exists a solution $x : [t_0, \beta) \to \mathbb{R}^n$ of the initial value problem (1.5), (1.4) for some $\beta > t_0$.

Remark 1.2. Condition 2 of Theorem 1.2 can be replaced by the condition:
   2'. For any $k = \pm 1, \pm 2, \ldots$ *and* $(t, x) \in \sigma_k$ *there exists the finite limit of* $f(s, y)$ *as* $(s, y) \to (t, x)$, $s > \tau_k(y)$.

Remark 1.3. The solution $x(t)$ of the initial value problem (1.5), (1.4) is unique, if the function $f(t, x)$ is such that the solution of the initial value problem (1.9) is unique. This requirement is met if, for instance, $f(t, x)$ is (locally) Lipschitz continuous with respect to x in a neighborhood of (t_0, x_0).

Now, we shall consider in more detail the system with fixed moments of impulsive effect:

$$\begin{cases} \dot{x} = f(t, x), \quad t \neq t_k, \\ \Delta x(t_k) = I_k(x(t_k)), \quad k = \pm 1, \pm 2, \ldots, \end{cases} \tag{1.10}$$

where $t_k < t_{k+1}$, $k = \pm 1, \pm 2, \ldots$ and $\lim\limits_{k \to \pm \infty} t_k = \pm \infty$.

Let us first establish some theorems.

Theorem 1.3 ([15]). *Let the following conditions hold:*

1. *The function* $f : \mathbb{R} \times \Omega \to \mathbb{R}^n$ *is continuous in the sets* $(t_k, t_{k+1}] \times \Omega$, $k = \pm 1, \pm 2, \ldots$.
2. *For each* $k = \pm 1, \pm 2, \ldots$ *and* $x \in \Omega$ *there exists the finite limit of* $f(t, y)$ *as* $(t, y) \to (t_k, x)$, $t > t_k$.
   Then for each $(t_0, x_0) \in \mathbb{R} \times \Omega$ *there exists* $\beta > t_0$ *and a solution* $x : [t_0, \beta) \to \mathbb{R}^n$ *of the initial value problem (1.10), (1.4).*
   If, moreover, the function $f(t, x)$ *is locally Lipschitz continuous with respect to* $x \in \Omega$ *then this solution is unique.*

Let us consider the problem of the continuability to the right of a given solution $\varphi(t)$ *of system (1.10).*

Theorem 1.4 ([15]). *Let the following conditions hold:*

1. *The function* $f : \mathbb{R} \times \Omega \to \mathbb{R}^n$ *is continuous in the sets* $(t_k, t_{k+1}] \times \Omega$, $k = \pm 1, \pm 2, \ldots$.
2. *For each* $k = \pm 1, \pm 2, \ldots$ *and* $x \in \Omega$ *there exists the finite limit of* $f(t, y)$ *as* $(t, y) \to (t_k, x)$, $t > t_k$.
3. *The function* $\varphi : (\alpha, \beta) \to \mathbb{R}^n$ *is a solution of (1.10).*

Then the solution $\varphi(t)$ *is continuable to the right of* β *if and only if there exists the limit*

$$\lim_{t \to \beta^-} \varphi(t) = \eta$$

and one of the following conditions hold:

(a) $\beta \neq t_k$ *for each* $k = \pm 1, \pm 2, \ldots$ *and* $\eta \in \Omega$.
(b) $\beta = t_k$ *for some* $k = \pm 1, \pm 2, \ldots$ *and* $\eta + I_k(\eta) \in \Omega$.

Theorem 1.5 ([15]). *Let the following conditions hold:*

1. *Conditions 1 and 2 of Theorem 1.1.11 hold.*
2. *The function f is locally Lipschitz continuous with respect to $x \in \Omega$.*
3. *$\eta + I_k(\eta) \in \Omega$ for each $k = \pm 1, \pm 2, \ldots$ and $\eta \in \Omega$.*
   Then for any $(t_0, x_0) \in \mathbb{R} \times \Omega$ there exists a unique solution of the initial value problem (1.10), (1.4) which is defined in an interval of the form $[t_0, \omega)$ and is not continuable to the right of ω.

Let the conditions of Theorem 1.5 be satisfied and let $(t_0, x_0) \in \mathbb{R} \times \Omega$. Denote by $J^+ = J^+(t_0, x_0)$ the maximal interval of the form $[t_0, \omega)$ in which the solution $x(t; t_0, x_0)$ is defined.

Theorem 1.6 ([15]). *Let the following conditions hold:*

1. *Conditions 1, 2 and 3 of Theorem 1.1.12 are met.*
2. *$\varphi(t)$ is a solution of the initial value problem (1.10), (1.4).*
3. *There exists a compact $Q \subset \Omega$ such that $\varphi(t) \in Q$ for $t \in J^+(t_0, x_0)$.*

Then $J^+(t_0, x_0) = (t_0, \infty)$.

Let $\varphi(t) : (\alpha, \omega) \to \mathbb{R}^n$ be a solution of system (1.10) and consider the question of the continuability of this solution to the left of α.

If $\alpha \neq t_k$, $k = \pm 1, \pm 2, \ldots$ then the problem of continuability to the left of α is solved in the same way as for ordinary differential equations without impulses [45]. In this case such an extension is possible if and only if there exists the limit

$$\lim_{t \to \sigma^+} \varphi(t) = \eta \tag{1.11}$$

and $\eta \in \Omega$.

If $\alpha = t_k$, for some $k = \pm 1, \pm 2, \ldots$, then the solution $\varphi(t)$ will be continuable to the left of t_k when there exists the limit (1.11), $\eta \in \Omega$, and the equation $x + I_k(x) = \eta$ has a unique solution $x_k \in \Omega$. In this case the extension $\psi(t)$ of $\varphi(t)$ for $t \in (t_{k-1}, t_k]$ coincides with the solution of the initial value problem

$$\begin{cases} \dot{\psi} = f(t, \psi), & t_{k-1} < t \leq t_k, \\ \psi(t_k) = x_k. \end{cases}$$

If the solution $\varphi(t)$ can be continued up to t_{k-1}, then the above procedure is repeated, and so on. Under the conditions of Theorem 1.5 for each $(t_0, x_0) \in \mathbb{R} \times \Omega$ there exists a unique solution $x(t; t_0, x_0)$ of the initial value problem (1.10), (1.4) which is defined in an interval of the form (α, ω) and is not continuable to the right of ω and to the left of α. Denote by $J(t_0, x_0)$ this maximal interval of existence of the solution $x(t; t_0, x_0)$ and set $J^- = J^-(t_0, x_0) = (\alpha, t_0]$. A straightforward verification shows that the solution $x(t) = x(t; t_0, x_0)$ of the initial value problem (1.10), (1.4) satisfies the following integro-summary equation

$$x(t) = \begin{cases} x_0 + \displaystyle\int_{t_0}^{t} f(s, x(s))ds + \sum_{t_0 < t_k < t} I_k(x(t_k)), & \text{for } t \in J^+, \\[3mm] x_0 + \displaystyle\int_{t_0}^{t} f(s, x(s))ds - \sum_{t \le t_k \le t_0} I_k(x(t_k)), & \text{for } t \in J^-. \end{cases} \qquad (1.12)$$

Now, we consider the linear system impulsive equations

$$\begin{cases} \dot{x} = A(t)x, \quad t \ne t_k, \\ \Delta x(t_k) = B_k x(t_k), \quad k = \pm 1, \pm 2, \ldots, \end{cases} \qquad (1.13)$$

under the assumption that the following conditions hold:

H1.1. $t_k < t_{k+1}$, $k = \pm 1, \pm 2, \ldots$ and $\lim_{k \to \pm\infty} t_k = \pm\infty$.

H1.2. $A \in PC[\mathbb{R}, \mathbb{R}^{n \times n}]$, $B_k \in \mathbb{R}^{n \times n}$, $k = \pm 1, \pm 2, \ldots$.

Theorem 1.7 ([15]). *Let conditions H1.1 and H1.2 hold. Then for any* $(t_0, x_0) \in \mathbb{R} \times \mathbb{R}^n$ *there exists a unique solution* $x(t)$ *of system (1.13) with* $x(t_0^+) = x_0$ *and this solution is defined for* $t \ge t_0$.

If moreover, $\det(E + B_k) \ne 0$, $k = \pm 1, \pm 2, \ldots$, *then this solution is defined for all* $t \in \mathbb{R}$.

Let $U_k(t, s)$ $(t, s \in (t_{k-1}, t_k])$ be the Cauchy matrix [65] for the linear equation

$$\dot{x}(t) = A(t)x(t), \quad t_{k-1} < t \le t_k, \quad k = \pm 1, \pm 2, \ldots.$$

Then by virtue of Theorem 1.7 the solution of the initial problem (1.13), (1.4) can be decomposed as:

$$x(t; t_0, x_0) = W(t, t_0^+)x_0, \qquad (1.14)$$

where

$$W(t, s) = \begin{cases} U_k(t, s) \quad \text{as } t, s \in (t_{k-1}, t_k], \\ U_{k+1}(t, t_k^+)(E + B_k)U_k(t_k, s) \quad \text{as } t_{k-1} < s \le t_k < t \le t_{k+1}, \\ U_k(t, t_k)(E + B_k)^{-1}U_{k+1}(t_k^+, s) \quad \text{as } t_{k-1} < t \le t_k < s \le t_{k+1}, \\ U_{k+1}(t, t_k^+) \displaystyle\prod_{j=k}^{i+1} (E + B_j)U_j(t_j, t_{j-1}^+)(E + B_i)U_i(t_i, s) \\ \qquad \text{as } t_{i-1} < s \le t_i < t_k < t \le t_{k+1}, \\ U_i(t, t_i) \displaystyle\prod_{j=i}^{k-1} (E + B_j)^{-1}U_{j+1}(t_j^+, t_{j+1})(E + B_k)^{-1}U_{k+1}(t_k^+, s) \\ \qquad \text{as } t_{i-1} < t \le t_i < t_k < s \le t_{k+1}, \end{cases}$$

is the solving operator of the (1.13).

1.1.2 *Piecewise Continuous Lyapunov Functions*

The second method of Lyapunov is one of the universal methods for investigating the dynamical systems from a different type. The method is also known as a direct method of Lyapunov or a method of the Lyapunov functions. Put forward in the end of the nineteenth century by Lyapunov [111], this method hasn't lost its popularity today. It has been applied initially to ordinary differential equations, and in his first work Lyapunov standardized the definition for stability and generalized the Lagrange's work on potential energy. The essence of the method is the investigation of the qualitative properties of the solutions without an explicit formula. For this purpose we need auxiliary functions—the so-called Lyapunov functions.

Different aspects of the Lyapunov second method applications for differential equations are given in [25, 39, 51, 52, 64, 83, 85, 86, 113, 134, 147, 193].

Gradually, there has been an expansion both in the class of the studied objects and in the mathematical problems investigated by means of the method.

Consider the system (1.5), and introduce the following notations:

$$G_k = \Big\{(t,x) \in \mathbb{R} \times \Omega : \tau_{k-1}(x) < t < \tau_k(x)\Big\}, \ G = \bigcup_{k=\pm 1, \pm 2, \dots} G_k,$$

Definition 1.3. A function $V : \mathbb{R} \times \Omega \to \mathbb{R}^+$ belongs to the class V_0, if:

1. $V(t,x)$ is continuous in G and locally Lipschitz continuous with respect to its second argument on each of the sets G_k, $k = \pm 1, \pm 2, \dots$.
2. For each $k = \pm 1, \pm 2, \dots$ and $(t_0^*, x_0^*) \in \sigma_k$ there exist the finite limits

$$V(t_0^{*-}, x_0^*) = \lim_{\substack{(t,x) \to (t_0^*, x_0^*) \\ (t,x) \in G_k}} V(t,x), \ \ V(t_0^{*+}, x_0^*) = \lim_{\substack{(t,x) \to (t_0^*, x_0^*) \\ (t,x) \in G_{k+1}}} V(t,x)$$

and the equality $V(t_0^{*-}, x_0^*) = V(t_0^*, x_0^*)$ holds.

Let the function $V \in V_0$ and $(t,x) \in G$. We define the derivative

$$\dot{V}_{(1.5)}(t,x) = \lim_{\delta \to 0^+} \sup \frac{1}{\delta} \Big[V(t+\delta, x+\delta f(t,x)) - V(t,x) \Big].$$

Note that if $x = x(t)$ is a solution of system (1.5), then for $t \neq \tau_k(x(t))$, $k = \pm 1, \pm 2, \dots$ we have $\dot{V}_{(1.5)}(t,x) = D_{(1.5)}^+ V(t, x(t))$, where

$$D_{(1.5)}^+ V(t, x(t)) = \lim_{\delta \to 0^+} \sup \frac{1}{\delta} \Big[V(t+\delta, x(t+\delta)) - V(t, x(t)) \Big] \qquad (1.15)$$

is *the upper right-hand Dini derivative* of $V \in V_0$ (with respect to system (1.5)).

The class of functions V_0 is also used for investigation of qualitative properties (such as stability, boundedness, almost periodicity) of the solutions of impulsive differential equations with fixed moments of impulse effect (1.10). In this case, $\tau_k(x) \equiv t_k$, $k = \pm1, \pm2, \dots$, σ_k are hyperplanes in $\mathbb{R}^{n+1}$, the sets G_k are

$$G_k = \{(t, x) \in \mathbb{R} \times \Omega : \quad t_{k-1} < t < t_k\},$$

and the condition 2 of Definition 1.3 is substituted by the condition:

2'. For each $k = \pm1, \pm2, \dots$ and $x \in \Omega$, there exist the finite limits

$$V(t_k^-, x) = \lim_{\substack{t \to t_k \\ t < t_k}} V(t, x), \qquad V(t_k^+, x) = \lim_{\substack{t \to t_k \\ t > t_k}} V(t, x),$$

and the following equalities are valid

$$V(t_k^-, x) = V(t_k, x).$$

For $t \neq t_k$, $k = \pm1, \pm2, \dots$ the upper right-hand derivative of Lyapunov's function $V \in V_0$, with respect to system (1.10) is

$$D^+ V_{(1.10)}(t, x(t)) = \lim_{\delta \to 0^+} \sup \frac{1}{\delta} \big[V(t + \delta, x(t + \delta)) - V(t, x) \big].$$

In Chap. 3 we shall use the next classes of piecewise continuous Lyapunov's functions

$$V_1 = \Big\{ V : \mathbb{R} \times \Omega \times \Omega \to \mathbb{R}^+, \ V \text{ is continuous in } (t_{k-1}, t_k] \times \Omega \times \Omega \text{ and}$$
$$\lim_{\substack{(t,x,y) \to (t_k, x_0, y_0) \\ t > t_k}} V(t, x, y) = V(t_k^+, x_0, y_0) \Big\}.$$

Definition 1.4. A function $V \in V_1$ belongs to class V_2, if:

1. $V(t, 0, 0) = 0$, $t \in \mathbb{R}$.
2. The function $V(t, x, y)$ is locally Lipschitz continuous with respect to its second and third arguments with a Lipschitz constant $H_1 > 0$, i.e. for $x_1, x_2 \in \Omega$, $y_1, y_2 \in \Omega$ and for $t \in \mathbb{R}$ it follows

$$|V(t, x_1, x_2) - V(t, y_1, y_2)| \leq H_1 \big(||x_1 - x_2|| + ||y_1 - y_2|| \big)$$

on each of the sets G_k, $k = \pm1, \pm2, \dots$.

Let $V \in V_2$, $t \neq t_k$, $k = \pm 1, \pm 2, \ldots$, $x \in PC[\mathbb{R}, \Omega]$, $y \in PC[\mathbb{R}, \Omega]$.
We introduce

$$D^+ V(t, x(t), y(t))$$

$$= \lim_{\delta \to 0^+} \sup \frac{1}{\delta} [V(t + \delta, x(t) + \delta f(t, x(t)), y(t) + \delta f(t, y(t))) - V(t, x(t), y(t))].$$

1.1.3 Impulsive Differential Inequalities

In this section we shall present the main comparison results and integral inequalities we use. The essence of the comparison method is in studying the relations between the given system and a comparison system so that some properties of the solutions of comparison system should imply the corresponding properties of the solutions of system under consideration. These relations are obtained employing differential inequalities. The comparison system is usually of lower order and its right-hand side possesses a certain type of monotonicity, which considerably simplifies the study of its solutions.

Define the following class:

$PC^1[J, \Omega] = \{\sigma \in PC[J, \Omega] : \sigma(t)$ is continuously differentiable everywhere except the points t_k at which $\dot{\sigma}(t_k^-)$ and $\dot{\sigma}(t_k^+)$ exist and $\dot{\sigma}(t_k^-) = \dot{\sigma}(t_k)$, $k = \pm 1, \pm 2, \ldots\}$.

Together with the system (1.10) we shall consider the comparison equation

$$\begin{cases} \dot{u}(t) = g(t, u(t)), \ t \neq t_k, \\ \Delta u(t_k) = B_k(u(t_k)), \ k = \pm 1, \pm 2, \ldots, \end{cases} \tag{1.16}$$

where $g : \mathbb{R} \times \mathbb{R}^+ \to \mathbb{R}^+$, $B_k : \mathbb{R}^+ \to \mathbb{R}^+$, $k = \pm 1, \pm 2, \ldots$.

Let $t_0 \in \mathbb{R}^+$ and $u_0 \in \mathbb{R}^+$. Denote by $u(t) = u(t; t_0, u_0)$ the solution of (1.16) satisfying the initial condition $u(t_0^+) = u_0$ and by $J^+(t_0, u_0)$—the maximal interval of type $[t_0, \beta)$ in which the solution $u(t; t_0, u_0)$ is defined.

Definition 1.5. The solution $u^+ : J^+(t_0, u_0) \to \mathbb{R}$ of the (1.16) for which $u^+(t_0; t_0, u_0) = u_0$ is said to be a *maximal solution* if any other solution $u : [t_0, \tilde{\omega}) \to \mathbb{R}$, for which $u(t_0) = u_0$ satisfies the inequality $u^+(t) \geq u(t)$ for $t \in J^+(t_0, u_0) \cap [t_0, \tilde{\omega})$.

Analogously, the *minimal solution* of (1.16) is defined.

In the case when the function $g : \mathbb{R} \times \mathbb{R}^+ \to \mathbb{R}^+$ is continuous and monotone increasing, all solutions of (1.16) starting from the point $(\bar{t}_0, u_0) \in [t_0, \infty) \times \mathbb{R}^+$ lie between two singular solutions—the maximal and the minimal ones.

The next result follows directly from the similar results in [94].

Theorem 1.8. *Let the following conditions hold:*

1. *Condition H1.1 holds.*

2. *The function* $g : \mathbb{R} \times \mathbb{R}^+ \to \mathbb{R}^+$ *is continuous in each of the sets* $(t_{k-1}, t_k] \times \mathbb{R}^+$, $k = \pm 1, \pm 2, \ldots$.

3. $B_k \in C[\mathbb{R}^+, \mathbb{R}^+]$ *and* $\psi_k(u) = u + B_k(u) \geq 0$, $k = \pm 1, \pm 2, \ldots$ *are non-decreasing with respect to* u.

4. *The maximal solution* $u^+ : J^+(t_0, u_0) \to \mathbb{R}^+$ *of (1.16),* $u^+(t_0^+; t_0, u_0) = u_0$ *is defined in* $\mathbb{R}$.

5. *The function* $V \in V_0$ *is such that* $V(t_0^+, x_0) \leq u_0$,

$$V(t^+, x + I_k(x)) \leq \psi_k(V(t, x)), \quad x \in \Omega, \ t = t_k, \ k = \pm 1, \pm 2, \ldots,$$

and the inequality

$$D_{(1.10)}^+ V(t, x(t)) \leq g(t, V(t, x(t))), \quad t \neq t_k, \ k = \pm 1, \pm 2, \ldots$$

is valid for $t \in \mathbb{R}$.
Then

$$V(t, x(t; t_0, x_0)) \leq u^+(t; t_0, u_0), \ t \in \mathbb{R}. \tag{1.17}$$

In the case when $g(t, u) = 0$ for $(t, u) \in \mathbb{R} \times \mathbb{R}^+$ and $\psi_k(u) = u$ for $u \in \mathbb{R}^+$, $k = \pm 1, \pm 2, \ldots$, the following corollary holds.

Corollary 1.3. *Let the following conditions hold:*

1. *Condition H1.1 holds.*
2. *The function* $V \in V_0$ *is such that*

$$V(t^+, x + I_k(x)) \leq V(t, x), \quad x \in \Omega, \ t = t_k, \ k = +1, +2, \ldots,$$

and the inequality

$$D_{(1.10)}^+ V(t, x(t)) \leq 0, \ t \neq t_k, \ k = \pm 1, \pm 2, \ldots$$

is valid for $t \in \mathbb{R}$.
Then

$$V(t, x(t; t_0, x_0)) \leq V(t_0^+, x_0), \ t \in \mathbb{R}.$$

We shall next consider Bihari and Gronwall type of integral inequality in a special case with impulses.

Theorem 1.9. *Let the following conditions hold:*

1. *Condition H1.1 holds.*
2. *The functions* $m : \mathbb{R} \to \mathbb{R}^+$, $p : \mathbb{R} \to \mathbb{R}^+$ *are continuous in each of the sets* $(t_{k-1}, t_k]$, $k = \pm 1, \pm 2, \ldots$.
3. $C \geq 0$, $\beta_k \geq 0$ *and*

$$m(t) \leq C + \int_{t_0}^{t} p(s)m(s)ds + \sum_{t_0 < t_k < t} \beta_k m(t_k). \qquad (1.18)$$

Then

$$m(t) \leq C \prod_{t_0 < t_k < t} (1 + \beta_k)e^{\int_{t_0}^{t} p(s)ds}. \qquad (1.19)$$

Proof. We shall proof this theorem by the method of mathematical induction [138].

Let $t \in [t_0, t_1]$. Then the inequality (1.18) is in the form

$$m(t) \leq C + \int_{t_0}^{t} p(s)m(s)ds.$$

From the Gronwall–Bellman's inequality, it follows that

$$m(t) \leq Ce^{\int_{t_0}^{t} p(s)ds},$$

i.e. for $t \in [t_0, t_1]$ the inequality (1.19) holds.

Let the inequality (1.19) holds for $t \in [t_k, t_{k+1}]$, $k = 1, 2, \ldots$, $k = i - 1$. Then, for $t \in (t_i, t_{i+1}]$ it follows

$$m(t) \leq C + \sum_{k=1}^{i} \beta_k C \prod_{j=1}^{k-1} (1 + \beta_j)e^{\int_{t_0}^{t_k} p(s)ds}$$

$$+ \sum_{k=1}^{i} \int_{t_{k-1}}^{t_k} p(s)C \prod_{j=1}^{k-1} (1 + \beta_j)e^{\int_{t_0}^{s} p(\sigma)d\sigma}ds + \int_{t_k}^{t} p(s)m(s)ds$$

$$= C \Big[1 + \sum_{k=1}^{i}\prod_{j=1}^{k-1} (1 + \beta_j)e^{\int_{t_0}^{t_k} p(s)ds}(1 + \beta_k) - \sum_{k=1}^{i}\prod_{j=1}^{k-1} (1 + \beta_j)e^{\int_{t_0}^{t_{k-1}} p(s)ds} \Big]$$

$$= C \prod_{j=1}^{k} (1 + \beta_j)e^{\int_{t_0}^{t_k} p(s)ds} + \int_{t_k}^{t} p(s)m(s)ds.$$

Consequently, for the function $m(t)$, $t \in (t_k, t_{k+1}]$ the next inequality holds

$$m(t) \leq C_1 + \int_{t_k}^{t} p(s)m(s)ds,$$

where

$$C_1 = C \prod_{k=1}^{i} (1 + \beta_k)e^{\int_{t_0}^{t_i} p(s)ds}.$$

Then, from Gronwall–Bellman's inequality for $t \in (t_k, t_{k+1}]$ it follows that

$$m(t) \leq C e^{\int_{t_0}^{t} p(s)ds},$$

or finally

$$m(t) \leq C \prod_{t_0 < t_k < t_k} (1 + \beta_k) e^{\int_{t_0}^{t} p(s)ds}. \qquad \square$$

The functions of Lyapunov and differential inequalities defined above are used in the investigation of stability of the impulsive systems. We shall use the next definitions for stability of these systems.

Let $B_h = \{x \in \mathbb{R}^n, \ ||x|| \leq h\}, \quad h > 0.$

Definition 1.6 ([138]). The zero solution $x(t) \equiv 0$ of system (1.10) is said to be:

(a) *Stable*, if

$$(\forall \varepsilon > 0)(\forall t_0 \in \mathbb{R})(\exists \delta > 0)(\forall x_0 \in B_\delta)(\forall t \geq t_0) : ||x(t; t_0, x_0)|| < \varepsilon.$$

(b) *Uniformly stable*, if the number δ in (a) is independent of $t_0 \in \mathbb{R}$.
(c) *Attractive*, if

$$(\forall t_0 \in \mathbb{R})(\exists \lambda > 0)(\forall x_0 \in B_\lambda) : \lim_{t \to \infty} ||x(t; t_0, x_0)|| = 0.$$

(d) *Equi-attractive*, if

$$(\forall t_0 \in \mathbb{R})(\exists \lambda > 0)(\forall \varepsilon > 0)(\forall x_0 \in B_\lambda)$$
$$(\exists T > 0)(\forall t \geq t_0 + T) : ||x(t; t_0, \varphi_0)|| < \varepsilon.$$

(e) *Uniformly attractive*, if the numbers λ and T in (d) are independent of $t_0 \in \mathbb{R}$.
(f) *Asymptotically stable*, if it is stable and attractive.
(g) *Uniformly asymptotically stable*, if it is uniformly stable and uniformly attractive.

Let $\psi(t) = \psi(t; t_0, \psi_0), \ \psi(t_0^+) = \psi_0 \in \Omega$, be a solution of system (1.10).

Definition 1.7 ([138]). The solution $\psi(t)$ is said to be:

(a) *Stable*, if

$$(\forall \varepsilon > 0)(\forall t_0 \in \mathbb{R})(\exists \delta > 0)(\forall x_0 \in \Omega, \ ||x_0 - \psi(t_0^+)|| < \delta) :$$
$$||x(t; t_0, x_0) - \psi(t)|| < \varepsilon.$$

(b) *Uniformly stable*, if the number δ in (a) is independent of $t_0 \in \mathbb{R}$.
(c) *Attractive*, if

$$(\forall t_0 \in \mathbb{R})(\exists \lambda > 0)(\forall x_0 \in \Omega, \ ||x_0 - \psi(t_0^+)|| < \lambda)$$

$$\lim_{t \to \infty} x(t; t_0, \varphi_0) = \psi(t).$$

(d) *Equi-attractive*, if

$$(\forall t_0 \in \mathbb{R})(\exists \beta > 0)(\forall \varepsilon > 0)(\forall x_0 \in \Omega, \ ||x_0 - \psi(t_0^+)|| < \beta)$$

$$(\exists \sigma > 0)(\forall t \geq t_0 + \sigma) : ||x(t; t_0, x_0) - \psi(t)|| < \varepsilon.$$

(e) *Uniformly attractive*, if the numbers λ and T in (d) are independent of $t_0 \in \mathbb{R}$.
(f) *Asymptotically stable*, if it is stable and attractive.
(g) *Uniformly asymptotically stable*, if it is uniformly stable and uniformly attractive.
(h) *Exponentially stable*, if

$$(\exists \lambda > 0)(\forall \alpha > 0)(\forall t_0 \in \mathbb{R})(\exists \gamma = \gamma(\alpha) > 0)$$

$$(\forall x_0 \in \Omega, \ ||x_0 - \psi(t_0^+)|| < \alpha)(\forall t \geq t_0) :$$

$$||x(t; t_0, x_0) - \psi(t)|| < \gamma(\alpha)||x_0 - \psi(t_0^+)|| \exp\{-\lambda(t - t_0)\}.$$

1.2 Almost Periodic Sequences

In this part, we shall follow [138] and consider the main definitions and properties of almost periodic sequences.

We shall consider the sequence $\{x_k\}$, $x_k \in \mathbb{R}^n$, $k = \pm 1, \pm 2, \ldots$, and let $\varepsilon > 0$.

Definition 1.8. The integer number p is said to be an ε-*almost period* of $\{x_k\}$, if for each $k = \pm 1, \pm 2, \ldots$

$$||x_{k+p} - x_k|| < \varepsilon. \tag{1.20}$$

It is easy to see that, if p and q are ε-almost periods of $\{x_k\}$, then $p+q$, $p-q$ are 2ε-almost periods of the sequence $\{x_k\}$.

Definition 1.9. The sequence $\{x_k\}$, $x_k \in \mathbb{R}^n$, $k = \pm 1, \pm 2, \ldots$ is said to be *almost periodic*, if for an arbitrary $\varepsilon > 0$ there exists a relatively dense set of its ε-almost periods, i.e there exists a natural number $N = N(\varepsilon)$ such that for an arbitrary integer number k, between integer numbers in the interval $[k, k+N]$, there exists at least one integer number p for which the inequality (1.20) holds.

Theorem 1.10. *Let the following conditions hold:*

1. *The sequence $\{x_k\} \subset B_\alpha$, $k = \pm 1, \pm 2, \ldots$, is almost periodic.*
2. *The function $y = f(x)$ is uniformly continuous in B_α.*

Then:

1. *The sequence $\{x_k\}$, $k = \pm 1, \pm 2, \ldots$ is bounded.*
2. *The sequence $\{y_k\}$, $y_k = f(x_k)$, $k = \pm 1, \pm 2, \ldots$ is almost periodic.*

Proof of Assertion 1. Let $\varepsilon > 0$ and k be an arbitrary integer number. Then there exists a natural number N such that in the interval $[-k, -k+N]$ there exists an ε-almost period p of $\{x_k\}$.

From $-k \le p \le -k+N$, we get $0 \le p+k \le N$ and

$$||x_k|| \le ||x_k - x_{k+p}|| + ||x_{k+p}|| < \varepsilon + \max_{0 \le k \le N} ||x_k||.$$

Then the sequence $\{x_k\}$ is bounded.

Proof of Assertion 2. For $\varepsilon > 0$ there exists $\delta = \delta(\varepsilon) > 0$ such that

$$||f(x') - f(x'')|| < \varepsilon,$$

when $||x' - x''|| < \delta$, $x', x'' \in B_\alpha$.

If p is a δ-almost period of the sequence $\{x_k\}$, then we have

$$||y_{k+p} - y_k|| = ||f(x_{k+p}) - f(x_k)|| < \varepsilon, \ k = \pm 1, \pm 2, \ldots.$$

Therefore, p is an ε-almost period of $\{y_k\}$.   □

Let us now consider the set of all bounded sequences,

$$BS = \left\{ \{x_k\}, \ x_k \in \mathbb{R}^n, \ k = \pm 1, \pm 2, \ldots \right\}$$

with norm $|x_k|_\infty = \sup_{k = \pm 1, \pm 2, \ldots} ||x_k||$.

Clearly, BS is a Banach space, and the sequence of almost periodic sequences $\{x_k^m\}$, $k = \pm 1, \pm 2, \ldots$, $m = 1, 2, \ldots$, is convergent to the sequence $\{y_k\}$, if there exists the limit $\lim_{m \to \infty} |x_k^m - y_k|_\infty = 0$.

Then on the set BS the following theorems hold.

Theorem 1.11 ([138]). *Let the following conditions hold:*

1. *For each $m = 1, 2, \ldots$ the sequence $\{x_k^m\}$, $k = \pm 1, \pm 2, \ldots$, is almost periodic.*
2. *There exists a limit $\{y_k\}$, $k = \pm 1, \pm 2, \ldots$ of the sequence $\{x_k^m\}$, $k = \pm 1, \pm 2, \ldots$ as $m \to \infty$.*

Then the limit sequence $\{y_k\}$, $k = \pm 1, \pm 2, \ldots$ is almost periodic.

Theorem 1.12. *The sequence $\{x_k\}$, $k = \pm1, \pm2, \ldots$ is almost periodic if and only if for any sequence of integer numbers $\{m_i\}$, $i = \pm1, \pm2, \ldots$ there exists a subsequence $\{m_{i_j}\}$, such that $\{x_{k+m_{i_j}}\}$ is convergent for $j \to \infty$ uniformly on $k = \pm1, \pm2, \ldots$.*

Proof. First, let $\{x_k\}$ be almost periodic, $\{m_i\}$ $i = \pm1, \pm2, \ldots$ be an arbitrary sequence of integer numbers and let $\varepsilon > 0$. Then, there exists $N = N(\varepsilon)$ such that in the interval $[m_i - N, m_i]$ there exists an ε-almost period p_i of the sequence $\{x_k\}$. From $m_i - N < p_i \leq m_i$, it follows $0 \leq m_i - p_i < N$.

Let now $q_i = m_i - p_i$. The sequence $\{q_i\}$ has only finite numbers of elements and there exists a number q such that $q_i = q$ for unbounded numbers of indices i_j^1.

Then, from

$$\|x_{k+m_i} - x_{k+q_i}\| = \|x_{k+m_i} - x_{k+m_i - p_i}\| < \varepsilon$$

it follows that

$$\|x_{k+m_{i_j^1}} - x_{k+q}\| < \varepsilon$$

for $k = \pm1, \pm2, \ldots$.

Let us use the decreasing sequence $\{\varepsilon_j\}$, $\varepsilon_j > 0$, $j = \pm1, \pm2, \ldots$, and let the sequence $\{x_{k+m_{i_j}}\}$ be a subsequence of $\{x_{k+m_i}\}$, such that $\|x_{k+m_{i_j}} - x_{k+q^1}\| < \varepsilon_1$.

Then, we choice a subsequence $\{x_{k+m_{i_j^2}}\}$ from the sequence $\{x_{k+m_{i_j}}\}$ such that $\|x_{k+m_{i_j^2}} - x_{k+q^2}\| < \varepsilon_2$. If we continue, we will find a subsequence $\{x_{k+m_{i_j^r}}\}$ for which $\|x_{k+m_{i_j^r}} - x_{k+q^r}\| < \varepsilon_r$. Finally, we construct a diagonal sequence $\{x_k + m_{i_j^j}\}$ and we shall proof that it is convergent uniformly on k for $j \to \infty$.

Let $\varepsilon > 0$ and the number N' is such that $\varepsilon_{N'} < \frac{\varepsilon}{2}$. Then, for $r, s \geq N'$, we have

$$\|x_{k+m_{i_s^r}} - x_{k+m_{i_s^s}}\| \leq \|x_{k+m_{i_s^r}} - x_{k+q^{N'}}\|$$
$$+ \|x_{k+q^{N'}} - x_{k+m_{i_s^s}}\| < \varepsilon_{N'} + \varepsilon_{N'} < \varepsilon,$$

since $\{m_{i_j^r}\}$ and $\{m_{i_j^s}\}$, $j = \pm1, \pm2, \ldots$, are subsequences of the sequence $\{m_{i_j}^{N'}\}$, $j = \pm1, \pm2, \ldots$.

Therefore,

$$\|x_{k+m_{i_j^r}} - x_{k+m_{i_j^s}}\| < \varepsilon,$$

and the sequence $\{x_{k+m_{i_j^j}}\}$, $j = \pm1, \pm2, \ldots$ is uniformly convergent on k.

On the other hand, let us suppose that the sequence $\{x_k\}$ is not almost periodic. Then there exists a number $\varepsilon_0 > 0$ such that for any natural number N, we have N consecutive integer numbers and between them there is not an ε_0-almost period.

Let now L_N be such a set of consecutive integer numbers, and let we choice arbitrary m_1 and m_2, so that $m_1 - m_2 \in L_1$. Let $L_1 = L_{\nu_1}$ and we choice $\nu_2 > |m_1 - m_2|$ and m_3, such that $m_3 - m_1, m_3 - m_2 \in L_{\nu_2}$. This is possible because, if $l, l+1, \ldots, l+\nu_2-1$ are from L_{ν_2} and $m_2 \leq m_1$, we can take $m_3 = l + m_1$, such that $m_3 - m_1 \in L_{\nu_2}$ and from

$$m_3 - m_2 = m_3 - m_1 + m_1 - m_2 < l + \nu_2, \quad m_3 - m_2 \geq l$$

it follows that $m_3 - m_2 \in L_{\nu_2}$.

Now for an arbitrary k, we can choice $\nu_k > \max\limits_{1 \leq \mu < \nu \leq k} |m_\mu - m_\nu|$ and m_{k+1} such that $m_{k+1} - m_\mu \in L_{\nu_k}$ for $1 \leq \mu \leq k$. Then, for the sequence $\{m_k\}$, we have

$$\sup_{k=\pm 1, \pm 2, \ldots} \|x_{k+m_r} - x_{k+m_s}\| = \sup_{k=\pm 1, \pm 2, \ldots} \|x_{k+m_r-m_s} - x_k\|.$$

On the other hand, $m_r - m_s \in L_{\nu_{r-1}}$, where $r \geq s$ and it is not an ε_0-almost period. Then, there exists a number k for which $\|x_{k+m_r-m_s} - x_k\| \geq \varepsilon_0$, and we have that $\sup\limits_{k=\pm 1, \pm 2, \ldots} \|x_{k+m_r} - x_{k+m_s}\| \geq \varepsilon_0$, or $\|x_{k+m_r} - x_{k+m_s}\| \geq \varepsilon_0$.

Therefore, for the sequence $\{m_k\}$, there exists a subsequence $\{m_{i_j}\}$, such that the sequence $\{x_{k+m_{i_j}}\}$, $j = \pm 1, \pm 2, \ldots$ is uniformly convergent on $k = \pm 1, \pm 2, \ldots$. Then, there exists an index j_0, such that for $j, l \geq j_0$, we get $\|x_{k+m_{i_j}} - x_{k+m_{i_l}}\| < \varepsilon_0$, which is a contradiction. $\qquad\square$

From this theorem, we get the next corollary.

Corollary 1.4. *Let the sequences $\{x_k\}$, $\{y_k\}$, $x_k, y_k \in \mathbb{R}^n$ are almost periodic and the sequence $\{\alpha_k\}$, $k = \pm 1, \pm 2, \ldots$, of real numbers is almost periodic.*

Then the sequences $\{x_k + y_k\}$ and $\{\alpha_k x_k\}$, $k = \pm 1, \pm 2, \ldots$, are almost periodic.

From Theorem 1.12 and Corollary 1.4 it follows that the set of all almost periodic sequences $\{x_k\}$, $k = \pm 1, \pm 2, \ldots$, $x_k \in \mathbb{R}^n$ is a linear space, and equipped with the norm $|x_k|_\infty = \sup\limits_{k=\pm 1, \pm 2, \ldots} \|x_k\|$ is a Banach space.

Theorem 1.13. *Let the sequences $\{x_k\}$, $\{y_k\}$, $k = \pm 1, \pm 2, \ldots$, $x_k, y_k \in \mathbb{R}^n$, are almost periodic.*

Then for any $\varepsilon > 0$ there exists a relatively dense set of their common ε-almost periods.

Proof. Let $\varepsilon > 0$ be fixed. There exist integer numbers $N_1 = N_1(\varepsilon)$ and $N_2 = N_2(\varepsilon)$ such that between integers in the intervals $[i, i+N_1]$ and $[i, i+N_2]$ there exists at least one $\frac{\varepsilon}{2}$-almost period of the sequences $\{x_k\}$, $\{y_k\}$, respectively.

Let now $N_3 = N_3(\varepsilon) = max\{N_1, N_2\}$. Then there exists at least one $\frac{\varepsilon}{2}$-almost period p_1 of the sequence $\{x_k\}$ and one $\frac{\varepsilon}{2}$-almost period p_2 of the sequence $\{y_k\}$ from the integer numbers in the interval $[i, i + N_3]$. Since $|p_1 - p_2| \leq N_3$, then the difference $p_2 - p_1$ has only a finite number of values, for arbitrary choices of the N_3 consecutive integer numbers. The pairs (p_1, p_2) and $(p_1', p_2',)$ are said to be equivalent, if $|p_1 - p_2| = |p_1' - p_2'|$. Since $|p_1 - p_2|$ can take only a finite number of values, then there exist a finite number classes of equivalence with respect to this relation of equivalence.

Let we choice arbitrary representative pairs (p_1^r, p_2^r), $r = 1, 2, \ldots, s$, and let $N_4 = N_4(\varepsilon) = \max\limits_{1 \leq r \leq s} |p_1^r|$, $N = N_3 + 2N_4$.

We shall show that there exists an ε-almost period from integers in $[i, i+N_4]$, which is common for the sequences $\{x_k\}$ and $\{y_k\}$. Let i be an integer, p_1, p_2 be $\frac{\varepsilon}{2}$-almost periods belong to the interval $[i + N_4, i + N_4 + N_3]$, and the pair (p_1^r, p_2^r) be from the same class like $(p_1, p_2,)$, so we have $|p_1 - p_2| = |p_1^r - p_2^r|$. Then, from $p_1^r - p_2^r = p_1 - p_2$, or $p_1^r - p_2^r = p_2 - p_1$, it follows $p_1^r - p_1 = p_2^r - p_2 = -p$, or $p_1^r + p_1 = p_2^r + p_2 = p$.

On the other hand, from the inequality $|p_1^r| \leq N_4$, we have that $i < p_1 + p_1^r \leq i + N_3 + 2N_4 = i + N$ and $i < p_1 - p_1^r \leq i + N_3 + 2N_4 = i + N$, i.e. in the both cases $i < p \leq i + N$. Then, the number p is a common ε-almost period for the sequences $\{x_k\}$ and $\{y_k\}$.

Indeed,

$$||x_{k+p} - x_k|| = ||x_{k+p_1 \pm p_1^r} - x_k|| \leq ||x_{k+p_1 \pm p_1^r} - x_{k+p_1}||$$

$$+ ||x_{k+p_1} - x_k|| < \frac{\varepsilon}{2} + \frac{\varepsilon}{2} = \varepsilon,$$

$$||y_{k+p} - y_k|| = ||y_{k+p_2 \pm p_2^r} - y_k|| \leq ||y_{k+p_2 \pm p_2^r} - y_{k+p_2}||$$

$$+ ||y_{k+p_2} - y_k|| < \frac{\varepsilon}{2} + \frac{\varepsilon}{2} = \varepsilon. \qquad \square$$

Theorem 1.14 ([138]). *For any almost periodic sequence* $\{x_k\}$, $k = \pm 1, \pm 2, \ldots$, $x_k \in \mathbb{R}^n$ *there exists uniformly on* k *the average value*

$$\lim_{n \to \infty} \frac{1}{n} \sum_{j=k}^{k+n-1} x_k = M(x_k) < \infty.$$

Now we shall consider the set $\mathcal{B}$,

$$\mathcal{B} = \left\{ \{t_k\}, \ t_k \in \mathbb{R}, \ t_k < t_{k+1}, \ k = \pm 1, \pm 2, \ldots, \ \lim_{k \to \pm \infty} t_k = \pm \infty \right\} \text{ of}$$

all unbounded increasing sequences of real numbers, and let $i(t, t + A)$ is the number of the points t_k in the interval $(t, t + A]$.

Lemma 1.1. *Let* $\{t_k\} \subset \mathcal{B}$ *be such that the sequence* $\{t_k^1\}$, $t_k^1 = t_{k+1} - t_k$, $k = \pm 1, \pm 2, \ldots$ *is almost periodic.*

Then, uniformly on $t \in \mathbb{R}$ there exists the limit

$$\lim_{A \to \infty} \frac{i(t, t+A)}{A} = p < \infty. \tag{1.21}$$

Proof. We shall show that there exists a nonzero limit

$$\lim_{n \to \infty} \frac{t_n}{n} = \frac{1}{p} < \infty.$$

Without loss of generality, let $t_{-1} < 0$ and $t_1 \geq 0$. Then,

$$t_n = t_1 + \sum_{j=1}^{n-1} t_j^1,$$

and

$$\frac{t_n}{n} = \frac{t_1}{n} + \frac{1}{n} \sum_{j=1}^{n-1} t_j^1.$$

The sequence $\{t_k^1\}$ is almost periodic, and from Theorem 1.14 it follows that there exists an average value and the finite limit

$$\lim_{n \to \infty} \frac{1}{n} \sum_{j=1}^{n-1} t_j^1 \neq 0,$$

since $t_k^1 > 0$.
Then,

$$\lim_{n \to \infty} \frac{t_n}{n} = \lim_{n \to \infty} \left\{ \frac{t_1}{n} + \frac{1}{n} \sum_{j=1}^{n-1} t_j^1 \right\} = \frac{1}{p} \neq 0,$$

and there exists

$$\lim_{n \to \infty} \frac{i(0, t_n)}{t_n} = p. \tag{1.22}$$

Now, from

$$\frac{A}{i(0, A)} = \frac{t_k + \theta_k}{k}$$

for a natural number k and $0 \leq \theta_k \leq \sup_{k=\pm 1, \pm 2, \ldots} \{t_k^1\}$, we have

$$\frac{A}{i(0, A)} - \frac{t_k}{k} = o\left(\frac{1}{k}\right),$$

where $o\left(\frac{1}{k}\right)$ is the Landaw symbol.

Then, from (1.22) and from the last equality, we get

$$\lim_{n\to\infty} \frac{i(0,A)}{A} = p.$$

On the other hand, from the almost periodicity of $\{t_k^1\}$ it follows

$$\left| \sum_{j=\nu}^{\nu+k-1} t_j^1 - \sum_{j=1}^{k-1} t_j^1 \right| \leq \frac{\varepsilon k}{4} + 2N \sup_{j=\pm1,\pm2,\ldots} \{t_j^1\}. \qquad (1.23)$$

Using (1.23), we have

$$\left| \frac{1}{k} \sum_{j=(\nu-1)k}^{\nu k-1} t_j^1 - \frac{1}{k} \sum_{j=1}^{k-1} t_j^1 \right| \leq \frac{\varepsilon}{4} + \frac{2N}{k} \sup_{j=\pm1,\pm2,\ldots} \{t_j^1\},$$

where $\nu = 1, 2, \ldots$.
Therefore, for

$$\frac{1}{n} \sum_{\nu=1}^{n} \left[\frac{1}{k} \sum_{j=(\nu-1)k}^{\nu k-1} t_j^1 - \frac{1}{k} \sum_{j=1}^{k-1} t_j^1 \right] = \frac{1}{nk} \sum_{j=1}^{nk-1} t_j^1 - \frac{1}{k} \sum_{j=1}^{k-1} t_j^1,$$

we have

$$\left| \frac{1}{nk} \sum_{j=1}^{nk-1} t_j^1 - \frac{1}{k} \sum_{j=1}^{k-1} t_j^1 \right| \leq \frac{\varepsilon}{4} + \frac{2N}{k} \sup_{j=\pm1,\pm2,\ldots} \{t_j^1\}. \qquad (1.24)$$

Then, from (1.23) and (1.24) it follows

$$\left| \frac{1}{k} \sum_{j=\nu}^{\nu+k-1} t_j^1 - \frac{1}{p} \right| < \frac{\varepsilon}{2} + \frac{4N}{k} \sup_{j=\pm1,\pm2,\ldots} \{t_j^1\},$$

or

$$\lim_{n\to\infty} \frac{t_{k+n} - t_k}{n} = \frac{1}{p} \qquad (1.25)$$

uniformly on $k = \pm1, \pm2, \ldots$.
Let in the interval $[t, t + A]$ there exist i elements of the sequence $\{t_k\}$, $t_{\nu+1}, t_{\nu+2}, \ldots, t_{\nu+i}$.
Then, we get

$$\frac{A}{i(t,t+A)} = \frac{A}{i} = \frac{t_{\nu+i} - t_\nu + \theta_i}{i}, \quad |\theta_i| \leq \sup_{k=\pm1,\pm2,\ldots} \{t_k^1\},$$

Now, it follows that

$$\lim_{A\to\infty} \frac{i(t,t+A)}{A} = p < \infty$$

for all $t \in \mathbb{R}$.   □

We shall consider the sequences $\{t_k^j\}$, $t_k^j = t_{k+j} - t_k$, $k,j = \pm1,\pm2,\ldots$.
It is easy to see that

$$t_{k+i}^j - t_k^j = t_{k+j}^i - t_k^i, \quad t_k^j - t_k^i = t_{k+i}^{j-i}, \quad i,j,k = \pm1,\pm2,\ldots. \tag{1.26}$$

□

Definition 1.10. The set of sequences $\{t_k^j\}, t_k^j = t_{k+j}-t_k, k,j = \pm1,\pm2,\ldots$,
is said to be *uniformly almost periodic*, if for an arbitrary $\varepsilon > 0$ there exists
a relatively dense set of ε-almost periods, common for all sequences $\{t_k^j\}$.

Example 1.1 ([138]). Let $\{\alpha_k\}$, $\alpha_k \in \mathbb{R}$, $k = \pm1,\pm2,\ldots$ be an almost
periodic sequence such that

$$\sup_{k=\pm1,\pm2,\ldots} |\alpha_k| = \alpha < \frac{A}{2}, \ A > 0,$$

and let $t_k = kA + \alpha_k$, $k = \pm1,\pm2,\ldots$.
Then

$$t_{k+1} - t_k \geq A - 2\alpha > 0,$$

and $\lim\limits_{k\to\perp\infty} t_k = \pm\infty$.

Let $\varepsilon > 0$ and p be an $\dfrac{\varepsilon}{2}$-almost period of the sequence $\{\alpha_k\}$. Then, for all
integers k and j it follows

$$|t_{k+p}^j - t_k^j| = |\alpha_{k+j+p} - \alpha_{k+j}| + |\alpha_{k+p} - \alpha_k| < \varepsilon.$$

The last inequality shows that the set of sequences $\{t_k^j\}$ is uniformly almost
periodic.

Example 1.2 ([67]). Let $t_k = k + \alpha_k$, where

$$\alpha_k = \frac{1}{4}|cosk - cosk\sqrt{2}|, \ k = \pm1,\pm2,\ldots.$$

The sequence $\{t_k\}$ is strictly increasing, since we have

$$t_{k+1} - t_k = 1 + \frac{1}{4}|cos(k+1) - cos(k+1)\sqrt{2}| - \frac{1}{4}|cosk - cosk\sqrt{2}| \geq \frac{1}{2},$$

and it is easy to see that $\lim\limits_{k\to\pm\infty} t_k = \pm\infty$.

We shall prove that the set of sequences $\{t_k^j\}$ is uniformly almost periodic. Let $\varepsilon > 0$ and p be an $\dfrac{\varepsilon}{2}$-almost period of the sequence $\{\alpha_k\}$. Then for all integers k and j, we have

$$\left| t_{k+p}^j - t_k^j \right| = \left| t_{k+p+j} - t_{k+p} - t_{k+j} + t_p \right|$$

$$\leq \left| \alpha_{k+p+j} - \alpha_{k+j} \right| + \left| \alpha_{k+p} - \alpha_k \right| < \varepsilon,$$

and from Definition 1.10 it follows that the set of sequences $\{t_k^j\}$ is uniformly almost periodic.

We shall use the following properties of the uniformly almost periodic sequences.

Lemma 1.2 ([138]). *Let the set of sequences $\{t_k^j\}$, $t_k^j = t_{k+j} - t_k$, $k, j = \pm 1, \pm 2, \ldots$, be uniformly almost periodic. Then for each $p > 0$ there exists a positive integer N such that on each interval of a length p, there exist no more than N elements of the sequence $\{t_k\}$ and*

$$i(s, t) \leq N(t - s) + N, \qquad (1.27)$$

where $i(s, t)$ is the number of points t_k in the interval (s, t).

Lemma 1.3 ([138]). *Let the set of sequences $\{t_k^j\}$, $t_k^j = t_{k+j} - t_k$, $k, j = \pm 1, \pm 2, \ldots$, be uniformly almost periodic. Then for each $\varepsilon > 0$ there exists a positive number $l = l(\epsilon)$ such that for each interval A of a length l, there exist a subinterval $I \subset A$ of a length $\varepsilon > 0$, and an integer number q such that*

$$\left| t_k^q - r \right| < \varepsilon, \ k = \pm 1, \pm 2, \ldots, \ r \in I. \qquad (1.28)$$

Lemma 1.4 ([63]). *Let the set of sequences $\{t_k^j\}$, $t_k^j = t_{k+j} - t_k$, $k, j = \pm 1, \pm 2, \ldots$, be uniformly almost periodic, and let the function $\Phi(t)$ be almost periodic in sense of Bohr. Then, for each $\varepsilon > 0$ there exists a positive $l = l(\epsilon)$ such that for each interval A of a length l, there exists $r \in A$ and an integer number q such that*

$$\left| t_k^q - r \right| < \varepsilon, \ \left| \Phi(t + r) - \Phi(t) \right| < \varepsilon,$$

for all $k = \pm 1, \pm 2, \ldots$, and $t \in \mathbb{R}$.

Lemma 1.5 ([63]). *Let the set of sequences $\{t_k^j\}$, $t_k^j = t_{k+j} - t_k$, $k, j = \pm 1, \pm 2, \ldots$, be uniformly almost periodic, and let the function $\Phi(t)$ be almost periodic in sense of Bohr. Then the sequence $\{\Phi(t_k)\}$ is almost periodic.*

Definition 1.11 ([139]). The set $T \in B$ is almost periodic, if for every sequence of real numbers $\{s_m'\}$ there exists a subsequence $\{s_n\}$, $s_n = s_{m_n}'$

such that $T - s_n = \{t_k - s_n\}$ is uniformly convergent for $n \to \infty$ to the set $T_1 \in \mathcal{B}$.

Lemma 1.6. *The set of sequences $\{t_k^j\}$, $t_k^j = t_{k+j} - t_k$, $k, j = \pm 1, \pm 2, \dots$ is uniformly almost periodic if and only if for every sequence of real numbers $\{s_m'\}$ there exists a subsequence $\{s_n\}$, $s_n = s_{m_n}'$ such that $T - s_n = \{t_k - s_n\}$ is uniformly convergent for $n \to \infty$ on $\mathcal{B}$.*

Proof. The proof follows directly from Theorem 1 in [139]. □

In the investigation of the existence of almost periodic solutions of impulsive differential equations, the question for the separation from the origin of the sequences $\{t_k\} \in \mathcal{B}$ is very important. Hence, we always shall suppose that the following inequality

$$\inf_{k=\pm 1, \pm 2, \dots} t_k^1 = \theta > 0$$

holds.

We shall use, also, the set $UAPS$, $UAPS \subset \mathcal{B}$ for which the sequences $\{t_k^j\}$, $t_k^j = t_{k+j} - t_k$, $k, j = \pm 1, \pm 2, \dots$, form uniformly almost periodic set and $\inf_{k=\pm 1, \pm 2, \dots} t_k^1 = \theta > 0$.

1.3 Almost Periodic Functions

In this part we shall consider the main definitions and properties of almost periodic piecewise continuous functions.

Definition 1.12. The function $\varphi \in PC[\mathbb{R}, \mathbb{R}^n]$ is said to be *almost periodic*, if the following holds:

(a) $\{t_k\} \in UAPS$.
(b) For any $\varepsilon > 0$ there exists a real number $\delta = \delta(\varepsilon) > 0$ such that, if the points t' and t'' belong to one and the same interval of continuity of $\varphi(t)$ and satisfy the inequality $|t' - t''| < \delta$, then $||\varphi(t') - \varphi(t'')|| < \varepsilon$.
(c) For any $\varepsilon > 0$ there exists a relatively dense set $\overline{T}$ such that, if $\tau \in \overline{T}$, then $||\varphi(t+\tau) - \varphi(t)|| < \varepsilon$ for all $t \in \mathbb{R}$ satisfying the condition $|t - t_k| > \varepsilon$, $k = \pm 1, \pm 2, \dots$.

The elements of $\overline{T}$ are called ε-*almost periods*.

Example 1.3 ([67]). Let $\{\mu_k\}$, $\mu_k \in \mathbb{R}$, $k = \pm 1, \pm 2, \dots$, be an almost periodic sequence and $\{t_k\} \in UAPS$, be uniformly almost periodic. Then the function $\varphi(t) = \mu_k$, $t_k \leq t < t_{k+1}$ is almost periodic.

Now we shall consider some properties of almost periodic functions.

Theorem 1.15. *Every almost periodic function is bounded on the real line.*

Proof. Let the function $\varphi \in PC[\mathbb{R}, \mathbb{R}^n]$ be almost periodic, and let the dense coefficient of the set $\overline{T}_1$ be $l = l(1)$. This means, that between the integer numbers k and $k + l$ there exists an integer number p such that

$$|t_k^p - \tau| < \varepsilon.$$

Let

$$M = \max_{0 \le t \le l} ||\varphi(t)||, \quad ||\varphi(t') - \varphi(t'')|| \le M_1, \quad M_1 > 0$$

for $|t' - t''| \le 1$, where t', t'' belong to one and the same interval of continuity of $\varphi(t)$. From Definition 1.12 it follows that for all $t \in \mathbb{R}$ and $|t - t_k| > 1$, there exists an 1-almost period r such that $t + r \in [0, l]$ and $||\varphi(t+r) - \varphi(t)|| \le 1$. Now, for all $t \in \mathbb{R}$, we have $||\varphi(t)|| < M + M_1 + 1$, and the proof is complete. $\qquad\square$

Theorem 1.16. *If $\varphi \in PC[\mathbb{R}, \mathbb{R}^n]$ is an almost periodic function, then for any $\varepsilon > 0$ there exists a relative dense set of intervals with a fixed length γ, $0 < \gamma < \varepsilon$ which contains $\varepsilon-$almost periods of the function $\varphi(t)$.*

Proof. Let Γ be the set of all $\frac{\varepsilon}{2}$-almost periods of the function $\varphi(t)$, and let l be the dense coefficient of Γ. Let the number $\frac{\gamma}{2} = \delta(\frac{\varepsilon}{2})$ is defined by the uniform continuity of $\varphi(t)$. Hence, if t', t'' belong to one and the same interval of continuity of the function $\varphi(t)$ and $|t' - t''| < \frac{\gamma}{2}$, then

$$||\varphi(t') - \varphi(t'')|| < \frac{\varepsilon}{2}.$$

Let for the simplicity $\gamma < \frac{\varepsilon}{2}$. Set $L = l + \gamma$, and consider an arbitrary interval $[a, a + L]$. From the definition of the almost period of a function, it follows that there exists an $\frac{\varepsilon}{2}$-almost period $r \in [a + \frac{\gamma}{2}, a + \frac{\gamma}{2} + l]$, such that $[r - \frac{\gamma}{2}, r + \frac{\gamma}{2}] \subset [a, a + L]$.

Let now ξ be an arbitrary number from the interval $[r - \frac{\gamma}{2}, r + \frac{\gamma}{2}]$. Then, from the inequality $|\xi - r| < \delta$, if $t' = t - r + \xi$, $t \in \mathbb{R}$ is such that $|t - t_k| > \varepsilon$, we get $|t' - t_k| > \frac{\varepsilon}{2}$ and

$$||\varphi(t + \xi) - \varphi(t)|| \le ||\varphi(t' + r) - \varphi(t')|| + ||\varphi(t') - \varphi(t)||$$

$$< \frac{\varepsilon}{2} + \frac{\varepsilon}{2} = \varepsilon. \qquad\qquad\square$$

Theorem 1.17. *Let $\varphi \in PC[\mathbb{R}, \mathbb{R}^n]$ be an almost periodic function with a range $Y \subset \mathbb{R}^n$. If the function $F(y)$ is uniformly continuous with a domain Y, then the function $F(\varphi(t))$ is almost periodic.*

Proof. The proof is trivial and we omit the details in this book. $\qquad\square$

Theorem 1.18. *For every two almost periodic functions with points of discontinuity from the sequence* $\{t_k\} \in UAPS$ *and for arbitrary* $\varepsilon > 0$, *there exists a relatively dense set of their common* ε-almost periods.

Proof. Let $\varphi_1(t)$ and $\varphi_2(t)$ be almost periodic functions with a common sequence $\{t_k\} \in UAPS$ of points of discontinuity. From Theorem 1.16 it follows, that there exist numbers l_1 and l_2, such that each of the intervals $[a, a + l_1]$ and $[a, a + l_2]$ will contain corresponding $\frac{\varepsilon}{4}$-almost periods r_1 and r_2, which are factors of the number γ, $0 < \gamma < \frac{\varepsilon}{4}$. If we set $l = \max\{l_1, l_2\}$, then there exist a pair of $\frac{\varepsilon}{4}$-almost periods $r_1 = n'\gamma$ and $r_2 = n''\gamma$ for every interval of the form $[a, a + l]$, where n' and n'' are integer numbers.

Since $r_1 - r_2 = (n' - n'')\gamma = n\gamma$ and $|n\gamma| \leq l$, then the number $n\gamma$ takes only finite values, $n_1\gamma, \ n_2\gamma, \ldots, n_p\gamma$. Let they are represented by the pair of numbers $(r_1^1, r_2^1), (r_1^2, r_2^2), \ldots, (r_1^p, r_2^p)$ such that $r_1^s - r_2^s = n_s\gamma$, $s = 1, 2, \ldots, p$. Set $\max\limits_s |r_1^s| = A$ and let $[a, a + l + 2A]$ be an arbitrary interval with length $l + 2A$. We choice in the interval $[a + A, a + l + A]$ two $\frac{\varepsilon}{4}$-almost periods $r_1 = n'\gamma$ and $r_2 = n''\gamma$ of the functions $\varphi_1(t)$ and $\varphi_2(t)$, respectively, and let $r_1 - r_2 = n_s\gamma = r_1^s - r_2^s$. Then, we have

$$r = r_1 - r_1^s > r_2 - r_2^s, \ r \in [a, a + l + 2A]. \tag{1.29}$$

All numbers defined by (1.29), which are factors of γ, form a relatively dense set of numbers $\overline{T}$. Now we shall show, that there exists a relatively dense set $\overline{T}_0 \subset \overline{T}$, such that $r_0 \in \overline{T}_0$ and $|t - t_k| > \varepsilon$, $k = \pm 1, \pm 2, \ldots$ imply $|t + r - t_k| > \frac{\varepsilon}{2}$.

Let $l = l(\varepsilon)$ be the dense coefficient of Γ and let $l' = l'(\varepsilon)$ be the dense coefficient of Γ', Γ' is defined by Lemma 1.3, for $\frac{\varepsilon}{4}$. Apparently, the number γ can be choice so that $l < +\infty$, $l' < +\infty$. Set $l'' = \max(l, l')$. Then, for every interval $[a, a + l'']$, there exist integer numbers m, m' and q, such that $m\gamma, m\gamma' \in [a, a + l'']$ and

$$|t_i^q - m\gamma| < \frac{\varepsilon}{4}, \ \|\varphi_j(t + m'\gamma) - \varphi_j(t)\| < \frac{\varepsilon}{2}, \tag{1.30}$$

$j = 1, 2$, $i = \pm 1, \pm 2, \ldots$.

The differences $m - m'$ can take only finite numbers of values n_s, $s = 1, 2, \ldots, p$. For every n_s, there exist triples (m_s, m_s', q), which represent the class defined by the number n_s. Let $\lambda = \max\limits_{1 \leq s \leq p} |m_s'\gamma|$ and for the intervals $I = [a, a + l'' + 2\lambda]$, $I' = [a + \lambda, a + \lambda + l'']$, $I' \subset I$, there exist integer numbers m', m, q for which (1.30) holds and $m\gamma$, $m'\gamma \in A$.

Let now $m\gamma - m'\gamma = n_s\gamma$, or $m\gamma - m'\gamma = m_s\gamma - m_s'\gamma$, and $m - m_s = m' - m_s'$. If $r = (m - m_s)\gamma$, $h = q - q_s$, then $r \in A$ and for $i = \pm 1, \pm 2, \ldots$ from (1.26) it follows that

$$|t_i^h - r| = |t_i^{q-q_s} - r| = |t_{i-q-q_s}^{q-q_s} - r| = |t_{i-q_s}^q - t_{i-q_s}^{q_s} - m\gamma + m_s\gamma|$$

$$\leq |t_{i-q_s}^q - m\gamma| + |t_{i-q_s}^{q_s} - m_s\gamma| < \frac{\varepsilon}{4} + \frac{\varepsilon}{4} = \frac{\varepsilon}{2}.$$

Now, let $|t - t_k| > \varepsilon$ and $t_i + \varepsilon < t < t_{i+1} - \varepsilon$. Then $t_i + r + \varepsilon < t + r < t_{i+1} + r - \varepsilon$. Since $|t_k^h - r| < \frac{\varepsilon}{2}$ for $t = 0, \pm 1, \pm 2, \ldots$, then $t_k^h - \frac{\varepsilon}{2} < r < t_k^h + \frac{\varepsilon}{2}$, and hence,

$$t_{i+h} - t_i + t_i + \varepsilon - \frac{\varepsilon}{2} < t + r < t_{i+1} + t_{i+h+1} - t_{i+1} + \frac{\varepsilon}{2} - \varepsilon, t_{i+h}$$

$$+ \frac{\varepsilon}{2} < t + r < t_{i+h+1} - \frac{\varepsilon}{2},$$

i.e.

$$|t + r - t_k| > \frac{\varepsilon}{2}, \ k = \pm 1, \pm 2, \ldots.$$

Consequently, the set $\overline{T}_0$ is not empty, and it is a relatively dense set on $\mathbb{R}$. For $j = 1, 2$ and $r \in \Gamma_0$, we get

$$||\varphi_j(t + r) - \varphi_k(t)|| = ||\varphi_j(t + (m' - m_s')\gamma) - \varphi(t)||$$

$$\leq ||\varphi_j(t + (m' - m_s')\gamma) - \varphi_j(t + m'\gamma)||$$

$$+ ||\varphi_j(t) - \varphi_j(t + m'\gamma)|| < \frac{\varepsilon}{2} + \frac{\varepsilon}{2} = \varepsilon. \qquad \square$$

The proof of the next theorem is similar to the proof in the continuous case.

Theorem 1.19. *The sum of two almost periodic functions with points of discontinuity t_k, $k = \pm 1, \pm 2, \ldots$, $\{t_k\} \in UAPS$, is an almost periodic function.*

Theorem 1.20. *The quotient $\dfrac{\varphi(t)}{\psi(t)}$ between two almost periodic functions with points of discontinuity t_k, $k = \pm 1, \pm 2, \ldots$, $\{t_k\} \in UAPS$, is an almost periodic function if*

$$\inf_{t \in \mathbb{R}} ||\psi(t)|| > 0.$$

Now, let we consider the following system of impulsive differential equations

$$\begin{cases} \dot{x} = A(t)x + f(t), \ t \neq t_k, \\ \Delta x(t_k) = B_k x(t_k) + I_k(x(t_k)), \ k = \pm 1, \pm 2, \ldots, \end{cases} \qquad (1.31)$$

where the $A : \mathbb{R} \to \mathbb{R}^{n \times n}$ is an almost periodic matrix in the sense of Bohr.

Lemma 1.7. *Let the following conditions hold:*

1. $U(t, s)$ *is the fundamental matrix of the linear part of (1.31).*
2. $f(t)$ *is an almost periodic function.*
3. *The sequence of functions $\{I_k\}$ and the sequence of matrices $\{B_k\}$ are almost periodic.*
4. *The set of sequences $\{t_k^j\}$ is uniformly almost periodic.*

Then for every $\varepsilon > 0$ and every $\theta > 0$, there exist ε_1, $0 < \varepsilon_1 < \varepsilon$, a relatively dense set $\overline{T}$ of real numbers and a set P of integer numbers, such that the following relations hold:

(a) $\|U(t+\tau, s+\tau) - U(t, s)\| < \varepsilon$, $t \in \mathbb{R}$, $\tau \in \overline{T}$, $0 \le t - s \le \theta$.
(b) $\|f(t+\tau) - f(t)\| < \varepsilon$, $t \in \mathbb{R}$, $\tau \in \overline{T}$, $|t - t_k| > \varepsilon$, $k = \pm 1, \pm 2, \dots$.
(c) $\|B_{k+q} - B_k\| < \varepsilon$, $q \in P$, $k = \pm 1, \pm 2, \dots$.
(d) $\|I_{k+q} - I_k\| < \varepsilon$, $q \in P$, $k = \pm 1, \pm 2, \dots$.
(e) $|t_k^q - \tau| < \varepsilon_1$, $q \in P$, $\tau \in \overline{T}$, $k = \pm 1, \pm 2, \dots$.

Proof. From Theorem 1.15, Theorem 1.18 and [47] it follows that there exists a relatively dense set containing the common almost periods of the function $f(t)$ and matrix $A(t)$, which are the factors of the number ε_2, $0 < \varepsilon_2 < \varepsilon$, such that

$$\|A(t+\tau) - A(t)\| < \frac{\varepsilon}{2\theta} e^{-m\theta}, \quad \|f(t+\tau) - f(t)\| < \frac{\varepsilon}{2},$$

where $\tau = n\varepsilon_2$, $|t - t_k| > \varepsilon$, $k = \pm 1, \pm 2, \dots$.

From Theorem 1.18, we have that, there exist a relatively dense set $\overline{T}$ of real numbers and a set P of integer numbers, such that

$$\|A(t+\tau) - A(t)\| < \frac{\varepsilon}{2\theta} e^{-m\theta}, \quad \|f(t+\tau) - f(t)\| < \frac{\varepsilon}{2},$$

where $|t - t_i| > \varepsilon$, $|t_i^p - \tau| < \frac{\varepsilon_1}{2}$, $0 < \varepsilon_1 < \varepsilon$, $i = \pm 1, \pm 2, \dots$.

On the other hand, for the sequences $\{I_k\}$ and $\{B_k\}$ there exists a relatively dense sets of their common almost periods. Then there exists a natural number N between the neighbours factors of the numbers p and q, such that

$$|t_i^p - m\varepsilon_1| < \frac{\varepsilon_1}{2}, \quad \|I_{i+q} - I_i\| < \frac{\varepsilon}{2}, \quad \|B_{i+q} - B_i\| < \frac{\varepsilon}{2}, \quad i = \pm 1, \pm 2, \dots. \quad (1.32)$$

The difference $p - q$ takes only a finite number of values n_k, $k = 1, 2, \dots, r$. Then, for every n_k we consider the pair of integer numbers (p^k, q^k), for which the inequalities (1.32) hold. Set $M = \max_k |p^k|$, and let $l + 1, l + 2, \dots, l + N + 2M$ be $N + 2M$ arbitrary integer numbers. For any N integer numbers $l + M + 1, \dots, l + M + N$, there exists a pair (p, q), such that the inequalities (1.32) hold, and let $p - q = n_j$. Then, $p - p^j = q - q^j = \mu$, and μ is one of the integer numbers $l + 1, l + 2, \dots, l + N + 2M$.

On the other hand, for every integer number $k = \pm 1, \pm 2, \ldots$, we have

$$||I_{i+\mu} - I_i|| \le ||I_{i+p^j-p} - I_{i+p^j}|| + ||I_{i+p^j} - I_i|| < \frac{\varepsilon}{2} + \frac{\varepsilon}{2} = \varepsilon.$$

Analogously, we get $||B_{i+\mu} - B_i|| < \varepsilon$.

Now, let τ and τ_j are almost periods, which are cofactors of ε_2, corresponding to the pairs (p, q) and (p^j, q^j).

If we note $\tau_\mu = \tau - \tau_j$, then

$$||A(t + \tau_\mu) - A(t)|| \le ||A(t + \tau_\mu) - A(t - \tau_j)|| + ||A(t - \tau_j) - A(t)||$$
$$< \frac{\varepsilon}{2\theta}e^{-m\theta} + \frac{\varepsilon}{2\theta}e^{-m\theta} = \frac{\varepsilon}{\theta}e^{-m\theta}. \tag{1.33}$$

Since $t_k^j - t_k^i = t_{k+i}^{j-i}$, then

$$|t_i^\mu - \tau_\mu| \le |t_{i-p^j}^p - \tau| + |t_{i-p^j}^{p^j} - \tau_j| < \frac{\varepsilon_1}{2} + \frac{\varepsilon_1}{2} = \varepsilon_1.$$

On the other hand, from $|t_i^p - \tau| < \varepsilon_1 < \frac{\varepsilon}{2}$ and $|t - t_k| > \varepsilon$, we get

$$|t - \tau - t_k| > \frac{\varepsilon}{2}, \ k = \pm 1, \pm 2, \ldots.$$

Indeed, let $t_k + \varepsilon < t < t_{k+1} - \varepsilon$. Then, we have $t_k + \tau + \varepsilon < t + \tau < t_{k+1} - \varepsilon + \tau$, and $t_{k+p} < t_k + \tau + \varepsilon_1$, $t_{k+1} + \tau - \varepsilon_1 < t_{k+p+1}$. Hence $t_{i+p} + \frac{\varepsilon}{2} < t + \tau < t_{k+p+1} - \frac{\varepsilon}{2}$, or $|t + \tau - t_k| > \frac{\varepsilon}{2}$. Now, we have

$$||f(t + \tau_\mu) - f(t)|| \le ||f(t + \tau_\mu) - f(t + \tau)||$$
$$+ ||f(t + \tau) - f(t)|| < \frac{\varepsilon}{2} + \frac{\varepsilon}{2} = \varepsilon. \qquad \square$$

We shall consider the following definition for almost periodic piecewise continuous functions.

Let $T, P \in \mathcal{B}$, and let $s(T \cup P) : \mathcal{B} \to \mathcal{B}$ be a map such that the set $s(T \cup P)$ forms a strictly increasing sequence. For $D \subset \mathbb{R}$ and $\varepsilon > 0$, we introduce the notations $\theta_\varepsilon(D) = \{t + \varepsilon, \ t \in D\}$, $F_\varepsilon(D) = \cap \{\theta_\varepsilon(D)$.

By $\phi = (\varphi(t), T)$ we denote an element from the space $PC[\mathbb{R}, \mathbb{R}^n] \times \mathcal{B}$ and for every sequence of real number $\{s_n\}, n = 1, 2, \ldots$ with $\theta_{s_n}\phi$, we shall consider the sets $\{\varphi(t + s_n), T - s_n\} \subset PC \times \mathcal{B}$, where

$$T - s_n = \{t_k - s_n, \ k = \pm 1, \pm 2, \ldots, \ n = 1, 2, \ldots\}.$$

Definition 1.13. The sequence $\{\phi_n\}$, $\phi_n = (\varphi_n(t), T_n) \in PC[\mathbb{R}, \mathbb{R}^n] \times \mathcal{B}$ is convergent to ϕ, $\phi = (\varphi(t), T)$, $(\varphi(t), T) \in PC[\mathbb{R}, \mathbb{R}^n] \times \mathcal{B}$, if and only if, for any $\varepsilon > 0$ there exists $n_0 > 0$ such that $n \ge n_0$ implies

$$\rho(T, T_n) < \varepsilon, \ ||\varphi_n(t) - \varphi(t)|| < \varepsilon$$

uniformly for $t \in \mathbb{R} \setminus F_\varepsilon(s(T_n \cup T))$, $\rho(.,.)$ is an arbitrary distance in $\mathcal{B}$.

Definition 1.14. The function $\varphi \in PC[\mathbb{R}, \mathbb{R}^n]$ is said to be an *almost periodic piecewise continuous function* with points of discontinuity of the first kind from the set $T \in \mathcal{B}$, if for every sequence of real numbers $\{s'_m\}$ there exists a subsequence $\{s_n\}$, $s_n = s'_{m_n}$, such that $\theta_{s_n}\phi$ is uniformly convergent on $PC[\mathbb{R}, \mathbb{R}^n] \times \mathcal{B}$.

Now, let $\Omega \subseteq \mathbb{R}^n$ and consider the impulsive differential system (1.10). Introduce the following conditions:

H1.3. The function $f(t, x)$ is almost periodic in t uniformly with respect to $x \in \Omega$.

H1.4. The sequence $\{I_k(x)\}$, $k = \pm 1, \pm 2, \dots$ is almost periodic uniformly with respect to $x \in \Omega$.

H1.5. The set of sequences $\{t_k\} \in UAPS$.

Let the assumptions H1.3–H1.5 hold, and let $\{s'_m\}$ be an arbitrary sequence of real numbers. Then there exists a subsequence $\{s_n\}$, $s_n = s_{m_n}'$, such that the sequence $\{f(t + s_n, x)\}$ is convergent uniformly to the function $\{f^s(t, x)\}$, and from Lemma 1.6 it follows that the set of sequences $\{t_k - s_n\}$, $k = \pm 1, \pm 2, \dots$ is convergent to the sequence t_k^s uniformly with respect to $k = \pm 1, \pm 2, \dots$ as $n \to \infty$.

By $\{k_{n_i}\}$ we denote the sequence of integers, such that the subsequence $\{t_{k_{n_i}}\}$ is convergent to the sequence t_k^s uniformly with respect to $k = \pm 1, \pm 2, \dots$ as $i \to \infty$.

Then, for every sequence $\{s'_m\}$, the system (1.10) "moves" to the system

$$\begin{cases} \dot{x} = f^s(t, x), \ t \neq t_k^s, \\ \Delta x(t_k^s) = I_k^s(x(t_k^s)), \ k = \pm 1, \pm 2, \dots. \end{cases} \tag{1.34}$$

Remark 1.4. In many papers, the limiting systems (1.33) are called *Hull* of the system (1.10), and is denoted by $H(f, I_k, t_k)$.

In Chap. 3, we shall use the next class of piecewise Lyapunov functions, connected with the system (1.33).

Definition 1.15. A function $W : \mathbb{R} \times \Omega \to \mathbb{R}^+$ belongs to class W_0, if:

1. The function $W(t, x)$ is continuous on $(t, x) \in \mathbb{R} \times \Omega$, $t \neq t_k^s$, $k = \pm 1, \pm 2, \dots$ and $W(t, 0) = 0$, $t \in \mathbb{R}$.
2. The function $W(t, x)$ is locally Lipschitz continuous with respect to its second argument.

3. For each $k = \pm 1, \pm 2, \ldots$ and $x \in \Omega$ there exist the finite limits

$$W(t_k^{s-}, x) = \lim_{\substack{t \to t_k^s \\ t < t_k^s}} W(t, x), \quad W(t_k^{s+}, x) = \lim_{\substack{t \to t_k^s \\ t > t_k^s}} W(t, x)$$

and the equality $W(t_k^{s-}, x) = W(t_k^s, x)$ holds.

Let the function $W \in W_0$ and $x \in PC[\mathbb{R}, \Omega]$. We shall use the upper right-hand Dini derivative

$$D^+ W(t, x(t)) = \lim_{\delta \to 0^+} \sup \frac{1}{\delta} \left[W(t + \delta, x(t) + \delta f^s(t, x(t))) - W(t, x(t)) \right].$$

Chapter 2
Almost Periodic Solutions

In the present chapter, we shall state some basic existence and uniqueness results for almost periodic solutions of impulsive differential equations. Applications to real world problems will also be discussed.

Section 2.1 will offer the existence and uniqueness theorems for almost periodic solutions of hyperbolic impulsive differential equations.

In Sect. 2.2, using weakly non-linear integro-differential systems, the existence and exponential stability of almost periodic solutions of impulsive integro-differential equations will be discussed.

In Sect. 2.3, we shall study the existence of almost periodic solutions for forced perturbed impulsive differential equations. The example here, will state the existence criteria for impulsive differential equations of Lienard's type.

Section 2.4 will deal with sufficient conditions for the existence of almost periodic solutions of impulsive differential equations with perturbations in the linear part.

In Sect. 2.5, we shall consider the strong stability and almost periodicity of solutions of impulsive differential equations with fixed moments of impulse effect. The investigations are carried out by means of piecewise continuous Lyapunov functions.

Section 2.6 is devoted to the problem of the existence of almost periodic projektor-valued functions for dichotomous impulsive differential systems.

In Sect. 2.7, we shall investigate separated solutions of impulsive differential equations with variable impulsive perturbations and we shall give sufficient conditions for almost periodicity of these solutions.

Finally, in Sect. 2.8, the existence results for almost periodic solutions of abstract differential equations in Banach space will be given. Applications for impulsive predator–prey systems with diffusion will be considered.

G.T. Stamov, *Almost Periodic Solutions of Impulsive Differential Equations,*
Lecture Notes in Mathematics 2047, DOI 10.1007/978-3-642-27546-3_2,
© Springer-Verlag Berlin Heidelberg 2012

2.1 Hyperbolic Impulsive Differential Equations

In this paragraph, we shall consider the following systems of impulsive differential equations with impulses at fixed moments

$$\begin{cases} \dot{z} = A(t)z + f(t), \ t \neq t_k, \\ \Delta z(t_k) = b_k, \ k = \pm 1, \pm 2, \ldots, \end{cases} \tag{2.1}$$

and

$$\begin{cases} \dot{z} = A(t)z + F(t,z), \ t \neq t_k, \\ \Delta z(t_k) = I_k(z(t_k)), \ k = \pm 1, \pm 2, \ldots, \end{cases} \tag{2.2}$$

where $t \in \mathbb{R}$, $\{t_k\} \in \mathcal{B}$, $A : \mathbb{R} \to \mathbb{R}^{n \times n}$, $f : \mathbb{R} \to \mathbb{R}^n$, $b_k \in \mathbb{R}^n$, $F : \mathbb{R} \times \Omega \to \mathbb{R}^n$, $I_k : \Omega \to \mathbb{R}^n$.

By $z(t) = z(t; t_0, z_0)$, we denote the solution of (2.1) or (2.2) with initial condition $z(t_0^+) = z_0$, $t_0 \in \mathbb{R}$, $z_0 \in \mathbb{R}^n$. Together with the systems (2.1) and (2.2), we shall consider the corresponding homogeneous system

$$\dot{z} = A(t)z. \tag{2.3}$$

Definition 2.1 ([71]). The system (2.3) is said to be *hyperbolic*, if there exist constants $\alpha > 0$, $\lambda > 0$ and for each $t \in \mathbb{R}$ there exist linear spaces $M^+(t)$, and $M^-(t)$, whose external direct sum is $M^+(t) \oplus M^-(t) = \mathbb{R}^n$, such that if $z_0 \in M^+(t_0)$, then for all $t \geq t_0$ the inequality

$$||z(t; t_0, z_0)|| \leq a||z_0||e^{-\lambda(t-t_0)},$$

holds true, while if $z_0 \in M^-(t_0)$ then for all $t \leq t_0$, we have

$$||z(t; t_0, z_0)|| \leq a||z_0||e^{\lambda(t-t_0)}.$$

In this part, we shall investigate the existence of almost periodic solutions of systems (2.1) and (2.2), assuming that the corresponding homogeneous system is hyperbolic.

Introduce the following conditions:

H2.1.   The matrix function $A \in C[\mathbb{R}, \mathbb{R}^{n \times n}]$ is almost periodic in the sense of Bohr.

H2.2.   The function $f \in PC[\mathbb{R}, \mathbb{R}^n]$ is almost periodic.

H2.3.   The sequence $\{b_k\}$, $k = \pm 1, \pm 2, \ldots$, is almost periodic.

H2.4.   The set of sequences $\{t_k^j\}$, $t_k^j = t_{k+j} - t_k$, $k = \pm 1, \pm 2, \ldots$, $j = \pm 1, \pm 2, \ldots$, is uniformly almost periodic, and $\inf_k t_k^1 = \theta > 0$.

H2.5.   The function $F \in C[\mathbb{R} \times \Omega, \mathbb{R}^n]$ is almost periodic with respect to t uniformly in $z \in \Omega$.

H2.6.   The sequence of functions $\{I_k(x)\}$, $I_k \in C[\Omega, \mathbb{R}^n]$, $k = \pm 1, \pm 2, \ldots$, is almost periodic with respect to k uniformly in $z \in \Omega$.

We shall use the following lemmas:

Lemma 2.1. *Let conditions H2.1–H2.4 hold. Then for each $\varepsilon > 0$ there exist ε_1, $0 < \varepsilon_1 < \varepsilon$, a relatively dense set $\overline{T}$ of real numbers, and a set P of integer numbers, such that the following relations are fulfilled:*

(a) $\|A(t+\tau) - A(t)\| < \varepsilon$, $t \in \mathbb{R}$, $\tau \in \overline{T}$.
(b) $\|f(t+\tau) - f(t)\| < \varepsilon$, $t \in \mathbb{R}$, $\tau \in \overline{T}$.
(c) $\|b_{k+q} - b_k\| < \varepsilon$, $q \in P$, $k = \pm 1, \pm 2, \ldots$.
(d) $|t_k^q - \tau| < \varepsilon_1$, $q \in P$, $\tau \in \overline{T}$, $k = \pm 1, \pm 2, \ldots$.

The proof of Lemma 2.1 is analogous to the proof of Lemma 1.7.

Lemma 2.2. *Let the system (2.3) is hyperbolic and the condition H2.1 holds. Then there exists a non-singular transformation, defined by almost periodic matrix $S(t)$, $S \in C[\mathbb{R}, \mathbb{R}^{n \times n}]$, which reduces the system (2.1.3) into the next ones*

$$\dot{x} = Q^+(t)x \tag{2.4}$$

and

$$\dot{y} = Q^-(t)y \tag{2.5}$$

where $x \in \mathbb{R}^k$, $y \in \mathbb{R}^{n-k}$, $Q^+ \in C[\mathbb{R}, \mathbb{R}^{k \times k}]$, $Q^- \in C[\mathbb{R}, \mathbb{R}^{(n-k) \times (n-k)}]$ and the following assertions hold true:

1. *$Q^+(t)$ and $Q^-(t)$ are almost periodic matrix-valued functions.*
2. *If $\Phi^+(t, s)$ and $\Phi^-(t, s)$ are the corresponding fundamental matrices of the systems (2.4) and (2.5), then the following inequalities hold true:*

$$\|\Phi^+(t, s)\| \le \overline{a} e^{-\lambda(t-s)}, \ t \ge s, \tag{2.6}$$

$$\|\Phi^-(t, s)\| \le \overline{a} e^{\lambda(t-s)}, \ t \le s, \tag{2.7}$$

   where $s, t \in \mathbb{R}$, $\overline{a} > 0$.
3. *For each $\varepsilon > 0$, $t \in \mathbb{R}$, $s \in \mathbb{R}$ there exists relatively dense set $\overline{T}$ of ε–almost periods, such that for each $\tau \in \overline{T}$, fundamental matrices $\Phi^+(t, s)$ and $\Phi^-(t, s)$ satisfy the inequalities*

$$\|\Phi^+(t+\tau, s+\tau) - \Phi^+(t, s)\| \le \varepsilon K e^{-\frac{\lambda}{2}(t-s)}, \ t \ge s, \tag{2.8}$$

$$\|\Phi^-(t+\tau, s+\tau) - \Phi^-(t, s)\| \le \varepsilon K e^{\frac{\lambda}{2}(t-s)}, \ t \le s, \tag{2.9}$$

   where $\lambda > 0$, $K > 0$.

Proof. Assertions 1 and 2 are immediate consequences of Theorem 1 in [71]. In fact, following the ideas used in [71], we define the matrix $S(t)$ to be formed by the vector-columns, which are solutions of (2.3). It follows from the condition H2.1, that $S(t)$ consists of almost periodic functions. On the other hand, the transformation $z = S(t)u$ rewrites (2.3) in the form

$$\dot{u} = Q(t)u,$$

where

$$Q(t) = S^{-1}(t)\Big(A(t)S(t) - \dot{S}(t)\Big).$$

Hence, $Q(t)$ is an almost periodic function. The estimates (2.6) and (2.7) are direct consequences of Theorem 1 in [71].

To prove Assertion 3, let $\Phi^+(t,s)$ and $\Phi^-(t,s)$ be the fundamental matrices of systems (2.4) and (2.5), respectively. Then for each $\varepsilon > 0$ the following relations hold true

$$\frac{\partial \Phi^+(t,s)}{\partial t} = Q^+(t)\Phi^+(t+\tau, s+\tau) + \Big(Q^+(t+\tau) - Q^+(t)\Big)\Phi^+(t+\tau, s+\tau),$$

$$\frac{\partial \Phi^-(t,s)}{\partial t} = Q^-(t)\Phi^-(t+\tau, s+\tau) + \Big(Q^-(t+\tau) - Q^-(t)\Big)\Phi^-(t+\tau, s+\tau)$$

and

$$\Phi^+(t+\tau, s+\tau) = \Phi^+(t,s)$$
$$+ \int_s^t \Phi^+(t,v)\Big(Q^+(v+\tau) - Q^+(v)\Big)\Phi^+(v+\tau, s+\tau)dv,$$
$$\Phi^-(t+\tau, s+\tau) = \Phi^-(t,s)$$
$$+ \int_s^t \Phi^-(t,v)\Big(Q^-(v+\tau) - Q^-(v)\Big)\Phi^-(v+\tau, s+\tau)dv.$$

Therefore,

$$\|\Phi^+(t+\tau, s+\tau) - \Phi^+(t,s)\| \le \int_s^t \|\Phi^+(t,v)\|\|(Q^+(v+\tau)$$
$$- Q^+(v)\|\|\Phi^+(v+\tau, s+\tau)\|dv,$$
$$\|\Phi^-(t+\tau, s+\tau) - \Phi^-(t,s)\| \le \int_s^t \|\Phi^-(t,v)\|\|Q^-(v+\tau)$$
$$- Q^-(v)\|\|\Phi^-(v+\tau, s+\tau)\|dv.$$

It follows from (2.6), that

$$||\Phi^+(t+\tau, s+\tau) - \Phi^+(t,s)|| \leq \varepsilon K e^{-\frac{\lambda}{2}(t-s)}, \ t \geq s,$$

where in this case $K = (\bar{a})^2$.

The proof of (2.9) is analogous. □

From Lemma 2.2 it follows that, by a transformation with the matrix $S(t)$, system (2.1) takes on the form

$$\begin{cases} \dot{x} = Q^+(t)x + f^+(t), \ t \neq t_k, \\ \Delta x(t_k) = b_k^+, \ k = \pm 1, \pm 2, \ldots, \\ \dot{y} = Q^-(t)y + f^-(t), \ t \neq t_k, \\ \Delta y(t_k) = b_k^-, \ k = \pm 1, \pm 2, \ldots, \end{cases} \tag{2.10}$$

where $x \in \mathbb{R}^k$, $y \in \mathbb{R}^{n-k}$, $f^+ : \mathbb{R} \to \mathbb{R}^k$, $f^- : \mathbb{R} \to \mathbb{R}^{n-k}$, b_k^+ and b_k^- are k and $n-k$-dimensional constant vectors, respectively.

In an analogous way, the system (2.2) after a transformation with the matrix $S(t)$, goes to the form

$$\begin{cases} \dot{x} = Q^+(t)x + F^+(t,x,y), \ t \neq t_k, \\ \Delta x(t_k) = I_k^+(x(t_k), y(t_k)), \ k = \pm 1, \pm 2, \ldots, \\ \dot{y} = Q^-(t)y + F^-(t,x,y), \ t \neq t_k, \\ \Delta y(t_k) = I_k^-(x(t_k), y(t_k)), \ k = \pm 1, \pm 2, \ldots, \end{cases} \tag{2.11}$$

where $F^+ : \mathbb{R} \times \mathbb{R}^k \times \mathbb{R}^{n-k} \to \mathbb{R}^k$, $F^- : \mathbb{R} \times \mathbb{R}^k \times \mathbb{R}^{n-k} \to \mathbb{R}^{n-k}$, and $I_k^+ : \mathbb{R}^k \times \mathbb{R}^{n-k} \to \mathbb{R}^k$, $I_k^- : \mathbb{R}^k \times \mathbb{R}^{n-k} \to \mathbb{R}^{n-k}$.

Theorem 2.1. *Let the following conditions hold:*

1. *Conditions H2.1–H2.4 hold.*
2. *The system (2.3) is hyperbolic.*

Then for the system (2.1) there exists a unique almost periodic solution, which is exponentially stable.

Proof. We consider the following equations

$$x(t) = \int_{-\infty}^t \Phi^+(t,s)f^+(s)ds + \sum_{t_k < t} \Phi^+(t, t_k)b_k^+,$$

$$y(t) = -\int_t^\infty \Phi^-(t,s)f^-(s)ds + \sum_{t_k > t} \Phi^-(t, t_k)b_k^-,$$

which are equivalent to the (2.10).

Let $\varepsilon > 0$ be an arbitrary chosen constant. It follows from Lemma 2.1 that there exist sets $\overline{T}$ and P such that for each $\tau \in \overline{T}$ and $q \in P$, the following estimates hold true:

$$\|x(t+\tau) - x(t)\| = \int_{-\infty}^{t} \|\Phi^+(t+\tau, s+\tau) - \Phi^+(t,s)\| \|f^+(s+\tau)\| ds$$

$$+ \int_{-\infty}^{t} \|\Phi^+(t,s)\| \|f^+(s+\tau) - f^+(s)\| ds$$

$$+ \sum_{t_k < t} \|\Phi^+(t+\tau, t_{k+q}) - \Phi^+(t, t_k)\| \|b_{k+q}^+\|$$

$$+ \sum_{t_k < t} \|\Phi^+(t, t_k)\| \|b_{k+q}^+ - b_k^+\|, \tag{2.12}$$

and

$$\|y(t+\tau) - y(t)\| = \int_{t}^{\infty} \|\Phi^-(t+\tau, s+\tau) - \Phi^-(t,s)\| \|f^-(s+\tau)\| ds$$

$$+ \int_{t}^{\infty} \|\Phi^-(t,s)\| \|f^-(s+\tau) - f^-(s)\| ds$$

$$+ \sum_{t > t_k} \|\Phi^-(t+\tau, t_{k+q}) - \Phi^-(t, t_k)\| \|b_{k+q}^-\|$$

$$+ \sum_{t > t_k} \|\Phi^-(t, t_k)\| \|b_{k+q}^- - b_k^-\|. \tag{2.13}$$

From Lemma 2.2, (2.12) and (2.13), we have

$$\|x(t+\tau) - x(t)\| \le K_1 \varepsilon, \tag{2.14}$$

where

$$K_1 = \frac{2K}{\lambda} \sup_{t \in \mathbb{R}} \|f^+(t)\| + \frac{\overline{a}}{\lambda} + \frac{2N\overline{a}}{1 - e^{-\frac{\lambda}{2}}} \sup_{k=\pm 1, \pm 2, \ldots} \|b_k^+\| + \frac{2N\overline{a}}{1 - e^{-\lambda}}.$$

In the same manner, we obtain

$$\|y(t+\tau) - y(t)\| \le K_2 \varepsilon, \tag{2.15}$$

where

$$K_2 = \frac{2K}{\lambda} \sup_{t \in \mathbb{R}} \|f^-(t)\| + \frac{\overline{a}}{\lambda} + \frac{2N\overline{a}}{1 - e^{-\frac{\lambda}{2}}} \sup_{k=\pm 1, \pm 2, \ldots} \|b_k^-\| + \frac{2N\overline{a}}{1 - e^{-\lambda}}.$$

The number N, which is defined in the last inequalities, is from Lemma 1.2 Now, from (2.14) and (2.15), we conclude that the solution $z(t) = (x(t), y(t))$ of system (2.1) is almost periodic.

On the other hand, each solution $(x(t), y(t))$ of (2.1) can be written in the form

$$x(t) = \Phi^+(t, t_0)\chi + \int_{t_0}^t \Phi^+(t, s)f^+(s)ds + \sum_{s < t_k < t} \Phi^+(t, t_k)b_k^+,$$

$$y(t) = -\int_t^\infty \Phi^-(t, s)f^-(s)ds + \sum_{t_k > t} \Phi^-(t, t_k)b_k^-,$$

where χ is a constant k-dimensional vector.

It follows that, for two different solutions $z_1(t)$ and $z_2(t)$ of system (2.1) the estimate

$$||z_1(t) - z_2(t)|| \le \bar{a}e^{-\lambda(t-t_0)}||z_1(t_0) - z_2(t_0)|| \tag{2.16}$$

holds true.

Thus, (2.16) implies that the solution $z(t)$ of (2.1) is unique and exponentially stable. $\qquad\square$

Let $\Omega \equiv B_h$.

Theorem 2.2. *Let the following conditions hold:*

1. *Conditions H2.1, H2.4–H2.6 hold.*
2. *The system (2.3) is hyperbolic.*
3. *The functions $F(t, z)$, $I_k(z)$, $k = \pm 1, \pm 2, \ldots$, are Lipschitz continuous with respect to $z \in B_h$ with a Lipschitz constant $L > 0$, i.e.,*

$$||F(t, z_1) - F(t, z_2)|| + ||I_k(z_1) - I_k(z_2)|| \le L||z_1 - z_2||,$$

and they are bounded, i.e. there exists a constant $L_1 > 0$, such that

$$max\left(\sup_{t \in \mathbb{R}, z \in B_h} ||F(t, z)||, \sup_{k=\pm 1, \pm 2, \ldots, z \in B_h} ||I_k(z))||\right) = L_1 < \infty.$$

4. *The following inequalities hold*

$$L_1\left(\frac{\bar{a}}{\lambda} + \frac{2\bar{a}N}{1 - e^{-\lambda}}\right) < h,$$

$$L\left(\frac{\bar{a}}{\lambda} + \frac{2\bar{a}N}{1 - e^{-\lambda}}\right) < 1.$$

Then for the system (2.2) there exists a unique almost periodic solution.

Proof. Denote by AP the set of all almost periodic solutions $\varphi(t)$, $\varphi \in PC[\mathbb{R}, \Omega]$, such that $||\varphi|| < h$.

We define in AP the operator SAP, such that if $\varphi \in AP$, then $\varphi = (\varphi^+, \varphi^-)$, where $\varphi^+ : \mathbb{R} \to \mathbb{R}^k$, $\varphi^- : \mathbb{R} \to \mathbb{R}^{n-k}$, $SAP\varphi = (SAP\varphi^+, SAP\varphi^-)$, $u = SAP\varphi^+$ is the almost periodic solution of

$$\begin{cases} \dot{u} = Q^+(t)u + F^+(t, \varphi(t)), t \neq t_k, \\ \Delta u(t_k) = I_k^+(\varphi(t_k)), k = \pm 1, \pm 2, \ldots, \end{cases}$$

and $v = SAP\varphi^-$ is the almost periodic solution of

$$\begin{cases} \dot{v} = Q^-(t)v + F^-(t, \varphi(t)), t \neq t_k, \\ \Delta v(t_k) = I_k^-(\varphi(t_k)), k = \pm 1, \pm 2, \ldots. \end{cases}$$

The existence of almost periodic solutions $u(t)$ and $v(t)$, is guaranteed by Theorem 2.1. In fact, the almost periodicity of the sequence $\{\varphi(t_k)\}$, $k = \pm 1, \pm 2, \ldots$ follows from Lemma 1.5, and from the method for finding of common almost periods, we obtain that the sequence $\{I_k(\varphi(t_k))\}$, $k = \pm 1, \pm 2, \ldots$, is almost periodic, also. The almost periodicity of the function $F(t, \varphi(t))$ follows from Theorem 1.17. Further on, conditions 2 and 3 imply that $SAP(AP) \subset AP$.

Let $\varphi, \psi \in AP$. Then, the estimate

$$||SAP\varphi - SAP\psi|| \leq L\bar{a}\left(\frac{1}{\lambda} + \frac{2N}{1 - e^{-\lambda}}\right)|\varphi - \psi|_\infty,$$

where $|\varphi - \psi|_\infty = \sup_{t \in \mathbb{R}} ||\varphi(t) - \psi(t)||$ holds true.

It follows from condition 3, and from the last inequality, that SAP is a contracting operator on SAP. Hence, for the system (2.2) there exists a unique almost periodic solution.   □

2.2 Impulsive Integro-Differential Equations

In this section, we shall present the main results on the existence of almost periodic solutions of impulsive integro-differential systems.

Consider the following linear system of impulsive integro-differential equations

$$\begin{cases} \dot{x} = A(t)x(t) + \int_{t_0}^{t} K(t, s)x(s)ds + f(t), \; t \neq t_k, \\ \Delta x(t_k) = B_k x(t_k), \; k = \pm 1, \pm 2, \ldots, \end{cases} \qquad (2.17)$$

where $t \in \mathbb{R}$, $\{t_k\} \in \mathcal{B}$, $A \in PC[\mathbb{R}, \mathbb{R}^{n \times n}]$, $K \in PC[\mathbb{R}^2, \mathbb{R}^{n \times n}]$, $f \in PC[\mathbb{R}, \mathbb{R}^n]$, $B_k \in \mathbb{R}^{n \times n}, k = \pm 1, \pm 2, \ldots$.

The solution of (2.17), $x(t) = x(t; t_0, x_0)$ with initial condition $x(t_0^+) = x_0, t_0 \in \mathbb{R}$, $x_0 \in \mathbb{R}^n$, is characterized at the following way:

1. For $t \neq t_k$, $k = \pm 1, \pm 2, \ldots$, the mapping point $(t, x(t))$ moves along some of the integral curves of the system

$$\dot{x} = A(t)x(t) + \int_{t_0}^{t} K(t, s)x(s)ds + f(t).$$

2. At the moment $t = t_k$, $k = \pm 1, \pm 2, \ldots$, the system is subject to an impulsive effect, as a result of which the mapping point is transferred "instantly" from the position $(t_k, x(t_k))$ into a position $(t_k, x(t_k) + B_k x(t_k))$. Afterwards, for $t_k < t < t_{k+1}$ the solution $x(t)$ coincides with the solution $y(t)$ of the system

$$\begin{cases} \dot{y} = A(t)y(t) + \int_{t_0}^{t} K(t, s)y(s)ds + f(t), \ t \neq t_k, \\ y(t_k) = x(t_k) + B_k x(t_k), \ k = \pm 1, \pm 2, \ldots, \end{cases}$$

At the moment $t = t_{k+1}$, the solution is subject to a new impulsive effect.

We shall, also, consider weakly nonlinear impulsive integro-differential systems

$$\begin{cases} \dot{x}(t) = A(t)x(t) + \int_{t_0}^{t} K(t, s)x(s)ds + F(t, x(t)), \ t \neq t_k, \\ \Delta x(t_k) = B_k x(t_k), \ k = \pm 1, \pm 2, \ldots, \end{cases} \tag{2.18}$$

where $F(t, x) \in PC[\mathbb{R} \times \mathbb{R}^n, \mathbb{R}^n]$, and

$$\begin{cases} \dfrac{\partial R(t, s)}{\partial t} = A(t)x(t) + \int_{t_0}^{t} K(t, v)R(v, s)dv, \ s \neq t_k, \ t \neq t_k, \\ R(t_k^+, s) = (E + B_k)R(t_k, s), \ k = \pm 1, \pm 2, \ldots, \end{cases} \tag{2.19}$$

where $R(t, s)$ is an $n \times n$-dimensional matrix function and $R(s, s) = E$, E is the identity matrix in $\mathbb{R}^n$.

Lemma 2.3 ([131]). *If $R(t, s)$ is a solution of (2.19), then the unique solution $x(t) = x(t; t_0, x_0)$ of (2.17) is given by*

$$x(t) = R(t, t_0)x(t_0) + \int\limits_{t_0}^{t} R(t, s)f(s)ds, \ x(t_0^+) = x_0.$$

Introduce the following conditions:

H2.7. There exists an $n \times n$-dimensional matrix function $R(t, s)$, satisfying (2.19).

H2.8. $det(E + B_k) \neq 0, \ k = \pm1, \pm2, \ldots$.

H2.9. $\mu[A(t) - R(t, t)] \leq -\alpha, \ \alpha > 0, \ \mu[.]$ is the logarithmic norm.

Lemma 2.4 ([15]). *Let conditions H2.7–H2.9 hold.*
  Then

$$\|R(t, s)\| \leq K_1 e^{-\alpha(t-s)}, \qquad (2.20)$$

where $K_1 > 0, \ t > s$.

Remark 2.1. In the special case, when in (2.17), $K(t, s) \equiv 0$, we obtain the linear impulsive system

$$\begin{cases} \dot{x} = A(t)x + f(t), \ t \neq t_k, \\ \Delta x(t_k) = B_k x(t_k), \ k = \pm1, \pm2, \ldots. \end{cases}$$

Then, from Lemma 2.3, it follows, respectively, well known variation parameters formula [94], where $R(t, s)$ is the fundamental matrix and $R(t_0, t_0) = E$.

We shall investigate the existence of almost periodic solutions of systems (2.17), (2.18), and we shall use the following conditions:

H2.10. $A(t)$ is an almost periodic $n \times n$-matrix function.

H2.11. The sequence $\{B_k\}, \ k = \pm1, \pm2, \ldots$ is almost periodic.

H2.12. The set of sequences $\{t_k^j\}, \ k = \pm1, \pm2, \ldots, \ j = \pm1, \pm2, \ldots$ is uniformly almost periodic, and $inf_k t_k^1 = \theta > 0$.

H2.13. The matrix $K(t, s)$ is almost periodic along the diagonal line, i.e. for any $\varepsilon > 0$, the set $T(K, \varepsilon)$ composed from ε-almost periods τ, such that for $\tau \in T(K, \varepsilon)$, $K(t, s)$ satisfies the inequality

$$\|K(t + \tau, s + \tau) - K(t, s)\| \leq \varepsilon e^{-\frac{\alpha}{2}(t-s)},$$

$t > s$, is relatively dense in $\mathbb{R}$.

H2.14. The function $f(t), \ f \in PC[\mathbb{R}, \mathbb{R}^n]$ is almost periodic.

H2.15. The function $F(t, x)$ is almost periodic along t uniformly with respect to $x \in \Omega$.

We shall use the next lemma, which is similar to Lemma 1.7.

Lemma 2.5. *Let conditions H2.10–H2.12 and H2.14 hold.*

Then for each $\varepsilon > 0$ there exist ε_1, $0 < \varepsilon_1 < \varepsilon$, a relatively dense set $\overline{T}$ of real numbers and a set P of integer numbers, such that the following relations are fulfilled:

(a) $\|A(t+\tau) - A(t)\| < \varepsilon$, $t \in \mathbb{R}$, $\tau \in \overline{T}$.
(b) $\|f(t+\tau) - f(t)\| < \varepsilon$, $t \in \mathbb{R}$, $\tau \in \overline{T}$, $|t - t_k| > \varepsilon$, $k = \pm 1, \pm 2, \ldots$.
(c) $\|B_{k+q} - B_k\| < \varepsilon$, $q \in P$, $k = \pm 1, \pm 2, \ldots$.
(d) $|t_k^q - \tau| < \varepsilon_1$, $q \in P$, $\tau \in \overline{T}$, $k = \pm 1, \pm 2, \ldots$.

Lemma 2.6. *Let conditions H2.7–H2.13 hold.*

Then $R(t, s)$ is almost periodic along the diagonal line and the following inequality holds

$$\|R(t+\tau, s+\tau) - R(t, s)\| \leq \varepsilon \Gamma e^{-\frac{\alpha}{2}(t-s)}, \tag{2.21}$$

where $t > s$, $\Gamma > 0$, $\varepsilon > 0$, τ is an almost period.

Proof. Let $\varepsilon > 0$ and τ be a common ε-almost period of $A(t)$ and $K(t, s)$.
Then, for $s \neq t_k'$, $t \neq t_k'$, we have

$$\frac{\partial R(t+\tau, s+\tau)}{\partial t} = A(t)R(t+\tau, s+\tau) + \big(A(t+\tau) - A(t)\big)R(t+\tau, s+\tau)$$

$$\times \int_s^t \big(K(t+\tau, v+\tau) - K(t, v)\big)R(v+\tau, s+\tau)dv$$

$$+ \int_s^t K(t, v)R(v+\tau, s+\tau)dv,$$

and

$$R(t_k'+\tau, s+\tau) = (E + B_k)R(t_k'+\tau, s+\tau) + (B_{k+q} - B_k)R(t_k+\tau, s+\tau),$$

where $t_k' = t_k - \tau$ and τ, q are the numbers from Lemma 2.5.
Hence, from (2.19), we obtain

$$R(t+\tau, s+\tau) - R(t, s)$$

$$= \int_s^t R(t, u)\big(A(u+\tau) - A(u)\big)R(u+\tau, s+\tau)du$$

$$+ \int_s^t R(t,u)\Big(\int_s^u \big(K(u+\tau,v+\tau) - K(u,v)R(v+\tau,s+\tau)\big)dv\Big)du$$

$$+ \sum_{s \le t'_v < t} R(t,t_v'^+)(B_{v+q} - B_v)R(t_v' + \tau, s + \tau). \tag{2.22}$$

From Lemma 2.5, it follows that, if $|t - t'_k| > \varepsilon$, $t \in \mathbb{R}$, then $t'_{k+q} < t+\tau < t'_{k+q+1}$ and from (2.20), (2.22), we obtain

$$\|R(t+\tau,s+\tau) - R(t,s)\| \le K_1^2 \varepsilon \Big(e^{-\alpha(t-s)}(t-s) + \frac{4}{\alpha^2}e^{-\frac{\alpha}{2}(t-s)}i(s,t)e^{-\alpha(t-s)}\Big),$$

where $i(t,s)$ is the number of points t_k in the interval (t,s).

Now, from the condition H2.12 and Lemma 1.2, it follows that there exists a positive integer N, such that for any $t \in \mathbb{R}$, $s \in \mathbb{R}$ and $t > s$ the following inequality holds

$$i(s,t) \le (t-s)N + N.$$

Therefore,

$$\|R(t+\tau,s+\tau) - R(t,s)\| \le \varepsilon \Gamma e^{-\frac{\alpha}{2}(t-s)},$$

where $t > s$, $\Gamma = K_1^2 \frac{2}{\alpha}\Big(1 + \frac{2}{\alpha}N + \frac{N\alpha}{2}\Big)$. □

The next theorems are the main in this paragraph.

Theorem 2.3. *Let conditions H2.7–H2.14 hold.*
Then for the system (2.17), there exists a unique exponentially stable almost periodic solution $\varphi(t)$, such that

$$\|\varphi(t)\| \le \frac{2K_1}{\alpha} \max_{s<t} \|f\|. \tag{2.23}$$

Proof. Consider the function

$$\varphi(t) = \int_{-\infty}^t R(t,s)f(s)ds. \tag{2.24}$$

From (2.19), (2.24), and Fubini's theorem, it follows that

$$\dot{\varphi}(t) = \int_{-\infty}^t \frac{\partial R(t,s)}{\partial t}f(s)ds + f(t)$$

$$= \int_{-\infty}^t \Big(A(t)R(t,s) + \int_u^t K(t,u)R(t,u)du\Big)f(s)ds + f(t)$$

$$= A(t)\varphi(t) + \int\limits_u^t \Big(\int\limits_{-\infty}^t K(t,u)R(u,s)f(s)ds \Big) du$$

$$= A(t)\varphi(t) + \int\limits_s^t K(t,s)\varphi(s)ds + f(t), \qquad (2.25)$$

where $s < t$, $s \neq t_k$, $t \neq t_k$, $k = \pm 1, \pm 2, \ldots$.

On the other hand, for $t = t_k$, $k = \pm 1, \pm 2, \ldots$, we have

$$\Delta\varphi(t_k) = \varphi(t_k^+) - \varphi(t_k) = B_k\varphi(t_k). \qquad (2.26)$$

Then, from (2.25) and (2.26), it follows that $\varphi(t)$ is a solution of system (2.17).

From Lemma 2.4, we obtain

$$||\varphi(t)|| \leq \int\limits_{-\infty}^t ||R(t,s)||||f(s)||ds \leq \frac{2K_1}{\alpha} \max_{s<t} ||f(t)||.$$

Let $\tau \in \overline{T}$, $q \in P$, where $\overline{T}$ and P are determined in Lemma 2.5. From Lemma 2.6, it follows that

$$||\varphi(t+\tau) - \varphi(t)|| = \int\limits_{-\infty}^t ||R(t+\tau, s+\tau)f(s+\tau) - R(t,s)f(s)||ds$$

$$\leq \int\limits_{-\infty}^t ||R(t+\tau, s+\tau) - R(t,s)||||f(s+\tau)||ds$$

$$+ \int\limits_{-\infty}^t ||R(t,s)||||f(s+\tau) - f(s)||ds$$

$$\leq \varepsilon\Big(\frac{2\Gamma M}{\alpha} + \frac{K_1}{\alpha} \Big), \qquad (2.27)$$

where $M = \max\limits_{s<t} ||f(t)||$. The estimate (2.27) means that $\varphi(t)$ is an almost periodic function.

Let $\eta(t)$ is one other solution of (2.17). Then, from (2.20), it follows that

$$||\varphi(t) - \eta(t)|| \leq K_1 e^{-\alpha(t-t_0)}||\varphi(t_0) - \eta(t_0)||,$$

and we obtain that the solution $\varphi(t)$ is unique and exponentially stable. $\square$

Theorem 2.4. *Let the following conditions hold:*

1. *Conditions H2.7–H2.13, and H2.15 hold.*
2. *The function $F(t,x)$ is Lipschitz continuous with respect to $x \in B_h$ with a Lipschitz constant $L > 0$, i.e.*

$$\|F(t,x_1) - F(t,x_2)\| \le L\|x_1 - x_2\|, \ x_1, x_2 \in B_h,$$

   and $F(t,x)$ is uniformly bounded, i.e. there exists a constant $G > 0$, such that
$$\|F(t,x)\| \le G, \ \|x\| < h.$$

3. *The following inequalities hold*

$$\frac{K_1 G}{\alpha} < h, \ \frac{KL}{\alpha} < 1.$$

Then there exists a unique exponentially stable almost periodic solution of (2.18).

Proof. Let us denote by AP the set of all almost periodic functions $\varphi(t)$, $\varphi \in PC[\mathbb{R}, \mathbb{R}^n]$, satisfying the inequality $\|\varphi(t)\| < h$, and let $|\varphi(t)|_\infty = \sup_{t \in \mathbb{R}} \|\varphi(t)\|$.

In AP, we define an operator S

$$S\varphi = \int_{-\infty}^{t} R(t,s) F(t, \varphi(s)) ds. \tag{2.28}$$

Let $\varphi \in AP$. From (2.28), it follows that

$$\|S\varphi\| \le \int_{-\infty}^{t} \|R(t,s)\| \|F(t,\varphi(s))\| ds$$

$$\le K_1 \int_{-\infty}^{t} e^{-\alpha(t-s)} G ds \le \frac{K_1 G}{\alpha} < h. \tag{2.29}$$

On the other hand, from Theorem 1.17, it follows that the function $F(t, \varphi(t))$ is almost periodic, and let τ be the common almost period of $\varphi(t)$ and $F(t, \varphi(t))$.

Then,

$$\|S\varphi(t+\tau) - S\varphi(t)\|$$

$$\le \int_{-\infty}^{t} \|R(t+\tau, s+\tau) F(s, \varphi(s+\tau)) - R(t,s) F(s, \varphi(s))\| ds$$

$$\le \left(\frac{2G\Gamma}{\alpha} + \frac{K_1}{\alpha}\right) \varepsilon. \tag{2.30}$$

Hence, using (2.29) and (2.30), we obtain that $S(AP) \subset AP$.
Let $\varphi \in AP$, $\eta \in AP$. From (2.28) and Lemma 2.6, we have

$$
\begin{aligned}
||S\varphi(t) - S\eta(t)|| &\leq \int_{-\infty}^{t} ||R(t,s)|| ||F(s,\varphi(s)) - F(s,\eta(s))|| ds \\
&\leq \frac{K_1 L}{\alpha} |\varphi(t) - \eta(t)|_{\infty}.
\end{aligned} \tag{2.31}
$$

Therefore, the inequality (2.31) shows that S is a contracting operator in AP, and hence, there exists a unique almost periodic solution of system (2.18).

Now, let $\psi(t)$ is one other solution of (2.18). Then, Lemma 2.3 and (2.20) imply that

$$||\varphi(t) - \psi(t)||$$

$$
\leq K_1 ||\varphi(t_0) - \psi(t_0)|| e^{-\alpha(t-s)} + \int_{t_0}^{t} K_1 e^{-\alpha(t-s)} L ||\varphi(s) - \psi(s)|| ds. \tag{2.32}
$$

Set

$$v(t) = ||\varphi(t) - \psi(t)|| e^{\alpha(t)}.$$

From (2.32) and Gronwall–Belman's inequality, we have

$$
v(t) \leq K_1 v(t_0) exp\left(\int_{t_0}^{t} K_1 L ds \right).
$$

Consequently,

$$
||\varphi(t) - \psi(t)|| \leq K_1 ||\varphi(t_0) - \psi(t_0)|| e^{(K_1 L - \alpha)(t-t_0)}.
$$

From the last inequality, it follows that $\varphi(t)$ is exponentially stable.   □

2.3   Forced Perturbed Impulsive Differential Equations

In this part, we shall consider sufficient conditions for the existence of almost periodic solutions for forced perturbed systems of impulsive differential equations with impulsive effects at fixed moments.

We shall consider the system

$$\begin{cases} \dot{x} = A(t)x + g(t) + \mu X(t, x, \mu), \ t \neq t_k, \\ \Delta x(t_k) = B_k x(t_k) + g_k + \mu X_k(x(t_k), \mu), \ k = \pm 1, \pm 2, \ldots, \end{cases} \quad (2.33)$$

where $t \in \mathbb{R}$, $\{t_k\} \in \mathcal{B}$, $A : \mathbb{R} \to \mathbb{R}^{n \times n}$, $g : \mathbb{R} \to \mathbb{R}^n$, $\mu \in M \subset \mathbb{R}$, $X : \mathbb{R} \times \Omega \times M \to \mathbb{R}^n$, $B_k \in \mathbb{R}^{n \times n}$, $g_k \in \mathbb{R}^n$, $X_k : \Omega \times M \to \mathbb{R}^n$, $k = \pm 1, \pm 2, \ldots$.

Denote by $x(t, \mu) = x(t; t_0, x_0, \mu)$ the solution of (2.33) with initial condition $x(t_0^+, \mu) = x_0$, $x_0 \in \Omega$, $\mu \in M$.

We shall use the following definitions:

Definition 2.2. The system

$$\begin{cases} \dot{x} = A(t)x + g(t), \ t \neq t_k, \\ \Delta x(t_k) = B_k x(t_k) + g_k, \ k = \pm 1, \pm 2, \ldots, \end{cases} \quad (2.34)$$

is said to be *generating system* of (2.33).

Definition 2.3 ([56]). The matrix $A(t)$ is said to has a *column dominant with a parameter* $\alpha > 0$ *on* $[a, b]$, if

$$a_{ii}(t) + \sum_{j \neq i} |a_{ji}(t)| \leq -\alpha < 0,$$

for each $i, j = 1, \ldots, n$, and $t \in [a, b]$.

Introduce the following conditions:

H2.16. The matrix function $A \in C[\mathbb{R}, \mathbb{R}^{n \times n}]$ is almost periodic in the sense of Bohr.

H2.17. $\{B_k\}$, $k = \pm 1, \pm 2, \ldots$ is an almost periodic sequence.

H2.18. $det(E + B_k) \neq 0$, $k = \pm 1, \pm 2, \ldots$ where E is the identity matrix in $\mathbb{R}^{n \times n}$.

H2.19. The function $g \in PC[\mathbb{R}, \mathbb{R}^n]$ is almost periodic.

H2.20. $\{g_k\}$, $k = \pm 1, \pm 2, \ldots$ is an almost periodic sequence.

H2.21. The function $X \in C[\mathbb{R} \times \Omega \times M, \mathbb{R}^n]$ is almost periodic in t uniformly with respect to $(x, \mu) \in \Omega \times M$, and is Lipschitz continuous with respect to $x \in B_h$ with a Lipschitz constant $l_1 > 0$, such that

$$||X(t, x, \mu) - X(t, y, \mu)|| \leq l_1 ||x - y||, \ x, y \in B_h,$$

for any $t \in \mathbb{R}$ and $\mu \in M$.

H2.22. The sequence of functions $\{X_k(x, \mu)\}$, $k = \pm 1, \pm 2, \ldots$, $X_k \in C[\Omega \times M, \mathbb{R}^n]$ is almost periodic uniformly with respect to $(x, \mu) \in \Omega \times M$, and the functions X_k are Lipschitz continuous with respect to $x \in B_h$ with a Lipschitz constant $l_2 > 0$, such that

$$||X_k(x,\mu) - X_k(y,\mu)|| \le l_2||x - y||, \ x, y \in B_h,$$

for $k = \pm 1, \pm 2, \ldots, \ \mu \in M$.

H2.23. The set of sequences $\{t_k^j\}$, $t_k^j = t_{k+j} - t_k$, $k = \pm 1, \pm 2, \ldots, \ j = \pm 1, \pm 2, \ldots$ is uniformly almost periodic, and $inf_k t_k^1 = \theta > 0$.

We shall use the next lemma, which is similar to Lemma 1.7.

Lemma 2.7. *Let conditions H2.16, H2.17, H2.19, H2.20 and H2.23 hold. Then for each $\varepsilon > 0$ there exist ε_1, $0 < \varepsilon_1 < \varepsilon$, a relatively dense set $\overline{T}$ of real numbers, and a set P of integer numbers, such that the following relations are fulfilled:*

(a) $||A(t + \tau) - A(t)|| < \varepsilon$, $t \in \mathbb{R}$, $\tau \in \overline{T}$.
(b) $||g(t + \tau) - g(t)|| < \varepsilon$, $t \in \mathbb{R}$, $\tau \in \overline{T}$, $|t - t_k| > \varepsilon$, $k = \pm 1, \pm 2, \ldots$.
(c) $||B_{k+q} - B_k|| < \varepsilon$, $q \in P$, $k = \pm 1, \pm 2, \ldots$.
(d) $||g_{k+q} - g_k|| < \varepsilon$, $q \in P$, $k = \pm 1, \pm 2, \ldots$.
(e) $|t_k^q - \tau| < \varepsilon_1$, $q \in P$, $\tau \in \overline{T}$, $k = \pm 1, \pm 2, \ldots$.

Lemma 2.8. *Let conditions H2.19, H2.20 and H2.23 hold. Then there exists a positive constant C_1 such that*

$$max(\sup_{t \in \mathbb{R}} ||g(t)||, \sup_{k = \pm 1, \pm 2, \ldots} ||g_k||) \le C_1.$$

Proof. The proof follows from Lemma 1.7.   □

Lemma 2.9 ([138]). *Let the following conditions hold:*

1. *Conditions H2.16–H2.18 and H2.23 are met.*
2. *For the Cauchy matrix $W(t, s)$ of the system*

$$\begin{cases} \dot{x} = A(t)x, \ t \ne t_k, \\ \Delta x(t_k) = B_k x(t_k), \ k = \pm 1, \pm 2, \ldots, \end{cases}$$

there exist positive constants K and λ such that

$$||W(t, s)|| \le Ke^{-\lambda(t-s)},$$

where $t \ge s$, $t, s \in \mathbb{R}$.

Then for any $\varepsilon > 0$, $t \in \mathbb{R}$, $s \in \mathbb{R}$, $|t - t_k| > \varepsilon > 0$, $|s - t_k| > \varepsilon$, $k = \pm 1, \pm 2, \ldots$, there exists a relatively dense set $\overline{T}$ of ε-almost periods of matrix $A(t)$ and a positive constant Γ, such that for $\tau \in \overline{T}$ it follows

$$||W(t + \tau, s + \tau) - W(t, s)|| \leq \varepsilon \Gamma e^{-\frac{\lambda}{2}(t-s)}.$$

Now, we are ready to proof the main theorem.

Theorem 2.5. *Let the following conditions hold:*

1. *Conditions H2.16–H2.23 are met.*
2. *There exists a positive constant L_1, such that*

$$max\{ \sup_{\substack{t \in \mathbb{R} \\ (x,\mu) \in \Omega \times M}} ||X(t, x, \mu)||, \sup_{\substack{k=\pm 1, \pm 2, \ldots \\ (x,\mu) \in \Omega \times M}} ||X_k(x, \mu)||\} \leq L_1.$$

3. *For the generating system (2.34), there exists a unique almost periodic solution.*

Then there exists a positive constant μ_0, $\mu_0 \in M$ such that:

1. *For any μ, $|\mu| < \mu_0$ and $C < C_1$, where the constant C_1 is from Lemma 2.8, there exists a unique almost periodic solution of (2.33).*
2. *There exists a positive constant L such that*

$$||x(t, \mu_1) - x(t, \mu_2)|| \leq L|\mu_1 - \mu_2|,$$

where $t \in \mathbb{R}$, $|\mu_i| < \mu_0$, $i = 1, 2$.
3. *For $|\mu| \to 0$, $x(t, \mu)$ converges to the unique almost periodic solution of (2.34).*
4. *The solution $x(t, \mu)$ is exponentially stable.*

Proof of Assertion 1. Let we denote by AP, the set of all almost periodic functions $\varphi(t, \mu)$, $\varphi \in AP \in PC[\mathbb{R} \times M, \mathbb{R}^n]$ satisfying the inequality $||\varphi|| < C$, and let $|\varphi|_\infty = \sup_{t \in \mathbb{R}, \ \mu \in M} ||\varphi(t, \mu)||$.

In AP, we define the operator S,

$$S\varphi = \int_{-\infty}^{t} W(t, s)\Big(g(s) + \mu X(s, \varphi(s, \mu), \mu)\Big) ds$$

$$+ \sum_{t_k < t} W(t, t_k)\Big(g_k + \mu X_k(\varphi(t_k, \mu), \mu)\Big). \tag{2.35}$$

From Lemma 2.8 and Lemma 2.9, it follows

$$||S\varphi|| = \int_{-\infty}^{t} ||W(t, s)||\Big(||g(s)|| + |\mu|||X(s, \varphi(s, \mu), \mu)||\Big) ds$$

$$+ \sum_{t_k < t} ||W(t, t_k)||\Big(||g_k|| + |\mu|||X_k(\varphi(t_k, \mu), \mu)||\Big)$$

$$\leq (C_1 + |\mu|L_1)\Big(\frac{K}{\lambda} + \frac{KN}{1 - e^{-\lambda}}\Big).$$

Consequently, there exists a positive constant μ_1 such that for $\mu \in (-\mu_1, \mu_1)$ and $C = (C_1 + |\mu|L_1)\left(\frac{K}{\lambda} + \frac{KN}{1-e^{-\lambda}}\right) < C_1$, we obtain

$$||S\varphi|| \leq C. \tag{2.36}$$

Now, let $\tau \in \overline{T}$, $q \in P$, where the sets $\overline{T}$ and P are determined in Lemma 2.7. From Lemma 1.5 and Theorem 1.17, we have

$$||S\varphi(t+\tau, \mu) - S\varphi(t, \mu)||$$

$$\leq \int_{-\infty}^{t} ||W(t+\tau, s+\tau) - W(t,s)|| \Big(||g(s+\tau)||$$

$$+ |\mu| ||X(s+\tau, \varphi(s+\tau, \mu), \mu)|| \Big) ds$$

$$+ \int_{-\infty}^{t} ||W(t,s)|| \Big(||g(s+\tau) - g(s)||$$

$$+ |\mu| ||X(s+\tau, \varphi(s+\tau, \mu), \mu) - X(s, \varphi(s, \mu), \mu)|| \Big) ds$$

$$+ \sum_{t_k < t} ||W(t+\tau, t_{k+q}) - W(t, t_k)|| \Big(||g_{k+q}||$$

$$+ |\mu| ||X_{k+q}(\varphi(t_{k+q}, \mu), \mu)|| \Big)$$

$$+ \sum_{t_k < t} ||W(t, t_k)|| \Big(||g_{k+q} - g_k||$$

$$+ |\mu| ||X_{k+q}(\varphi(t_{k+q}, \mu), \mu) - X_k(\varphi(t_k, \mu), \mu)|| \Big)$$

$$\leq \varepsilon \Big((C_1 + |\mu|L_1)\Big(\frac{2\Gamma}{\lambda} + \frac{N\Gamma}{1-e^{-\lambda}}\Big) + (1+|\mu|)\Big(\frac{K}{\lambda} + \frac{NK}{1+e^{-\lambda}}\Big) \Big). \tag{2.37}$$

Thus, by (2.35) and (2.36), we obtain $S\varphi \in AP$.
Let $\varphi \in AP$, $\psi \in AP$. Then from (2.35), it follows

$$||S\varphi - S\psi|| \leq |\mu| \int_{-\infty}^{t} ||W(t,s)|| ||X(s, \varphi(s, \mu), \mu) - X(s, \psi(s, \mu), \mu)|| ds$$

$$+ |\mu| \sum_{t_k < t} ||W(t, t_k)|| ||X_k(\varphi(t_k, \mu), \mu) - X_k(\psi(t_k, \mu), \mu)||$$

$$\leq |\mu| K \Big(\frac{l_1}{\lambda} + \frac{l_2}{1-e^{-\lambda}}\Big) |\varphi - \psi|_\infty.$$

Since there exists a positive constant $\mu_0 < \mu_1$ such that

$$\mu_0 K\Big(\frac{l_1}{\lambda} + \frac{l_2}{1 - e^{-\lambda}}\Big) < 1,$$

we have that S is a contracting operator in AP.

Proof of Assertion 2. Let $\varphi_j = \varphi_j(t, \mu_j)$, $j = 1, 2$, and $|\mu_j| < \mu_0$.
Then,

$$||\varphi_1 - \varphi_2|| \le |\mu_1 - \mu_2|\Big(\int_{-\infty}^t ||W(t,s)||||X(s, \varphi_1(s, \mu_1), \mu_1)||ds$$

$$+ \sum_{t_k < t} ||W(t, t_k)||||X_k(\varphi_1(t_k, \mu_1), \mu_1)||\Big)$$

$$+ |\mu_2|\Big(\int_{-\infty}^t ||W(t,s)||||X(s, \varphi_1(s, \mu_1), \mu_1) - X(s, \varphi_2(s, \mu_2), \mu_2)||ds$$

$$+ \sum_{t_k < t} ||W(t, t_k)||||X_k(\varphi_1(t_k, \mu_1), \mu_1) - X_k(\varphi_2(t_k, \mu_2), \mu_2)||\Big)$$

$$\le L|\mu_1 - \mu_2|, \tag{2.38}$$

where

$$L = L_1 K\Big(\frac{l_1}{\lambda} + \frac{l_2}{1 - e^{-\lambda}}\Big)(1 - \mu_0 K)K\Big(\frac{l_1}{\lambda} + \frac{Nl_2}{1 - e^{-\lambda}}\Big).$$

Proof of Assertion 3. Let we denote by $x(t)$ the almost periodic solution of (2.33).
From (2.35) and Lemma 2.9, it follows

$$||x(t, \mu) - x(t)|| \le |\mu|\Big(\int_{-\infty}^t ||W(t,s)||||X(s, \varphi(s, \mu), \mu)||ds$$

$$+ \sum_{t_k < t} ||W(t, t_k)||||X_k(\varphi(t_k, \mu), \mu)||\Big)$$

$$\le |\mu| L_1 K\Big(\frac{1}{\lambda} + \frac{N}{1 - e^{-\lambda}}\Big).$$

Then $x(t, \mu) \to x(t)$ for $|\mu| \to 0$.

Proof of Assertion 4. Let $y(t)$ be an arbitrary solution of (2.34). Then using (2.35), we obtain

$$y(t) - x(t, \mu) = W(t, t_0)\big(y(t_0) - x(t_0, \mu)\big)$$

$$+ \mu\Big(\int_{t_0}^t W(t, s)\big(X(s, y(s), \mu) - X(s, x(s, \mu), \mu)\big)ds$$

$$+ \sum_{t_0 < t_k < t} W(t, t_k)\big(X_k(y(t_k), \mu) - X_k(x(t_k, \mu), \mu)\big)\Big).$$

Now, we have

$$||y(t) - x(t, \mu)|| \leq K e^{-\lambda(t-t_0)} ||y(t_0) - x(t_0, \mu)||$$

$$+ |\mu| \left(\int_{t_0}^{t} K l_1 e^{-\lambda(t-s)} ||y(s) - x(s, \mu)|| ds \right.$$

$$\left. + \sum_{t_0 < t_k < t} K l_2 e^{-\lambda(t-t_k)} ||y(t_k) - x(t_k, \mu)|| \right).$$

Set $u(t) = ||y(t) - x(t, \mu)|| e^{-\lambda t}$ and from Gronwall–Bellman's inequality, it follows

$$||y(t) - x(t, \mu)|| \leq K ||y(t_0) - x(t_0, \mu)|| (1 + |\mu| K l_1)^{i(t_0, t)} e^{(-\lambda + |\mu| K l_2)(t-t_0)},$$

where $i(t, s)$ is the number of points t_k in the interval (t, s). Obviously, if there exists $\mu \in M$ such that $N \ln(1 + |\mu| K l_1) + |\mu| K l_2 < \lambda$, then the solution $x(t, \mu)$ is exponentially stable. $\qquad \Box$

Lemma 2.10. *Let the following conditions hold:*

1. *Conditions H2.16, H2.17 are met.*
2. *The matrix-valued function $A(t)$ has a column dominant with a parameter $\alpha > 0$ for $t \in \mathbb{R}$.*

Then for the Cauchy's matrix $W(t, s)$ it follows

$$||W(t, s)|| \leq K e^{-\alpha(t-s)},$$

where $t \in \mathbb{R}$, $s \in \mathbb{R}$, $t \geq s$, $K > 0$.

Proof. The proof follows from the definition of matrix $W(t, s)$. $\qquad \Box$

Example 2.1. We consider the following system of impulsive differential equations of Lienard's type:

$$\begin{cases} \ddot{x} + f(t)\dot{x} + q(t) = \mu h(t, x, \dot{x}, \mu), \ t \neq t_k, \\ \Delta x(t_k) = b_k^1 x(t_k) + g_k^1 + \mu X_k^1(x(t_k), \dot{x}(t_k), \mu), \\ \Delta \dot{x}(t_k) = b_k^2 x(t_k) + g_k^2 + \mu X_k^2(x(t_k), \dot{x}(t_k), \mu), \ k = \pm 1, \pm 2, \ldots, \end{cases} \qquad (2.39)$$

where $t \in \mathbb{R}$, $x \in \mathbb{R}$, $\mu \in M$, $\{t_k\} \in \mathcal{B}$, the functions $f \in PC[\mathbb{R}, \mathbb{R}]$, $q \in PC[\mathbb{R}, \mathbb{R}]$ are almost periodic, the function $h \in C[\mathbb{R}^3 \times M, \mathbb{R}]$ is almost periodic in t uniformly with respect to $x, \dot{x}$ and μ, $b_k^m \in \mathbb{R}$, $g_k^m \in \mathbb{R}$, the sequences $\{b_k^m\}$, $\{g_k^m\}$ are almost periodic, $X_k^m \in C[\mathbb{R}^2 \times M, \mathbb{R}]$ and the sequences $\{X_k^m\}$, $k = \pm 1, \pm 2, \ldots$, $m = 1, 2$, are almost periodic uniformly with respect to $x, \dot{x}$ and μ.

Set

$$\dot{x} = y - (f(t) - a)x,$$
$$\dot{y} = \big(af(t) - a^2 - \dot{f}(t)\big)x - ay - q(t) + \mu h(t, x, \dot{x}, \mu),$$
$$z = \begin{pmatrix} x \\ y \end{pmatrix}, \quad A(t) = \begin{pmatrix} -f(t) + a & 1 \\ af(t) - a^2 - \dot{f}(t) & -a \end{pmatrix}, \quad X = \begin{pmatrix} 0 \\ h \end{pmatrix},$$
$$X_k = \begin{pmatrix} X_k^1 \\ \big(f(t_k^+) - a\big)X_k^1 + X_k^2 \end{pmatrix},$$
$$B_k = \begin{pmatrix} b_k^1 & 0 \\ \big(f(t_k^+) - a\big)b_k^1 - b_k^2(f(t_k) - a) & b_k^2 \end{pmatrix},$$
$$g_k = \begin{pmatrix} g_k^1 \\ \big(f(t_k^+) - a\big)g_k^1 \end{pmatrix}, \quad g(t) = \begin{pmatrix} 0 \\ -q(t) \end{pmatrix}.$$

Then, we can rewrite system (2.39) in the form

$$\begin{cases} \dot{z} = A(t)z + g(t) + \mu X(t, z, \mu), \quad t \neq t_k, \\ \Delta z(t_k) = B_k z(t_k) + g_k + \mu X_k(z(t_k), \mu), \quad k = \pm 1, \pm 2, \ldots. \end{cases}$$

Now, the conditions for the column dominant of the matrix $A(t)$ are

$$1 < a < \frac{1}{2}\Big(f(t) - 1 + \sqrt{\big(f(t) - 1\big)^2 + 4f(t) - 4\dot{f}(t)}\Big),$$
$$a - f(t) + \big|af(t) - a^2 - \dot{f}(t)\big| < 0,$$

i.e.

$$\big(f(t) - 1\big)^2 < 4\dot{f}(t) < \big(f(t) + 1\big)^2, \tag{2.40}$$
$$2f(t) - \dot{f}(t) - 2 > 0.$$

Theorem 2.6. *Let the following conditions hold:*

1. *Condition H2.23 and the inequalities (2.40) are met.*
2. $b_k^1 b_k^2 + b_k^1 + b_k^2 + 1 \neq 0, \ k = \pm 1, \pm 2, \ldots.$
3. *The functions* $h(t, x, \dot{x}, \mu)$, $X_k(x, \dot{x}, \mu)$ *are Lipschitz continuous with respect to* x *and* $\dot{x}$ *uniformly for* $t \in \mathbb{R}$, $k = \pm 1, \pm 2, \ldots$, *and* $\mu \in M$ *respectively.*

Then there exists a positive constant μ_0, $\mu_0 \in M$ *such that:*

1. *For any* $\mu, |\mu| < \mu_0$ *the system* (2.39) *has a unique almost periodic solution.*
2. *The almost periodic solution is exponentially stable.*
3. *For* $|\mu| \to 0$ *the solution is convergent to the unique almost periodic solution of the system*

$$\begin{cases} \dot{z} = A(t)z + g(t), \ t \neq t_k, \\ \Delta z(t_k) = B_k z(t_k) + g_k, \ k = \pm 1, \pm 2, \dots. \end{cases}$$

Proof. The proof follows directly from Theorem 2.5. □

Now, we shall consider the following systems

$$\begin{cases} \dot{x} = f(t, x), \ t \neq t_k, \\ \Delta x(t_k) = I_k(x(t_k)), \ k = \pm 1, \pm 2, \dots, \end{cases} \tag{2.41}$$

and

$$\begin{cases} \dot{x} = f(t, x) + g(t) + \mu X(t, x, \mu), \ t \neq t_k, \\ \Delta x(t_k) = I_k(x(t_k)) + g_k + \mu X_k(x(t_k), \mu), \ k = \pm 1, \pm 2, \dots. \end{cases} \tag{2.42}$$

Introduce the following conditions:

H2.24. The function $f \in C[\mathbb{R} \times \Omega, \mathbb{R}^n]$ is almost periodic in t uniformly with respect to $x \in \Omega$ and it is Lipschitz continuous with respect to $x \in B_h$ with a Lipschitz constant $l_3 > 0$, such that uniformly in $t \in \mathbb{R}$

$$\|f(t, x) - f(t, y)\| \leq l_3 \|x - y\|, \ x, y \in B_h.$$

H2.25. The sequence of functions $\{I_k\}$, $I_k \in C[\Omega, \mathbb{R}^n]$, $k = \pm 1, \pm 2, \dots$ is almost periodic uniformly with respect to $x \in \Omega$, and the functions I_k are Lipschitz continuous with respect to $x, y \in B_h$ with a Lipschitz constant $l_4 > 0$, such that

$$\|I_k(x) - I_k(y)\| \leq l_4 \|x - y\|,$$

where $x, y \in B_h$, $k = \pm 1, \pm 2, \dots$.

We shall suppose that for the system (2.42) there exists an almost periodic solution $\varphi(t)$, and consider the system

$$\begin{cases} \dot{x} = \dfrac{\partial f}{\partial x}(t, \varphi(t))x, \ t \neq t_k, \\ \\ \Delta x(t_k) = \dfrac{\partial I_k}{\partial x}(\varphi(t_k)), \ k = \pm 1, \pm 2, \dots. \end{cases} \tag{2.43}$$

Let

$$L_1(\delta) = \sup_{t \in \mathbb{R}, \ z \in B_\delta} \|f(t, \varphi(t) + z) - f(t, \varphi(t))\|,$$

$$L_2(\delta) = \sup_{k = \pm 1, \pm 2, \dots, \ z \in B_\delta} \|I_k(\varphi(t_k) + z) - I_k(\varphi(t_k))\|.$$

Theorem 2.7. *Let the following conditions hold:*

1. *Conditions H2.19–H2.25 are met.*
2. *Condition 2 of Theorem 2.5 holds.*
3. *For the Cauchy's matrix $W_1(t,s)$ of the system (2.43), conditions of Lemma 2.3.5 are met.*
4. *There exist positive constants C_0, C_1, C_2 and μ_0 such that*

$$\frac{K}{\lambda}\left(l_3 + \mu_0 l_1 + \sup_{t \in \mathbb{R}}||\frac{\partial f}{\partial x}(t, \varphi(t))||\right) + \frac{K}{1 - e^{-\lambda}}\left(l_3 + \mu_0 l_2\right)$$

$$+ \sup_{k=\pm 1, \pm 2, \dots} ||\frac{\partial I_k}{\partial x}(\varphi(t_k))||\right) < 1,$$

$$\frac{K}{\lambda}\left(C_1 + \mu_0 L_1 + \sup_{t \in \mathbb{R}}||\frac{\partial f}{\partial x}(t, \varphi(t))||\right)$$

$$+ \frac{K}{1 - e^{-\lambda}}\left(C_2 + \mu_0 L_1 + \sup_{k=\pm 1, \pm 2, \dots}||\frac{\partial I_k}{\partial x}(\varphi(t_k))||\right) < C_0.$$

Then there exists a positive constant $\mu_0 \in M$, and for any μ, $|\mu| < \mu_0$, system (2.42) has a unique almost periodic solution, such that:

1. $||x(t, \mu) - \varphi(t)|| \leq C_0.$
2. $\lim_{|\mu| \to 0} x(t, \mu) = x(t, 0).$
3. *The solution $x(t, \mu)$ is exponentially stable.*

Proof. Set $x = z + \varphi(t)$ and from (2.43), it follows the equation

$$\begin{cases} \dot{z} = \dfrac{\partial f}{\partial x}(t, \varphi(t))z + R(t, z) + \mu X(t, z + \varphi(t), \mu), \ t \neq t_k, \\ \Delta z(t_k) = \dfrac{\partial I_k}{\partial z}(\varphi(t_k)) + R_k(z(t_k)) + \mu X(z(t_k) + \varphi(t_k, \mu), \mu), \\ \qquad k = \pm 1, \pm 2, \dots, \end{cases} \qquad (2.44)$$

where

$$R(t, z) = f(t, \varphi(t) + z) - f(t, \varphi(t)) + g(t) - \frac{\partial f}{\partial z}(t, \varphi(t))z,$$

$$R_k(z) = I_k(\varphi(t_k) + z) - I_k(\varphi(t_k)) + g_k - \frac{\partial I_k}{\partial z}(\varphi(t_k)).$$

Let AP, $AP \subset PC[\mathbb{R} \times M, \mathbb{R}^n]$ is the set of all almost periodic functions $\varphi(t, \mu)$, satisfying the inequality $||\varphi|| < C_0$.
Let us define in AP an operator S_μ,

$$S_\mu z = \int_{-\infty}^{t} W_1(t, s)\Big(R(t, z(s)) + \mu X(s, z(s) + \varphi(s), \mu)\Big) ds$$

$$+ \sum_{t_k < t} W_1(t, t_k)\Big(R_k(z(t_k)) + \mu X_k(z(t_k) + \varphi(t_k))\Big). \qquad (2.45)$$

From (2.45), Lemma 1.5, Theorem 1.17, Lemma 2.8 and the conditions of Theorem 2.7 it follows that the operator S_μ is contracting in AP. Hence, there exists a unique almost periodic solution $z(t, \mu)$ of system (2.44). Moreover, $x(t, \mu) = z(t, \mu) + \varphi(t)$ is an almost periodic solution of (2.42). The proof of Assertions 1–3 are analogous to the proof of Theorem 2.5.   □

2.4   Perturbations in the Linear Part

In this paragraph, sufficient conditions for the existence of almost periodic solutions of differential equations with perturbations in the linear part, are obtained.

We shall consider the system of impulsive differential equations

$$\begin{cases} \dot{x} = A(t)x + f(t), \ t \neq t_k, \\ \Delta x(t_k) = A_k x(t_k) + l_k, \ k = \pm 1, \pm 2, \dots, \end{cases} \tag{2.46}$$

where $t \in \mathbb{R}$, $\{t_k\} \in \mathcal{B}$, $A : \mathbb{R} \to \mathbb{R}^{n \times n}$, $f : \mathbb{R} \to \mathbb{R}^n$, $A_k \in \mathbb{R}^{n \times n}$, $l_k \in \mathbb{R}^n$, $k = \pm 1, \pm 2, \dots$. By $x(t) = x(t; t_0, x_0)$ we denote the solution of (2.46) with initial condition $x(t_0^+) = x_0$, $t_0 \in \mathbb{R}$, $x_0 \in \Omega$.

Together with the system (2.46), we shall consider the following systems of impulsive differential equations with perturbations in the linear part:

$$\begin{cases} \dot{x} = \big(A(t) + B(t)\big)x + f(t), \ t \neq t_k, \\ \Delta x(t_k) = \big(A_k + B_k\big)x(t_k) + l_k, \ k = \pm 1, \pm 2, \dots, \end{cases} \tag{2.47}$$

and

$$\begin{cases} \dot{x} = \big(A(t) + B(t)\big)x + F(t, x), \ t \neq t_k, \\ \Delta x(t_k) = \big(A_k + B_k\big)x(t_k) + I_k(x(t_k)), \ k = \pm 1, \pm 2, \dots, \end{cases} \tag{2.48}$$

where $B : \mathbb{R} \to \mathbb{R}^{n \times n}$, $F : \mathbb{R} \times \Omega \to \mathbb{R}^n$, $B_k \in \mathbb{R}^{n \times n}$, and $I_k : \Omega \to \mathbb{R}^n$, $k = \pm 1, \pm 2, \dots$.

Introduce the following conditions:

H2.26. The matrix function $A \in C[\mathbb{R}, \mathbb{R}^{n \times n}]$ is almost periodic in the sense of Bohr.

H2.27. $det(E + A_k) \neq 0$, where E is the identity matrix in $\mathbb{R}^n$, and the sequence $\{A_k\}$, $k = \pm 1, \pm 2, \dots$ is almost periodic.

H2.28. The set of sequences $\{t_k^j\}$, $t_k^j = t_{k+j} - t_k$, $k = \pm 1, \pm 2, \dots$, $j = \pm 1, \pm 2, \dots$ is uniformly almost periodic, and $\inf_k t_k^1 = \theta > 0$.

H2.29. The function $f \in PC[\mathbb{R}, \mathbb{R}^n]$ is almost periodic.

H2.30. The sequence $\{l_k\}$, $k = \pm 1, \pm 2, \dots$ is almost periodic.

H2.31. The matrix function $B \in C[\mathbb{R}, \mathbb{R}^{n \times n}]$ is almost periodic in the sense of Bohr.

H2.32. The sequence $\{B_k\}$, $k = \pm 1, \pm 2, \ldots$ is almost periodic.

Let us denote with $W(t, s)$ the Cauchy matrix for the linear impulsive system

$$\begin{cases} \dot{x} = A(t)x, \ t \neq t_k, \\ \Delta x(t_k) = A_k x(t_k), \ k = \pm 1, \pm 2, \ldots, \end{cases} \tag{2.49}$$

and with $Q(t, s)$ the Cauchy matrix for the linear perturbed impulsive system

$$\begin{cases} \dot{x} = \big(A(t) + B(t)\big)x, \ t \neq t_k, \\ \Delta x(t_k) = \big(A_k + B_k\big)x(t_k), \ k = \pm 1, \pm 2, \ldots. \end{cases}$$

In this part, we shall use the following lemmas:

Lemma 2.11 ([138]). *For the system (2.46) there exists only one almost periodic solution, if and only if:*

1. *Conditions H2.26–H2.30 hold.*
2. *The matrix $W(t, s)$ satisfies the inequality*

$$||W(t, s)|| \leq K e^{-\alpha(t-s)}, \tag{2.50}$$

where $s < t$, $K \geq 1$, $\alpha > 0$.

Lemma 2.12 ([148]). *Let the following conditions hold:*

1. *Conditions H2.26–H2.28, H2.31 and H2.32 hold.*
2. *For $K \geq 1$, $\alpha > 0$ and $s < t$, it follows*

$$||W(t, s)|| \leq K e^{-\alpha(t-s)}.$$

Then:

1. *If there exists a constant $d > 0$ such that*

$$\sup_{t \in (t_0, \infty)} ||B(t)|| < d, \quad \sup_{t_k \in (t_0, \infty)} ||B_k|| < d,$$

then

$$||Q(t, s)|| < K e^{-(\alpha - Kd)(t-s) + i(s,t)}, \tag{2.51}$$

where $s < t$.

2. *If there exists a constant $D > 0$ such that*

$$\int_{t_0}^{\infty} ||B(\sigma)||d\sigma + \sum_{t_0 \le t_k} ||B_k|| \le D,$$

then

$$||Q(t,s)|| \le Ke^{KD}e^{-\alpha(t-s)}, \qquad (2.52)$$

where $s < t$.

The proof of the next lemma is similar to the proof of Lemma 1.7.

Lemma 2.13. *Let the conditions H2.26–H2.32 hold. Then for each $\varepsilon > 0$ there exist ε_1, $0 < \varepsilon_1 < \varepsilon$, a relatively dense set $\overline{T}$ of real numbers and a set P of integer numbers, such that the following relations are fulfilled:*

(a) $||A(t+\tau) - A(t)|| < \varepsilon$, $t \in \mathbb{R}$, $\tau \in \overline{T}$.
(b) $||B(t+\tau) - B(t)|| < \varepsilon$, $t \in \mathbb{R}$, $\tau \in \overline{T}$.
(c) $||f(t+\tau) - f(t)|| < \varepsilon$, $t \in \mathbb{R}$, $\tau \in \overline{T}$.
(d) $||A_{k+q} - A_k|| < \varepsilon$, $q \in P$, $k = \pm 1, \pm 2, \ldots$.
(e) $||B_{k+q} - B_k|| < \varepsilon$, $q \in P$, $k = \pm 1, \pm 2, \ldots$.
(f) $||l_{k+q} - l_k|| < \varepsilon$, $q \in P$, $k = \pm 1, \pm 2, \ldots$.
(g) $|t_k^q - \tau| < \varepsilon_1$, $q \in P$, $\tau \in \overline{T}$, $k = \pm 1, \pm 2, \ldots$.

Lemma 2.14 ([148]). *Let the conditions H2.31 and H2.32 hold. Then there exist positive constants d_1, and d_2, such that*

$$\sup_{t \in (t_0, \infty)} ||B(t)|| < d_1, \qquad \sup_{t_k \in (t_0, \infty)} ||B_k|| < d_2.$$

Lemma 2.15. *Let the following conditions hold:*

1. *Conditions H2.26–H2.28, H2.31 and H2.32 are met.*
2. *The following inequalities hold*

   (a) $||W(t,s)|| \le Ke^{-\alpha(t-s)}$, *where $s < t$, $K \ge 1$ and $\alpha > 0$,*
   (b) $\nu = -\alpha - Kd - N(1 + Kd) > 0$,

where $d = max(d_1, d_2)$, d_1 and d_2 are from Lemma 2.14, N is the number of the points t_k lying in the interval (s,t).

   Then for each $\varepsilon > 0$, $t \in \mathbb{R}$, $s \in \mathbb{R}$ there exists a relatively dense set T of ε-almost periods, common for $A(t)$ and $B(t)$ such that for each $\tau \in T$ the following inequality holds

$$||Q(t+\tau, s+\tau) - Q(t,s)|| < \varepsilon \Gamma e^{-\frac{\nu}{2}(t-s)}, \qquad (2.53)$$

where $\Gamma = \frac{1}{\nu}2Ke^{N \ln(1+Kd)}(1 + N + \frac{Nd}{2})$.

Proof. Let $\overline{T}$ and P be the sets, defined in Lemma 2.13.

Then for $\tau \in \overline{T}$ and $q \in P$ the matrix $Q(t + \tau, s + \tau)$ is a solution of the system

$$
\begin{cases}
\dfrac{\partial Q}{\partial t} = \big(A(t) + B(t)\big)Q(t + \tau, s + \tau) \\
\quad + \big(A(t + \tau) + B(t + \tau) - A(t) - B(t)\big)Q(t + \tau, s + \tau),\ t \neq t'_k, \\
\Delta Q(t'_k) = \big(A_k + B_k\big)\big(Q(t'_k + \tau, s + \tau)\big) \\
\quad + \big(A_{k+q} + B_{k+q} - A_k - B_k\big)Q(t'_k + \tau, s + \tau),
\end{cases}
$$

where $k = \pm 1, \pm 2, \ldots,\ t'_k = t_k - \tau$.
    Then

$$
Q(t + \tau, s + \tau) - Q(t, s) = \int_s^t Q(t, s)\big(A(\sigma + \tau) + B(\sigma + \tau) - A(\sigma)
$$

$$
- B(\sigma)\big)Q(\sigma + \tau, s + \tau)d\sigma + \sum_{s \leq t'_\nu < t} Q(t, t'^+_\nu)
$$

$$
\times \big(A_{k+q} + B_{k+q} - A_k - B_k\big)Q(t'_\nu + \tau, s + \tau).
$$

From Lemmas 1.2 and 2.13, we have

$$
\|Q(t + \tau, s + \tau) - Q(t, s)\| \leq \varepsilon K e^{N \ln(1 + Kd)}\big(e^{-\nu(t-s)}(t - s)
$$

$$
+ i(s, t)e^{-\nu(t-s)}\big) \leq \varepsilon \Gamma e^{-\frac{\nu}{2}(t-s)}. \qquad \qquad \square
$$

The proof of the next lemma is analogously.

Lemma 2.16. *Let the following conditions hold:*

1. *Conditions H2.26–H2.28, H2.31 and H2.32 are met.*
2. *The following inequalities hold*

    (a) $\|W(t, s)\| \leq K e^{-\alpha(t-s)}$, *where* $s < t$, $K \geq 1$, $\alpha > 0$,

    (b) $\displaystyle\int_{t_0}^{\infty} \|B(\sigma)\|d\sigma + \sum_{t_0 < t_k} \|B_k\| \leq D$, $D > 0$, *where* $s < t$, $D > 0$.

Then for each $\varepsilon > 0$, $t \in \mathbb{R}$, $s \in \mathbb{R}$ *there exists a relatively dense set* $\overline{T}$ *of* ε-*almost periods, common for* $A(t)$ *and* $B(t)$ *such that for each* $\tau \in \overline{T}$ *the following inequality holds*

$$
\|Q(t + \tau, s + \tau) - Q(t, s)\| < \varepsilon \overline{\Gamma} e^{-\frac{\alpha}{2}(t-s)}, \tag{2.54}
$$

where $\overline{\Gamma} = K e^{KD}\dfrac{2}{\alpha}\Big(1 + N + \dfrac{2N}{\alpha}\Big)$.

Now, we are ready to proof the main results in this paragraph.

Theorem 2.8. *Let the following conditions hold:*

1. *Conditions H2.26–H2.32 are met.*
2. *For the system (2.46), there exists a unique almost periodic solution.*

Then there exists a constant d_0 such that for $d \in (0, d_0]$ for the system (2.47) there exists a unique almost periodic solution $\varphi(t)$, and

$$||\varphi(t)|| \leq C max \left(\sup_{t \in \mathbb{R}} ||f||, \sup_{k=\pm 1, \pm 2, \ldots} ||l_k|| \right), \tag{2.55}$$

where $C > 0$.

Proof. Let the inequalities (2.50) and (2.51) hold, and let we consider the function

$$\varphi(t) = \int_{-\infty}^{t} Q(t, s) f(s) ds + \sum_{t_k < t} Q(t, t_k^+) l_k.$$

A straightforward verification yields, that $\varphi(t)$ is a solution of (2.47).   □

Then, from Lemma 2.15 it follows that there exists a constant $d_0 > 0$ such that for any $d \in (0, d_0]$, we have

$$\nu = \alpha - Kd - N \ln(1 + Kd) > 0.$$

Now, we obtain

$$||\varphi(t)|| \leq \frac{K}{\nu} \sup_{t \in \mathbb{R}} ||f(t)|| + K e^{N \ln(1 + K_1 d)} \sup_{k=\pm 1, \pm 2, \ldots} ||l_k|| \sum_{t_k < t} e^{-\nu(t - t_k)}. \tag{2.56}$$

Then, from the relations

$$\sum_{t_k < t} e^{-\nu(t - t_k)} = \sum_{k=0}^{\infty} \sum_{t-k-1 < t_k < t-k} e^{-\nu(t - t_k)} \leq \frac{2N}{1 - e^{-\nu}},$$

and (2.56), we obtain

$$||\varphi(t)|| \leq C max \left(\sup_{t \in \mathbb{R}} ||f(t)||, \sup_{k=\pm 1, \pm 2, \ldots} ||l_k|| \right),$$

where $C = K e^{N \ln(1 + Kd)} \left(\frac{1}{\nu} + \frac{2N}{1 - e^{-\nu}} \right)$.

Let $\varepsilon > 0$ be an arbitrary chosen constant. It follows from Lemma 2.13, that there exist sets $\overline{T}$ and P, such that for each $\tau \in \overline{T}$, $q \in P$, and $d \in (0, d_0]$

the following estimates hold:

$$||\varphi(t+\tau) - \varphi(t)|| \le \int_{-\infty}^{t} ||Q(t+\tau, \sigma+\tau) - Q(t,\sigma)|| \, ||f(\sigma+\tau)|| d\sigma$$

$$+ \int_{-\infty}^{t} ||Q(t,\sigma)|| \, ||f(\sigma+\tau) - f(\sigma)|| d\sigma$$

$$+ \sum_{t_k < t} ||Q(t+\tau, t_{k+q}^+) - Q(t, t_k^+)|| \, ||l_{k+q}||$$

$$+ \sum_{t_k < t} ||Q(t, t_k^+)|| \, ||l_{k+q} - l_k|| \le M\varepsilon,$$

where $M > 0$, $|t - t_k| > \varepsilon$.

The last inequality implies, that the function $\varphi(t)$ is almost periodic.

The uniqueness of this solution follows from the fact that the homogeneous part of system (2.47) has only the zero bounded solution under conditions H2.26, H2.27, H2.31 and H2.32, and from the estimate (2.50). □

Theorem 2.9. *Let the following conditions hold:*

1. *Conditions H2.26–H2.32 are met.*
2. *For the system (2.46), there exists a unique almost periodic solution.*
3. *There exists a constant $D_0 > 0$, such that*

$$\int_{t_0}^{\infty} ||B(\sigma)|| d\sigma + \sum_{t_0 < t_k} ||B_k|| < D_0.$$

Then, for $D \in (0, D_0]$ for the system (2.47), there exists a unique almost periodic solution $\varphi(t)$ such that

$$||\varphi(t)|| \le C max\Big(\sup_{t \in \mathbb{R}} ||f||, \sup_{k=\pm1,\pm2,\dots} ||l_k|| \Big),$$

where $C > 0$.

Proof. Using Lemma 2.16 and (2.52), the proof of Theorem 2.9 is carried out in the same way as the proof of Theorem 2.8. □

Theorem 2.10. *Let the following conditions hold:*

1. *Conditions H2.26–H2.30 are met.*
2. *For the system (2.46), there exists a unique almost periodic solution.*
3. *$B(t) = B$, $B_k = \Lambda$, where B and Λ are constant matrices such that*

$$||B|| + ||\Lambda|| \le d_1, \ d_1 > 0.$$

Then there exists a constant $d_0 > 0$, $d_0 \le d_1$, such that for $d \in (0, d_0]$ for the system (2.47), there exists a unique almost periodic solution.

Proof. The proof of Theorem 2.10 is carried out in the same way as the proof of Theorem 2.8. □

Example 2.2. We shall consider the systems

$$\begin{cases} \dot{x} = -x + f(t), \ t \neq t_k, \\ \Delta x(t_k) = l_k, \ k = \pm 1, \pm 2, \ldots, \end{cases} \tag{2.57}$$

and

$$\begin{cases} \dot{x} = \big(b(t) - 1\big)x + f(t), \ t \neq t_k, \\ \\ \Delta x(t_k) = l_k + g_k, \ k = \pm 1, \pm 2, \ldots, \end{cases} \tag{2.58}$$

where $t \in \mathbb{R}$, $x \in \mathbb{R}$, $\{t_k\} \in \mathcal{B}$, the function $b \in C[\mathbb{R}, \mathbb{R}]$ is almost periodic in the sense of Bohr, the function $f \in PC[\mathbb{R}, \mathbb{R}]$ is almost periodic, $b_k \in \mathbb{R}$, $l_k \in \mathbb{R}$ and $\{b_k\}$, $\{l_k\}$, $k = \pm 1, \pm 2, \ldots$, are almost periodic sequences.

Let condition H2.28 holds. From [138] it follows that for the system (2.57) there exists a unique almost periodic solution.

Then, the conditions of Theorem 2.8. are fulfilled, and hence, there exists a constant d_0 such that for any $d \in (0, d_0]$ for the system (2.58), there exists a unique almost periodic solution in the form

$$x(t) = \int_{-\infty}^{t} Q(t, \sigma) f(\sigma) d\sigma + \sum_{t_k < t} Q(t, t_k^+) l_k,$$

where

$$Q(t, s) = \prod_{s \leq t_k < t} (1 + b_k) exp\Big\{ \int_{s}^{t} b(\sigma) d\sigma - (t - s) \Big\}.$$

Now, we shall investigate the existence of almost periodic solutions for the system (2.48).

Introduce the following conditions:

H2.33. The function $F \in C[\mathbb{R} \times \Omega, \mathbb{R}^n]$ is almost periodic in t uniformly with respect to $x \in \Omega$, and it is Lipschitz continuous with respect to $x \in B_h$ with a Lipschitz constant $L > 0$,

$$||F(t, x) - F(t, y)|| \leq L||x - y||, \ x, y \in B_h, \ t \in \mathbb{R}.$$

H2.34. The sequence of functions $\{I_k(x)\}$, $I_k \in C[\Omega, \mathbb{R}^n]$ is almost periodic uniformly with respect to $x \in \Omega$, and the functions $I_k(x)$ are Lipschitz continuous with respect to $x \in B_h$ with a Lipschitz constant $L > 0$,

$$||I_k(x) - I_k(y)|| \leq L||x - y||, \ x, y \in B_h, \ k = \pm 1, \pm 2, \ldots.$$

Theorem 2.11. *Let the following conditions hold:*

1. Conditions H2.26–H2.28, H2.31–H2.34 are met.

2. *For the functions* $F(t,x)$ *and* $I_k(x)$, $k = \pm 1, \pm 2, \ldots$, *there exists a constant* $L_1 > 0$ *such that*

$$max\left(\sup_{t \in \mathbb{R}, x \in B_h} ||F(t,x)||, \quad \sup_{k = \pm 1, \pm 2, \ldots, \ x \in B_h} ||I_k(x))|| \right) \leq L_1.$$

3. *The inequalities (2.50) and*

$$CL_1 < h, \quad CL < 1. \tag{2.59}$$

   hold.

Then there exists a constant $d_0 > 0$ *such that for any* $d \in (0, d_0]$, *for the system (2.48) there exists a unique almost periodic solution.*

Proof. Let we denote by AP the set of all almost periodic solutions $\varphi(t)$, $\varphi \in PC[\mathbb{R}, \mathbb{R}^n]$, satisfy the inequality $||\varphi|| < h$, and let $|\varphi|_\infty = \sup_{t \in \mathbb{R}} ||\varphi(t)||$.

We define in AP the operator S, such that if $\varphi \in AP$, then $y = S\varphi(t)$ is the almost periodic solution of the system

$$\begin{cases} \dot{y} = (A(t) + B(t))y + F(t, \varphi(t)), \ t \neq t_k, \\ \Delta y(t_k) = (A_k + B_k)y(t_k) + I_k(\varphi(t_k)), \ k = \pm 1, \pm 2, \ldots, \end{cases}$$

determined by Theorem 2.8.

We shall note that the almost periodicity of the sequence $\{\varphi(t_k)\}$, the function $F(t, \varphi(t))$ and the sequence $\{I_k(\varphi(t_k))\}$ follows from Lemma 1.5 and Theorem 1.17.

On the other hand, there exists a positive constant $d_0 > 0$ such that for any $d \in (0, d_0]$,

$$\alpha - Kd - N \ln(1 + Kd) > 0.$$

From the last inequality and (2.59), it follows that (2.51) and conditions of Lemma 2.15 hold.

Then $S(AP) \subset AP$.

If $\varphi \in AP$, $\psi \in AP$, then from (2.51) and condition 2 of Theorem 2.11, we get

$$||S\varphi(t) - S\psi(t)|| \leq CL|\varphi - \psi|_\infty. \tag{2.60}$$

Finally, from (2.59) and (2.60,) it follows that S is contracting in AP, i.e. there exists a unique almost periodic solution of system (2.48). □

2.5 Strong Stable Impulsive Differential Equations

In this section, conditions for strong stability and almost periodicity of solutions of impulsive differential equations with impulsive effect at fixed moments will be proved. The investigations are carried out by means of piecewise continuous Lyapunov functions.

We shall consider the system of impulsive differential equations

$$\begin{cases} \dot{x} = f(t,x), \ t \neq t_k, \\ \Delta x(t_k) = I_k(x(t_k)), \ k = \pm 1, \pm 2, \ldots, \end{cases} \tag{2.61}$$

where $t \in \mathbb{R}$, $\{t_k\} \in \mathcal{B}$, $f : \mathbb{R} \times \Omega \to \mathbb{R}^n$, $I_k : \Omega \to \mathbb{R}^n$, $k = \pm 1, \pm 2, \ldots$.

Set

$$\rho(x,y) = ||x - y||, \ x,y \in \mathbb{R}^n,$$

$$B_h(a) = \{x \in \mathbb{R}^n, \ ||x - a|| < h\}, \ h > 0, \ a \in \mathbb{R}^n,$$

$$\Psi_h = \{(t,x) \in \mathbb{R} \times B_h, \ x \in B_h, \text{ if } (t,x) \in G \text{ and } x + I_k(x) \in B_h,$$

$$\text{if } t = t_k\},$$

where G is the set from Sect. 1.1.

Introduce the following conditions:

H2.35. The function $f \in C[\mathbb{R} \times B_h, \mathbb{R}^n]$, and has continuous partial derivatives of the first order with respect to all components of $x \in B_h$.

H2.36. The functions $I_k \in C[B_h, \mathbb{R}^n]$, $k = \pm 1, \pm 2, \ldots$ and have continuous partial derivatives of the first order with respect to all components of $x \in B_h$.

H2.37. There exists h_0, $0 < h_0 < h$ such that if $x \in B_{h_0}$, then $x + I_k(x) \in B_h$, $k = \pm 1, \pm 2, \ldots$.

H2.38. The functions $L_k(x) = x + I_k(x)$, $k = \pm 1, \pm 2, \ldots$ are such that $L_k^{-1}(x) \in B_h$ for $x \in B_h$.

From [138] if the conditions H2.35–H2.38 are satisfied, then for each point $(t_0, x_0) \in \mathbb{R} \times B_h$, there exists a unique solution $\bar{x}(t) = \bar{x}(t; t_0, x_0)$ of system (2.61), which satisfies the initial condition $x(t_0^+) = x_0$.

We need the following condition in our subsequent analysis:

H2.39. $f(t,0) = 0$, $I_k(0) = 0$ for $t \in \mathbb{R}$ and $k = \pm 1, \pm 2, \ldots$, respectively.

If the conditions H2.35–H2.39 hold, then there exists a zero solution for system (2.61).

Definition 2.4 ([90]). The zero solution $x(t) \equiv 0$ of system (2.61) is said to be *strongly stable*, if

$$(\forall \varepsilon > 0)(\exists \delta > 0)(\forall t_0 \in \mathbb{R})(\forall x_0 \in B_\delta \; : \; (t_0, x_0) \in \Psi_\delta)$$

$$(\forall t \in \mathbb{R}) : \rho(\overline{x}(t; t_0, x_0), 0) < \varepsilon.$$

Definition 2.5 ([90]). An arbitrary solution $\overline{x}(t) = \overline{x}(t; t_0, x_0)$ of (2.61) is said to be *strongly stable*, if

$$(\forall \varepsilon > 0)(\forall \eta > 0)(\exists \delta > 0)(\forall \tau_1 \in \mathbb{R}, \; \forall \tau_2 \in \mathbb{R}, \; \rho(\overline{x}(\tau_1), \overline{x}(\tau_2)) < \delta)$$

$$(\forall t \in \mathbb{R}) : \rho(\overline{x}(t + \tau_1), \overline{x}(t + \tau_2)) < \varepsilon.$$

Definition 2.6. The function $V \in V_0$ belongs to the class V_0^*, if V has continuous partial derivatives on the sets G_k.

For each function $V \in V_0^*$, we define the function

$$\dot{V}(t, x) = \frac{\partial V(t, x)}{\partial t} + \sum_{i=1}^{n} \frac{\partial V(t, x)}{\partial x_i} f_i(t, x)$$

for $(t, x) \in G$.

If $\overline{x}(t)$ is a solution of system (2.61), then

$$\frac{d}{dt} V(t, \overline{x}(t)) = \dot{V}(t, \overline{x}(t)), \; t \in \mathbb{R}, \; t \neq t_k.$$

Definition 2.7. The function $V \in V_0$ belongs to the class V_0^{**}, if V has continuous partial derivatives of the second order in the sets G_k.

Let $V \in V_0^{**}$. If the function $f(t, x)$ satisfies condition H2.35 and has a continuous partial derivative with respect to t, we can define the function

$$\ddot{V}(t, x) = \frac{\partial \dot{V}(t, x)}{\partial t} + \sum_{i=1}^{n} \frac{\partial \dot{V}(t, x)}{\partial x_i} f_i(t, x)$$

for $(t, x) \in G$.

In the further considerations, we shall use the next class K of functions

$$K = \big\{ a \in C[\mathbb{R}, \mathbb{R}^+], \text{ a is strictly increasing and } a(0) = 0 \big\}.$$

Introduce the following conditions:

H2.40. The function $f(t, x)$ is almost periodic in t uniformly with respect to x, $x \in B_h$.
H2.41. The sequence $\{I_k(x)\}$, $k = \pm 1, \pm 2, \ldots$, is almost periodic uniformly with respect to x, $x \in B_h$.

H2.42. The set of sequences $\{t_k^j\}$, $t_k^j = t_{k+j} - t_k$, $k = \pm 1, \pm 2, \ldots$, $j = \pm 1, \pm 2, \ldots$, is uniformly almost periodic, and $\inf_k t_k^1 = \theta > 0$.

Definition 2.8 ([114]). The set S, $S \subset \mathbb{R}$ is said to be:

(a) $\Delta - m$ set, if from every $m + 1$ real numbers $\tau_1, \tau_2, \ldots, \tau_{m+1}$ one can find $i \neq j$, such that $\tau_i - \tau_j \in S$.

(b) *symmetric* $\Delta - m$ set, if S is $\Delta - m$ set symmetric with respect to the number 0.

Lemma 2.17 ([114]). *Every symmetric* $\Delta - m$ *set is relatively dense.*

Theorem 2.12. *Let conditions H2.35–H2.42 hold. Then any strongly stable bounded solution of (2.61) is almost periodic.*

Proof. Let $\bar{x} = \bar{x}(t; t_0, x_0)$ be a unique bounded solution of system (2.61) with initial condition $\bar{x}(t_0) = x_0$. Let $\varepsilon > 0$ be given, $\delta(\varepsilon) > 0$, and the points $a_1, a_2, \ldots, a_{N+1}$, $a_l \in \mathbb{R}^n$, $l = 1, 2, \ldots, N + 1$, are such that for $t \in \mathbb{R}$, $t \geq t_0$, it follows that $\bar{x}(t) \in B_{\frac{\delta}{2}}(a_l)$. If $t_0, \ldots, t_{N+1}$ are given real numbers, then for some $i \neq j$ and some $l \in \{1, \ldots, N + 1\}$, we get

$$\rho(\bar{x}(\tau_i), a_l) < \frac{\delta(\varepsilon)}{2}, \quad \rho(\bar{x}(\tau_j), a_l) < \frac{\delta(\varepsilon)}{2}.$$

Consequently, $\rho(\bar{x}(t_i), \bar{x}(t_j)) < \delta(\varepsilon)$.

On the other hand, the solution $\bar{x}(t)$ is strongly stable, i.e. it follows that $\rho(\bar{x}(t + \tau_i), \bar{x}(t + \tau_j)) < \varepsilon$, where $t \in \mathbb{R}$.

Then, for $t \in \mathbb{R}$ we have $\rho(\bar{x}(t + \tau_i - \tau_j), \bar{x}(t)) < \varepsilon$ and consequently, $\tau_i - \tau_j$ is an ε-almost period of the solution $\bar{x}(t)$.

Let $\overline{T}$ be the set of all ε-almost periods of $x(t)$. Then, for any sequence of numbers $\tau_0, \ldots, \tau_N$ from above, it follows that there exists $i \neq j$, such that $\tau_i - \tau_j \in \overline{T}$.

From Definition 2.8, we get that $\overline{T}$ is a symmetric $\Delta - N$ set, and from Lemma 2.17, it follows that $\overline{T}$ is a relatively dense set. Then, $\bar{x}(t)$ is an almost periodic function. $\square$

Let $\bar{x}(t)$ be a solution of the system (2.61). Set $z = x - \bar{x}(t)$, and consider the system

$$\begin{cases} \dot{z} = g(t, z), \ t \neq t_k, \\ \Delta z(t_k) = J_k(z(t_k)), \ k = \pm 1, \pm 2, \ldots, \end{cases} \quad (2.62)$$

where $g(t, z) = f(t, z + \bar{x}(t)) - f(t, \bar{x}(t))$, $J_k(z) = I_k(z + \bar{x}) - I_k(\bar{x})$.

Theorem 2.13. *Let the following conditions hold:*

1. *Conditions H2.35–H2.42 are met.*
2. *There exist functions $V \in V_0^*$ and a, $b \in K$ such that:*

(a) $a(||z||) \leq V(t, z) \leq b(||z||)$, $(t, z) \in \mathbb{R} \times B_h$.
(b) $\dot{V}(t, z) \equiv 0$, for $(t, z) \in \mathbb{R} \times B_h$, $t \neq t_k$.
(c) $V(t_k^+, z + I_k(z)) = V(t_k, z)$, $k = \pm 1, \pm 2, \ldots$, $z \in B_h$.

Then the solution $\overline{x}(t)$ of (2.61) is almost periodic.

Proof. Let $0 < \varepsilon < h$, $0 < \mu < h$ be given, and let

$$\delta = \delta(\varepsilon) < min\{\varepsilon, \ b^{-1}(a(\varepsilon)), \ b^{-1}(a(\mu))\},$$

where $a, b \in \mathcal{K}$. If $z(t) = z(t; t_0, z_0)$ be a solution of (2.62) such that $t_0 \in \mathbb{R}$, $(t_0, x_0) \in S_\delta$, then from condition 2 of Theorem 2.13, it follows that

$$a(||z||) \leq V(t, z(t)) = V(t_0^+, z_0) \leq b(||z_0||) < b(\delta(\epsilon)) < min\{a(\varepsilon), a(\mu)\}.$$

Consequently, $||z(t; t_0, z_0)|| < min(\varepsilon, \mu)$ for $t \in \mathbb{R}$, i.e. the zero solution of (2.62) is strongly stable. Then, $\overline{x}(t)$ is strongly stable, and from conditions H2.40–H2.42, and Theorem 2.12, it follows that $\overline{x}(t)$ is almost periodic. $\square$

Definition 2.9 ([90]). The zero solution of system (2.62) is said to be *uniformly stable to the right (to the left)*, if for any $\varepsilon > 0$ there exists $\delta(\varepsilon) > 0$, such that if $t_0 \in \mathbb{R}$ and $(t_0, z_0) \in \mathbb{R} \times B_{\delta(\varepsilon)}$, then $||z(t; t_0, z_0)|| < \varepsilon$ for all $t \geq t_0$ (for all $t \leq t_0$), where $z(t; t_0, z_0)$ is a solution of (2.62) such that $z(t_0^+) = z_0$.

Lemma 2.18 ([90]). *The zero solution of system (2.62) is uniformly stable to the left if and only if for any $\varepsilon > 0$ the following inequality holds:*

$$\gamma(\varepsilon) = inf\{||z(t; t_0, z_0)|| \ : \ t_0 \in \mathbb{R}, \ ||z_0|| \geq \varepsilon\} > 0.$$

Lemma 2.19 ([90]). *The zero solution of system (2.62) is strongly stable if and only if it is stable to the left and to the right at the same time.*

Example 2.3. We shall consider the linear impulsive system of differential equations

$$\begin{cases} \dot{x} = A(t)x, \ t \neq t_k, \\ \Delta x(t_k) = B_k x(t_k), \ k = \pm 1, \pm 2, \ldots, \end{cases} \qquad (2.63)$$

where $A(t)$ is a square matrix, the elements of which are almost periodic continuous functions for $t \in \mathbb{R}$, $\{B_k\}$ is an almost periodic sequence of constant matrices such that $det(E + B_k) \neq 0$, and for the points t_k the condition H2.42 is fulfilled. Let $W(t, s)$ be the Cauchy matrix of system (2.63).

Since the nontrivial solution of (2.63) is given by the formula $x(t; t_0, x_0) = W(t, t_0)x_0$, then $x_0 = W^{-1}(t, t_0)x(t; t_0, x_0)$. Hence, for any $\varepsilon > 0$ and $||x_0|| \geq \varepsilon$, we have

$$\varepsilon \leq ||x_0|| \leq ||W^{-1}(t, t_0)|| ||x(t; t_0, x_0)||,$$

and

$$||x(t; t_0, x_0)|| \geq \varepsilon ||W^{-1}(t, t_0)||^{-1}.$$

However, for $t = t_0$ and $||x_0|| = \varepsilon$, we have

$$||x(t; t_0, x_0)|| = \varepsilon ||W^{-1}(t, t_0)||^{-1}.$$

Hence,

$$\gamma(\varepsilon) = inf\left\{ \varepsilon ||W^{-1}(t, t_0)||^{-1} \ : \ t \geq t_0 \right\} > 0$$

and, applying Lemma 2.18, we conclude that the zero solution of system (2.63) is uniformly stable to the left if and only if the function $||W^{-1}(t, s)||$ is bounded on the set $s \leq t < \infty$. Moreover, it is clear that the zero solution of (2.63) is uniformly stable to the right if and only if the function $||W(t, s)||$ is bounded on the set $s \leq t < \infty$. Then, by virtue of Lemma 2.18, the zero solution of system (2.63) is strongly stable if and only if the functions $||W(t, t_0)||$ and $||W^{-1}(t, t_0)||$ are bounded for $t \in \mathbb{R}$. Consequently, an arbitrary solution $x(t)$ of the system (2.63) is bounded and strongly stable. From Theorem 2.7, it follows that the solution $z(t)$ is almost periodic.

Now, we consider the following scalar impulsive differential equations:

$$\begin{cases} \dot{u} = \omega_1(t, u), \ t \neq t_k, \\ \Delta u(t_k) = P_k(u(t_k)), \ k = \pm 1, \pm 2, \ldots, \end{cases} \tag{2.64}$$

where $\omega_1 : [t_0 - T, t_0] \times \chi \to \mathbb{R}$, χ is an open interval in $\mathbb{R}$, and t_0 and T are constants such that $t_0 > T$, $P_k : \chi \to \chi$;

$$\begin{cases} \dot{v} = \omega_2(t, v), \ t \neq t_k, \\ \Delta v(t_k) = P_k(v(t_k)), \ k = \pm 1, \pm 2, \ldots, \end{cases} \tag{2.65}$$

where $\omega_2 : [t_0, t_0 + T] \times \chi \to \mathbb{R}$;

$$\begin{cases} \ddot{u} = \omega(t, u, \dot{u}), \ t \neq t_k, \\ \Delta u(t_k) = A_k(u(t_k)), \ k = \pm 1, \pm 2, \ldots, \\ \Delta \dot{u}(t_k) = B_k(u(t_k), \dot{u}(t_k)), \ k = \pm 1, \pm 2, \ldots, \end{cases} \tag{2.66}$$

where $\omega : [t_0 - T, t_0 + T] \times \chi_1 \times \chi_2 \to \mathbb{R}$, $A_k : \chi_2 \to \chi_1$, $B_k : \chi_1 \times \chi_2 \to \chi_2$, χ_1 and χ_2 are open intervals in $\mathbb{R}$.

Theorem 2.14. *Let the following conditions hold:*

1. *Conditions H2.35–H2.42 are met.*
2. *The zero solution $u(t) \equiv 0$, $(v(t) \equiv 0)$ of (2.64), (2.65) is uniformly stable to the left (to the right).*
3. *The functions $u + P_k(u)$, $k = \pm 1, \pm 2, \ldots$, are monotone increasing in $\mathbb{R} \times B_h$.*

4. *There exist functions $V \in V_0^*$ and $a, b \in K$ such that*

    (a) $a(||z||) \leq V(t, z) \leq b(||z||)$, $(t, z) \in \mathbb{R} \times B_h$.
    (b) $\omega_1(t, V(t, z)) \leq \dot{V}(t, z) \leq \omega_2(t, V(t, z))$ $(t, z) \in \mathbb{R} \times B_h$.
    (c) $V(t_k^+, z + J_k(z)) = V(t_k, z) + P_k(V(t_k, z))$, $k = \pm 1, \pm 2, \dots$.

5. *The solution $\bar{x}(t)$ of system (2.61) is bounded.*

Then the solution $\bar{x}(t)$ of system (2.61) is almost periodic.

Proof. From conditions of the theorem and [90], it follows that the zero solution of system (2.61) is strongly stable, i.e. the solution $\bar{x}(t)$ is strongly stable. Then, from H2.40–H2.42 and Theorem 2.12, it follows that $\bar{x}(t)$ is almost periodic.   □

Definition 2.10 ([90]). The zero solution $x(t) \equiv 0$ of (2.66) is said to be *u-strongly stable*, if

$$(\forall \varepsilon > 0)(\exists \delta > 0)(\forall t_0 \in \mathbb{R})(\forall u_0 : 0 \leq u_0 < \delta(\varepsilon))(\forall \dot{u}_0 \in \mathbb{R} : |\dot{u}_0| < \delta(\varepsilon))$$

$$(\forall t \in \mathbb{R}) : 0 \leq u(t; t_0, u_0, \dot{u}_0) < \varepsilon.$$

Theorem 2.15. *Let the following conditions hold:*

1. *Conditions H2.35–H2.42 are met.*
2. *The function $g(t, x)$ has continuous partial derivative of the first kind with respect to t.*
3. *There exist functions $V \in V_0^{**}$ and $a, b \in K$, such that*

    (a) $a(||z||) \leq V(t, z) \leq b(||z||)$, $(t, z) \in \mathbb{R} \times B_h$.

    (b) $\dot{V}(t, z) \leq c||z||$, $c = const > 0$, $(t, z) \in G$.

    (c) $\ddot{V}(t, z) \leq \omega(t, V(t, z), \dot{V}(t, z))$ *for* $(t, z) \in \mathbb{R} \times B_h$, $t \neq t_k$,
        where $\omega(t, u_1, u_2)$, $\omega : \mathbb{R}^3 \to \mathbb{R}^+$ *is continuous and monotone increasing on u_1 and $\omega(t, 0, 0) = 0$ for $t \in \mathbb{R}$.*
    (d) $V(t_k^+, z + J_k(z)) \leq V(t_k, z) + A_k(\dot{V}(t_k, z))$.

    (e) $\dot{V}(t_k^+, z + J_k(z)) \leq \dot{V}(t_k, z) + B_k(V(t_k, z), \dot{V}(t_k, z))$, $k = \pm 1, \pm 2, \dots, z \in B_h$.

4. *The following inequalities hold*

$$u_1 + A_k(v_1) \leq u_2 + A_k(v_2),$$
$$v_1 + B_k(u_1, v_1) \leq v_2 + B_k(u_2, v_2)$$

for $u_1 \leq u_2$, $v_1 \leq v_2$, where $u_1, u_2 \in \chi_1$, $v_1, v_2 \in \chi_2$, $k = \pm 1, \pm 2, \dots$.

5. *The zero solution of equation (2.66) is strongly u-stable.*
6. *The solution $\overline{x}(t)$ of system (2.61) is bounded.*

Then the solution $\overline{x}(t)$ of system (2.61) is almost periodic.

Proof. The proof of Theorem 2.15 is analogous to the proof of Theorem 2.14.

$\square$

2.6   Dichotomies and Almost Periodicity

In this part, the existence of an almost periodic projector-valued function of dichotomous impulsive differential systems with impulsive effects at fixed moments is considered.

First, we shall consider the linear system of impulsive differential equations

$$\begin{cases} \dot{x} = A(t)x, \ t \neq t_k, \\ \Delta x(t_k) = B_k x(t_k), \ k = \pm 1, \pm 2, \ldots, \end{cases} \tag{2.67}$$

where $t \in \mathbb{R}$, $\{t_k\} \in \mathcal{B}$, $A : \mathbb{R} \to \mathbb{R}^{n \times n}$, $B_k \in \mathbb{R}^{n \times n}$, $k = \pm 1, \pm 2, \ldots$.

By $x(t) = x(t; t_0, x_0)$ we denote the solution of (2.67) with initial condition $x(t_0^+) = x_0$, $x_0 \in \mathbb{R}^n$.

Introduce the following conditions:

H2.43. The matrix-valued function $A \in PC[\mathbb{R}, \mathbb{R}^{n \times n}]$ is almost periodic.

H2.44. $\{B_k\}$, $k = \pm 1, \pm 2, \ldots$ is an almost periodic sequence.

H2.45. $det(E + B_k) \neq 0$, $k = \pm 1, \pm 2, \ldots$ where E is the identity matrix in $\mathbb{R}^n$.

H2.46. The set of sequences $\{t_k^j\}$, $t_k^j = t_{k+j} - t_k$, $k = \pm 1, \pm 2, \ldots$, $j = \pm 1, \pm 2, \ldots$ is uniformly almost periodic, and $inf_k t_k^1 = \theta > 0$.

Let $W(t, s)$ be the Cauchy matrix of system (2.67). From conditions H2.43–H6.46, it follows that the solutions $x(t)$ are written down in the form

$$x(t; t_0, x_0) = W(t, t_0)x_0.$$

It is easy to verify, that the equalities $W(t, t) = E$ and $W(t, t_0) = X(t)X^{-1}(t_0)$ are valid, $X(t) = (x_1(t), x_2(t), \ldots, x_n(t))$ is some non degenerate matrix solution of (2.67).

Definition 2.11. The linear system (2.67) is said to has an *exponential dichotomy in* $\mathbb{R}$, if there exist a projector P and positive constants K, L, α, β such that

$$\begin{aligned} ||X(t)PX^{-1}(s)|| &\leq K \ e^{-\alpha(t-s)}, \ t \geq s, \\ ||X(t)(E - P)X^{-1}(s)|| &\leq L \ e^{-\beta(t-s)}, \ s \geq t. \end{aligned} \tag{2.68}$$

Lemma 2.20. *Let the system (2.67) has an exponential dichotomy in* $\mathbb{R}$. *Then any other fundamental matrix of the form $X(t)C$ satisfies inequalities*

(2.68) with the same projector P if and only if the constant matrix C commutes with P.

Proof. The proof of this lemma does not use the particular form of the matrix $X(t)$, and is analogous to the proof of a similar lemma in [46].   □

Definition 2.12. The functions $f \in PC[\mathbb{R}, \Omega]$, $g \in PC[\mathbb{R}, \Omega]$ are said to be *ε-equivalent*, and denoted $f \overset{\varepsilon}{\sim} g$, if the following conditions hold:

(a) The points of possible discontinuity of these functions can be enumerated t_k^f, t_k^g, admitting a finite multiplicity by the order in $\mathbb{R}$, so that $|t_k^f - t_k^g| < \varepsilon$.
(b) There exist strictly increasing sequences of numbers $\{t_k'\}$, $\{t_k''\}$, $t_k' < t_{k+1}'$, $t_k'' < t_{k+1}''$, $k = \pm 1, \pm 2, \ldots$, for which we have

$$\sup_{t \in (t_k', t_{k+1}'), \; t' \in (t_k'', t_{k+1}'')} \|f(t) - g(t)\| < \varepsilon, \; |t_k' - t_k''| < \varepsilon, \; k = \pm 1, \pm 2, \ldots.$$

By $\rho(f, g) = \inf \varepsilon$ we denote the distance between functions $f \in PC[\mathbb{R}, \Omega]$ and $g \in PC[\mathbb{R}, \Omega]$, and by $PC\varphi$ the set of all functions $\varphi \in PC[\mathbb{R}, \Omega]$, for which $\rho(f, \varphi)$ is a finite number. It is easy to verify, that $PC\varphi$ is a metric space.

Definition 2.13 ([9]). The function $\varphi \in PC[\mathbb{R}, \Omega]$ is said to be *almost periodic*, if for any ε the set

$$\overline{T} = \{\tau : \; \rho(\varphi(t+\tau), \varphi(t)) < \varepsilon, \; t, \tau \in \mathbb{R}\}$$

is relatively dense in $\mathbb{R}$.

By $D = \{M_i\}, i \in I$, we denote the family of countable sets of real numbers unbounded below an above and not having limit points, where I is a countable index set. Let M_1 and M_2 be sets of D.

Lemma 2.21 ([9]). *The function $\varphi \in PC[\mathbb{R}, \Omega]$ is almost periodic if and only if for an arbitrary sequence $\{s_n\}$ the sequence $\{\varphi(t + s_n)\}$ is compact in $PC\varphi$.*

Definition 2.14. The sets M_1 and M_2 are said to be *ε-equivalent*, if their elements can be renumbered by integers m_k^1, m_k^2, admitting a finite multiplicity by their order in $\mathbb{R}$, so that

$$\sup_{k = \pm 1, \pm 2, \ldots} |m_k^1 - m_k^2| < \varepsilon.$$

Definition 2.15. The number $\rho_D(M_1, M_2) = \inf_{M_1 \overset{\varepsilon}{\sim} M_2} \varepsilon$ is said to be a *distance* in D.

Throughout the rest of this paragraph, the following notation will be used:

Let conditions H2.43–H2.46 hold and let $\{s'_m\}$ be an arbitrary sequence of real numbers. Analogously to the process from Chap. 1, it follows that there exists a subsequence $\{s_n\}$, $s_n = s'_m$ such that the system (2.67) moves to the system

$$\begin{cases} \dot{x} = A^s(t)x, \ t \neq t^s_k, \\ \Delta x(t^s_k) = B^s_k x(t^s_k), \ k = \pm 1, \pm 2, \ldots. \end{cases} \tag{2.69}$$

The systems of the form (2.69), we shall denote by E^s, and in this meaning we shall denote (2.67) by E_0. From [127], it follows that, each sequence of shifts E^{s_n} of system E_0 is compact, and let denote by $H(A, B_k, t_k)$ the set of shifts of E_0 for an arbitrary sequence $\{s_n\}$.

Now, we shall consider the following scalar impulsive differential equation

$$\begin{cases} \dot{v} = p(t)v, \ t \neq t_k, \\ \Delta v(t_k) = b_k v(t_k), \ k = \pm 1, \pm 2, \ldots, \end{cases} \tag{2.70}$$

where $p \in PC[\mathbb{R}, \mathbb{R}]$, $b_k \in \mathbb{R}$.

Lemma 2.22. *Let the following conditions hold:*

1. *Condition H2.46 holds.*
2. *The function $p(t)$ is almost periodic.*
3. *The sequence b_k is almost periodic.*
4. *The function $v(t)$ is a nontrivial almost periodic solution of (2.70).*

Then $\inf\limits_{t \in \mathbb{R}} |v(t)| > 0$ and the function $1/v(t)$ is almost periodic.

Proof. Suppose that $\inf\limits_{t \in \mathbb{R}} |v(t)| = 0$. Then, there exists a sequence $\{s'_m\}$ of real numbers such that $\lim\limits_{n \to \infty} v(s_n) = 0$. From the almost periodicity of $p(t)$ and $v(t)$ it follows that, the sequences of shifts $p(t + s_n)$ and $v(t + s_n)$ are compact in the sets PC_p and PC_v, respectively. Hence, from Ascoli's diagonal process, it follows that there exists a subsequence $\{s_{n_k}\}$, common for $p(t)$ and $v(t)$ such that the limits

$$\lim_{k \to \infty} p(t + s_{n_k}) = p^s(t),$$

and

$$\lim_{k \to \infty} v(t + s_{n_k}) = v^s(t)$$

exist uniformly for $t \in \mathbb{R}$. Analogously, it is proved that for the sequences of shifts $\{t_k + n_k\}$ and $\{b_k + n_k\}$ there exists a subsequence of $\{n_k\}$, for which there exist the limits $\{t^s_k\}$ and $\{b^s_k\}$. Consequently, for the system

$$\begin{cases} \dot{v}^s = p^s(t)v^s, \ t \neq t^s_k, \\ \Delta v^s(t^s_k) = b^s_k v^s(t^s_k), \ k = \pm 1, \pm 2, \ldots, \end{cases}$$

with initial condition $v^s(0) = 0$ it follows that there exists only the trivial solution.

Then,

$$v(t) = \lim_{k \to \infty} v^{\alpha}(t - s_{n_k}) = 0$$

for all $t \in \mathbb{R}$, which contradicts the conditions of Theorem 1.20. Hence, $\inf_{t \in \mathbb{R}} |v(t)| > 0$, and from Lemma 2.21 it follows that $1/v(t)$ is an almost periodic solution. $\square$

Theorem 2.16. *Let the following conditions hold:*

1. *Conditions H2.43–H2.46 are met.*
2. *The fundamental matrix $X(t)$, $X \in PC[\mathbb{R}, \mathbb{R}^n]$ is almost periodic.*

Then $X^{-1}(t)$ is an almost periodic matrix-valued function.

Proof. From the representation of $W(t, s)$ in Sect. 1.1, we have that $X(t) = W(t, t_0)X(t_0)$, hence

$$X^{-1}(t) = X^{-1}(t_0)W^{-1}(t, t_0)$$
$$= X^{-1}(t_0)\Big(\det W(t, t_0) \Big)^{-1} \Big(adj\ W(t, t_0) \Big)^{T},$$

where by $adj\ W(t, t_0)$ we denote the matrix of cofactors of matrix $W(t, t_0)$.

Then, $X(t)$ will be almost periodic when the following function

$$\Big(v(t) \Big)^{-1} = \Big(\det W(t, t_0) \Big)^{-1}$$

is almost periodic.

From

$$\det W(t, t_0) = \begin{cases} \prod_{t_0 \leq t_k < t} \det(E + B_k) exp\Big(\int_{t_0}^{t} Tr\ A(s)ds \Big), & t > t_0, \\ \prod_{t \leq t_k < t_0} \det(E + B_k) exp\Big(\int_{t_0}^{t} Tr\ A(s)ds \Big), & t \leq t_0, \end{cases}$$

where $Tr A(t)$ is the trace of the matrix A, and a straightforward verification, it follows that the function $v(t) = \det W(t, t_0)$ is a nontrivial almost periodic solution of the system

$$\begin{cases} \dot{v} = Tr\ A(t)v, & t \neq t_k, \\ \Delta v(t_k) = b_k v(t_k), & k = \pm 1, \pm 2, \ldots. \end{cases}$$

Then, from Lemma 2.22 it follows that $1/v(t)$ is an almost periodic function. $\square$

Theorem 2.17. *Let the following conditions hold:*

1. *Conditions H2.43–H2.46 are met.*
2. *The fundamental matrix $X(t)$ satisfies inequalities (2.68).*

Then the fundamental matrix $X^s(t)$ of system (2.70) also satisfies inequalities (2.68).

Proof. Let we denote by H the square root of the positively definite Hermite matrix

$$H^2 = PX^*XP + (E - P)X^*X(E - P).$$

Since P commutes with H^2, then P commutes with H and H^{-1}.

The matrix $X(t)$ is continuously differentiable for $t \neq t_k$ and with points of discontinuity at the first kind at $t = t_k$. Hence, the matrices H, XH^{-1}, HX^{-1} enjoy the properties of $X(t)$, and let $\{s_n\}$ be an arbitrary sequence of real numbers. By a straightforward verification we establish that the matrix $X_n = x(t + s_n)H^{-1}(s_n)$ is a fundamental matrix of system (2.69).

On the other hand, the matrix $H^{-1}(s_n)$ commutes with P, consequently, from Lemma 2.20 it follows that the matrix $X_n(t)$ satisfies inequalities (2.68).

Hence, the matrices $X_n(0)$, $X_n^{-1}(0)$ are bounded, and then there exists a subsequence, common for both matrix sequences such that $X_n(0) \to X_0^s$, where X_0^s is invertible. Then, from the continuous dependence of the solution on initial condition and on parameter, it follows that $X_n(t)$ tends, uniformly on each compact interval, to the matrix solution $X^s(t)$ of (2.69). Since $n \to \infty$, we obtain that $X(t)$ satisfies (2.68). □

Theorem 2.18. *Let the following conditions hold:*

1. *Conditions H2.43–H2.46 are met.*
2. *For the system (2.67) there exists an exponential dichotomy with an hermitian projector P and fundamental matrix $X(t)$.*

Then, the projector-valued function $P(t) = X(t)X^{-1}(t)$ is almost periodic.

Proof. Let $\{s'_m\}$ be an arbitrary sequence of real numbers, which moves the system (2.67) to the system (2.69).

Since the function $P(t) = X(t)X^{-1}(t)$ is bounded and uniformly continuous in the intervals of the form $(t_k, t_{k+1}]$, hence the sequence $\{P(t + s'_m)\}$ is uniformly bounded and uniformly continuous on the intervals $(t_k - s'_m, t_{k+1} - s'_m]$. From Ascoli's diagonal process it follows that there exists a subsequence $\{s_n\}$ of the sequence $\{s'_m\}$ such that the sequence $\{P(t + s_n)\}$ is convergent at each compact interval, and let we denote its limit by $Y(t)$. If $\{s_n\}$ is a subsequence of $\{s'_m\}$, such that $X(s_n)H^{-1}(s_n) \to X_0^s$ is invertible, then from Theorem 2.17 it follows that the sequence $\{X(t + s_n)H^{-1}(s_n)\}$ tends uniformly in each compact interval to the fundamental matrix $X^s(t)$ of system (2.69) and $X^s(t)$ satisfies $Y(t) = X^s(t)P\Big(X^s(t)\Big)^{-1}$.

From Theorem 2.17 it follows that each uniformly convergent in a compact interval subsequences of $\{P(t + s_n)\}$ tends to one and the same limit. Thus, the sequence $\{P(t + s_n)\}$ tends uniformly to $Y(t)$ on each compact interval.

Further on, we shall show that this convergence is uniform in $\mathbb{R}$. Suppose that this is not true. Then, for some $\gamma > 0$ there exists a sequence $\{h_n\}$ of real numbers and a subsequence $\{s'_n\}$ of $\{s_n\}$ such that

$$\|P(h_n + s'_n) - Y(h_n)\| \geq \gamma, \tag{2.71}$$

for each n. It is easily to verify that $E^{h_n+s'_n}$ and E^{h_n} are uniformly convergent in $H(A, B_k, t_k)$. From the almost periodicity and from the process of the construction of E^s it follows that the limit of such system in $H(A, B_k, t_k)$ is one and the same, and let we denote it by E^r. Analogously, $\{P(t + h_n + s'_n)\}$ tends uniformly on each compact interval to $Z(t)PZ^{-1}(t)$, where $Z(t)$ is the fundamental matrix of system E^r, for which there exists an exponential dichotomy with a projector P. Hence, $Y(t+h_n)$ tends to $Z(t)PZ^{-1}(t)$. Then

$$\|P(h_n + s'_n) - Y(h_n)\| \to 0,$$

which contradicts the assumption (2.71).   □

2.7 Separated Solutions and Almost Periodicity

In the present paragraph, by using the notion of separated solutions, sufficient conditions for the existence of almost periodic solutions of impulsive differential equations with variable impulsive perturbations are obtained. Amerio, formulated in [12] the concept of separated solutions, in order to give sufficient conditions for the existence of almost periodic solutions to ordinary differential equations.

The objective of this section is to extend the notion of separated solutions for impulsive differential equations.

Consider the system of impulsive differential equations with variable impulsive perturbations

$$\begin{cases} \dot{x} = f(t, x), \; t \neq \tau_k(x), \\ \Delta x = I_k(x), \; t = \tau_k(x), \; k = \pm 1, \pm 2, \dots, \end{cases} \tag{2.72}$$

where $t \in \mathbb{R}$, $f : \mathbb{R} \times \Omega \to \mathbb{R}^n$, $\tau_k : \Omega \to \mathbb{R}$, and $I_k : \Omega \to \mathbb{R}^n$, $k = \pm 1, \pm 2 \dots$.
Introduce the following conditions:

H2.47. The function $f \in C^1[\mathbb{R} \times \Omega, \mathbb{R}^n]$.
H2.48. The functions $I_k \in C^1[\Omega, \mathbb{R}^n]$, $k = \pm 1, \pm 2 \dots$.

H2.49. If $x \in \Omega$, then $x + I_k(x) \in \Omega$, $L_k(x) = x + I_k(x)$ are invertible on Ω
and $L_k^{-1}(x) \in \Omega$ for $k = \pm 1, \pm 2 \ldots$.

H2.50. $\tau_k(x) \in C^1(\Omega, \mathbb{R})$ and $\lim\limits_{k \to \pm\infty} \tau_k(x) = \pm\infty$ uniformly on $x \in \Omega$.

H2.51. The following inequalities hold:

$$sup\Big\{ \|f(t,x)\| \; : \; (t,x) \in \mathbb{R} \times \Omega \Big\} \leq A < \infty,$$

$$sup\Big\{ \|\tfrac{\partial \tau_k(x)}{\partial x}\| \; : \; x \in \Omega, \; k = \pm 1, \pm 2, \ldots \Big\} \leq B < \infty, \; AB < 1,$$

$$sup\Big\{ \langle \tfrac{\partial \tau_k}{\partial x}(x + sI_k(x)), I_k(x) \rangle \; : \; s \in [0,1], \; x \in \Omega, \; k = \pm 1, \pm 2, \ldots \Big\} \leq 0.$$

From Chap. 1, it follows that, if conditions H2.47–H2.51 are satisfied, then system (2.72) has a unique solution $x(t) = x(t; t_0, x_0)$ with the initial condition

$$x(t_0^+) = x_0.$$

Assuming that conditions H2.48–H2.51 are fulfilled, we consider the hypersurfaces:

$$\sigma_k = \Big\{ (t,x) : \; t = \tau_k(x), \; x \in \Omega \Big\}, \quad k = \pm 1, \pm 2, \ldots.$$

Let t_k be the moments in which the integral curve $(t, x(t; t_0, x_0))$ meets the hypersurfaces σ_k, $k = \pm 1, \pm 2, \ldots$.

Introduce the following conditions:

H2.52. The function $f(t,x)$ is almost periodic in t uniformly with respect to $x \in \Omega$.

H2.53. The sequences $\{I_k(x)\}$ and $\{\tau_k(x)\}$, $k = \pm 1, \pm 2, \ldots$, are almost periodic uniformly with respect to $x \in \Omega$.

H2.54. The set of sequences $\{t_k^j\}$, $t_k^j = t_{k+j} - t_k$, $k = \pm 1, \pm 2, \ldots$, $j = \pm 1, \pm 2, \ldots$, is uniformly almost periodic, and $inf_k t_k^1 = \theta > 0$.

Let conditions H2.47–H2.54 hold, and let $\{s_m'\}$ be an arbitrary sequence of real numbers. Then, there exists a subsequence $\{s_n\}$, $s_n = s_{m_n}'$, so that analogous to the process in Chap. 1, the system (2.72) moves to the system

$$\begin{cases} \dot{x} = f^s(t,x), \; t \neq \tau_k^s, \\ \Delta x = I_k(x), \; t = \tau_k^s, \; k = \pm 1, \pm 2, \ldots, \end{cases} \qquad (2.73)$$

and in this case, the set of systems in the form (2.73) we shall denote by $H(f, I_k, \tau_k)$.

We shall introduce the following operator notation. Let $\alpha = \{\alpha_n\}$ be a subsequence of the sequence $\alpha' = \{\alpha_n\}_{n=0}^{\infty}$, and denote $\alpha \subset \alpha'$. Also with $\alpha + \beta$ we shall denote $\{\alpha_n + \beta_n\}$ of the sequences $\{\alpha_n\}$ and $\{\beta_n\}$.

By $\alpha > 0$ we mean $\alpha_n > 0$ for each n. If $\alpha \subset \alpha'$ and $\beta \subset \beta'$, then α and β are said to have matching subscripts, if $\alpha = \{\alpha_{n_k}'\}$ and $\beta = \{\beta_{n_k}'\}$.

Let we denote by $S_{\alpha+\beta}\phi$ and $S_\alpha S_\beta\phi$ the limits $\lim\limits_{n\to\infty}\theta_{\alpha_n+\beta_n}(\phi)$ and $\lim\limits_{n\to\infty}\theta_{\alpha_n}(\lim\limits_{m\to\infty}\theta_{\beta_m}\phi)$, respectively, where the number θ_{α_n} is defined in Chap. 1, and $\phi = (\varphi(t), T)$, $\phi \in PC[\mathbb{R}, \Omega] \times UAPS$.

Lemma 2.23. *The function $\varphi(t)$ is almost periodic if and only if from every pair of sequences α', β' one can extracts common subsequences $\alpha \subset \alpha'$, $\beta \subset \beta'$ such that*

$$S_{\alpha+\beta}\varphi = S_\alpha S_\beta\varphi, \qquad (2.74)$$

exists pointwise.

Proof. Let (2.74) exists pointwise, γ' be a sequence, such that for $\gamma \subset \gamma'$, $S_\gamma\varphi$ exists. If $S_\gamma\phi$ is uniform, we are done. If not, we can find $\varepsilon > 0$ and sequences $\beta \subset \gamma$, $\beta' \subset \gamma$ such that

$$\rho(T_n^\beta, T_n^{\beta'}) < \varepsilon,$$

but

$$\sup_{t\in\mathbb{R}\setminus F_\varepsilon(s(T_n^\beta \cup T_n^{\beta'}))} ||\varphi(t + \beta_n) - \varphi(t + \beta'_n)|| \geq \varepsilon > 0,$$

where T_n^β and $T_n^{\beta'}$ are the points of discontinuity of functions $\varphi(t + \beta_n)$, $\varphi(t + \beta'_n)$, $n = 0, 1, 2, \ldots$, respectively.

From the intermediate value theorem for the common intervals of continuity of functions $\varphi(t + \beta_n)$ and $\varphi(t + \beta'_n)$, and the fact that

$$\lim_{n\to\infty} ||\varphi(\beta_n) - \varphi(\beta'_n)|| = 0,$$

it follows that there exists a sequence α such that

$$\sup_{t\in\mathbb{R}\setminus F_\varepsilon(s(T_n^\beta \cup T_n^{\beta'}))} ||\varphi(\alpha_n + \beta_n) - \varphi(\alpha_n + \beta'_n)|| \geq \varepsilon > 0. \qquad (2.75)$$

Then, for the sequence α there exist common subsequences $\alpha_1 \subset \alpha$, $\beta_1 \subset \beta$, $\beta_2 \subset \beta$ such that

$$S_{\alpha_1+\beta_1}\phi = R_1, \quad S_{\alpha_1+\beta_2}\phi = R_2,$$

where $R_j = (r_j(t), P_j)$, $r_j \in PC$, $P_j \in UAPS$, $j = 1, 2$, exist pointwise.

From (2.74), we get

$$R_1 = S_{\alpha_1+\beta_1}\phi = S_{\alpha_1} S_{\beta_1}\phi = S_{\alpha_1} S_\gamma\phi$$

$$= S_{\alpha_1} S_{\beta_2}\phi = S_{\alpha_1+\beta_2}\phi = R_2, \qquad (2.76)$$

for $t \in \mathbb{R} \setminus F_\varepsilon(s(P_1 \cup P_2))$.

On the other hand, from (2.75) it follows that

$$||r_1(0) - r_2(0)|| > 0,$$

which is a contradiction of (2.76).

Let $\varphi(t)$ be almost periodic and if α' and β' are given, we take subsequences $\alpha \subset \alpha'$, $\beta \subset \beta'$ successively, such that they are common subsequences and $S_\alpha \phi = \phi_1$, $S_\beta \phi_1 = \phi_2$ and $S_{\alpha+\beta}\phi = \phi_3$, where $\phi_j = (\phi_j, T_j)$, $\phi_j \in PC[\mathbb{R}, \Omega] \times UAPS$, $j = 1, 2, 3$, exist uniformly for $t \in \mathbb{R} \setminus F_\varepsilon(s(T_1 \cup T_2 \cup T_3))$.

If $\varepsilon > 0$ is given, then

$$||\varphi(t + \alpha_n + \beta_n) - \varphi_3(t)|| < \frac{\varepsilon}{3},$$

for n large and for all $t \in \mathbb{R} \setminus F_\varepsilon(s(T_{n,n} \cup T_3))$, where $T_{n,n}$ is the set of points of discontinuity of functions $\varphi(t + \alpha_n + \beta_n)$.

Also,

$$||\varphi(t + \alpha_n + \beta_m) - \varphi_1(t + \beta_n)|| < \frac{\varepsilon}{3},$$

for n, m large and for all $t \in \mathbb{R} \setminus F_\varepsilon(s(T_{n,m} \cup T_{1,n}))$, where $T_{n,m}$ is the set of points of discontinuity of functions $\varphi(t + \alpha_n + \beta_m)$ and $T_{1,n}$ is formed by the points of discontinuity of functions $\varphi_1(t + \beta_n)$.

Finally,

$$||\varphi_1(t + \beta_m) - \varphi_2(t)|| < \frac{\varepsilon}{3},$$

for m large and all $t \in \mathbb{R} \setminus F_\varepsilon(s(T_{1,m} \cup T_2))$, where $T_{1,m}$ is the set of points of discontinuity of functions $\varphi_1(t + \beta_m)$.

By the triangle inequality for $n = m$ large, we have $||\varphi_2(t) - \varphi_3(t)|| < \varepsilon$ for all $t \in \mathbb{R} \setminus F_\varepsilon(s(T_2 \cup T_3))$.

Since ε is arbitrary, we get $\varphi_2(t) = \varphi_3(t)$ for all $t \in \mathbb{R} \setminus F_\varepsilon(s(T_{1,m} \cup T_2))$, i.e. (2.74) holds.   □

Definition 2.16. The function $\varphi(t)$, $\varphi \in PC[\mathbb{R}, \Omega]$, is said to *satisfy the condition SG*, if for a given sequence γ', $\lim\limits_{n \to \infty} \gamma'_n = \infty$ there exist $\gamma \subset \gamma'$ and a number $d(\gamma) > 0$ such that $S_\gamma \phi$, $\phi = (\varphi(t), T)$, $T \in UAPS$ exists pointwise for each $\varepsilon > 0$. If α is a sequence with $\alpha > 0$, $\beta' \subset \gamma$ and $\beta'' \subset \gamma$ are such that $S_{\alpha+\beta'}\phi = (r_1(t), P_1)$, $S_{\alpha+\beta''}\phi = (r_2(t), P_2)$, then either $r_1(t) = r_2(t)$ or $||r_1(t) - r_2(t)|| > d(\gamma)$ hold for $t \in \mathbb{R} \setminus F_\varepsilon(s(P_1 \cup P_2))$.

Definition 2.17. Let $K \subset \Omega$ be a compact. The solution $x(t)$ of system (2.72) with points of discontinuity in the set T is said to be *separated in K*, if for any other solution $y(t)$ of (2.72) in Ω with points of discontinuity in the set T there exists a number $d(y(t))$ such that $||x(t) - y(t)|| > d(y(t))$ for $t \in \mathbb{R} \setminus F_\varepsilon(s(T))$. The number $d(y(t))$ is said to be a *separated constant*.

Theorem 2.19. *The function $\varphi(t), \varphi \in PC[\mathbb{R}, \Omega]$, is almost periodic if and only if φ satisfies the condition SG.*

Proof. Let φ satisfies the condition SG, and let γ' be a sequence such that $\lim_{n \to \infty} \gamma'_n = \infty$. Then there exists $\gamma \subset \gamma'$ such that $S_\gamma \phi$, $\phi = (\varphi(t), T)$ exists pointwise. If the convergence is not uniformly in $\mathbb{R}$, then there exist sequences $\delta' > 0$, $\alpha' \subset \gamma$, $\beta' \subset \gamma$, and a number $\varepsilon > 0$ such that $\|\varphi(\alpha'_n + \delta'_n) - \varphi(\beta'_n + \delta'_n)\| \geq \varepsilon$, where we may pick $\varepsilon < d(\gamma)$. Since $S_\gamma(\varphi(0), T)$ exists, we have

$$\|\varphi(\alpha'_n) - \varphi(\beta'_n)\| < d(\gamma), \tag{2.77}$$

for large n.

Consequently, $k(t) = \varphi(t + \alpha'_n) - \varphi(t + \beta'_n)$ satisfies $\|k(0)\| < d(\gamma)$ and $\|k(\delta'_n)\| \geq \varepsilon$ for large n. Hence, there exists δ''_n such that $\delta''_n \subset \delta'_n$ and $\varepsilon \leq \|k(\delta''_n)\| < d(\gamma)$.

We shall consider the sequences $\alpha' + \delta''$ and $\beta' + \delta''$. By SG there exist sequences $\alpha + \delta \subset \alpha' + \delta''$ and $\beta + \delta \subset \beta' + \delta''$ with matching subscripts such that $S_{\alpha+\delta}\phi = \phi_1$, $S_{\alpha+\delta}\phi = \phi_2$, $\phi_j = (\varphi_j, T_j)$ exist pointwise, and $\varphi_1(t) = \varphi_2(t)$ or $\|\varphi_1(t) - \varphi_2(t)\| > 2d(\gamma)$, for $t \in \mathbb{R} \setminus F_\varepsilon(s(T_1 \cup T_2))$.

On the other hand,

$$\|\varphi_1(0) - \varphi_2(0)\| = \lim_{n \to \infty} \|\varphi(\alpha_n + \delta_n) - \varphi(\beta_n + \delta_n)\|,$$

and from (2.77), it follows that $\|\varphi_1(0) - \varphi_2(0)\| \leq d(\gamma)$. The contradiction shows that $S_\gamma \varphi$ exists uniformly on $t \in \mathbb{R} \setminus F_\varepsilon(s(T))$.

Conversely, if $\varphi(t)$ is an almost periodic function, and γ' be given with $\lim_{n \to \infty} \gamma'_n = \infty$ then, there exists $\gamma \subset \gamma'$ such that $S_\gamma \phi$ exists uniformly on $t \in \mathbb{R} \setminus F_\varepsilon(s(T))$ and $S_\gamma \varphi = (k(t), Q)$, $(k(t), Q) \in PC[\mathbb{R}, \Omega] \times UAPS$.

Let the subsequences $\beta' \subset \gamma$, $\beta'' \subset \gamma$, and $\alpha > 0$ be such that $S_{\alpha+\beta'}\phi = (r_1(t), P_1)$, $S_{\alpha+\beta''}\phi = (r_2(t), P_2)$, $(r_j(t), P_j) \in PC[\mathbb{R}, \Omega] \times UAPS$.

From Lemma 2.23 it follows that there exist $\alpha' \subset \alpha$, $\overline{\beta}' \subset \beta'$, $\overline{\beta}'' \subset \beta''$ such that

$$(r_1(t), P_1) = S_{\alpha' + \overline{\beta}'}(p(t), T) = S_{\alpha'} S_{\overline{\beta}'}(p(t), T) = S_{\alpha'} S_\gamma(p(t), T)$$

$$= S_{\alpha'}(k(t), Q) = S_\alpha(k(t), Q), \tag{2.78}$$

$$(r_2(t), P_2) = S_{\alpha' + \overline{\beta}''}(p(t), T) = S_{\alpha'} S_{\overline{\beta}''}(p(t), T) = S_{\alpha'} S_\gamma(p(t), T)$$

$$= S_{\alpha'}(k(t), Q) = S_\alpha(k(t), Q). \tag{2.79}$$

Hence, from (2.78) and (2.79), we get $r_1(t) = r_2(t)$ for $t \in \mathbb{R} \setminus F_\epsilon(s(P_1 \cup P_2))$. Then, $(\varphi(t), T)$ satisfies SG. $\square$

Now, let $K \subset \Omega$ be a compact. We shall consider the system of impulsive differential equations

$$\begin{cases} \dot{x} = g(t, x), \ t \neq \sigma_k(x), \\ \Delta x = G_k(x), \ t = \sigma_k(x), \ k = \pm 1, \pm 2, \dots, \end{cases} \tag{2.80}$$

where $(g, G_k, \sigma_k) \in H(f, I_k, \tau_k)$.

Theorem 2.20. *Let the following conditions hold:*

1. *Conditions H2.47–H2.54 are met.*
2. *Every solution of system (2.80) in K is separated.*

Then every system in $H(f, I_k, \tau_k)$ has only a finite number of solutions and the separated constant d may be picked to be independent of solutions.

Proof. The fact that each system has only a finite number solutions in K is a consequence of a compactness of K and the resulting compactness of the solutions in K. But no solution can be a limit of others by the separated condition. Consequently, the number of solutions of any system from $H(f, I_k, \tau_k)$ is finite and d may be picked as a function of the system.

Let $(h, L_k, l_k) \in H(f, I_k, \tau_k)$ and $S_{\alpha'}(g, G_k, \sigma_k) = (h, L_k, l_k)$, with $\lim_{n \to \infty} \alpha'_n = \infty$.

Let $(\varphi(t), T)$, $(\varphi_0(t), T_0)$ be two solutions in K, and let $\alpha \subset \alpha'$ be such that $S_\alpha(\varphi(t), T)$ and $S_\alpha(\varphi_0(t), T_0)$ exist uniformly on K, and are solutions of (2.80).

Then,

$$\|S_\alpha(\varphi(t), T) - S_\alpha(\varphi_0(t), T_0)\| \geq d(g, G_k, \sigma_k).$$

So, if $\varphi_1, \dots, \varphi_n$ are solutions of (2.80) in K, then $S_\alpha(\varphi_j(t), T_j)$, $j = 1, 2, \dots, n$, are distinct solutions of (2.80) in K such that

$$\|S_\alpha(\varphi_j(t), T_j) - S_\alpha(\varphi_i(t), T_i)\| \geq d(g, G_k, \sigma_k), \ i \neq j.$$

Hence, the number of solutions of (2.80) in K is greater or equal than n. By "symmetry" arguments the reverse is true, hence each system has the same number of solutions.

On the other hand, $S_\alpha(\varphi_i, T_i)$ exhaust the solutions of (2.80) in K, so that $d(g, G_k, \sigma_k) \leq d(h, L_k, l_k)$. Again by symmetry, $d(h, L_k, l_k) \geq d(g, G_k, \sigma_k)$. $\square$

Theorem 2.21. *Let the following conditions hold:*

1. *Conditions H2.47–H2.54 are met.*
2. *For every system in $H(f, I_k, \tau_k)$ there exist only separated solutions on K.*

Then:

1. *If for some system in $H(f, I_k, \tau_k)$ there exists a solution in K, then for every system in $H(f, I_k, \tau_k)$ there exists a solution in K.*

2. *All such solutions in K are almost periodic and for every system in $H(f, I_k, \tau_k)$ there exists an almost periodic solution in K.*

Proof. The first statement has been proved in Theorem 2.20. Let $\varphi(t)$ be a solution of system (2.80) in K and δ be the separation constant.

Let γ' be a sequence such that $\lim\limits_{n\to\infty} \gamma' = \infty$ and $\gamma \subset \gamma'$, $S_\gamma(g, G_k, \sigma_k) = (h, L_k, l_k)$, and $S_\gamma(\varphi(t), T)$ exists.

Let $\beta' \subset \gamma$, $\beta'' \subset \gamma$ and $\alpha > 0$ are such that

$$S_{\alpha+\beta'}(\varphi(t), T) = (\varphi_1(t), T_1),$$
$$S_{\alpha+\beta''}(\varphi(t), T) = (\varphi_2(t), T_2).$$

Again, take further subsequences with matching subscripts, so that (without changing notations)

$$S_{\alpha+\beta'}(g, G_k, \sigma_k) = S_\alpha S_{\beta'}(g, G_k, \sigma_k)$$
$$= S_\alpha S_\gamma(g, G_k, \sigma_k) = S_\alpha(h, L_k, l_k),$$

and

$$S_{\alpha+\beta''}(g, G_k, \sigma_k) = S_\alpha(h, L_k, l_k).$$

Consequently, $\varphi_1(t)$ and $\varphi_2(t)$ are solutions of the same system and for $\varepsilon > 0$, $\varphi_1 \equiv \varphi_2$, for $\mathbb{R} \setminus F_\varepsilon(s(T_1 \bigcup T_2))$ or $\|\varphi_1(t) - \varphi_2(t)\| \geq \delta = 2d$ on $\mathbb{R} \setminus F_\varepsilon(s(T_1 \bigcup T_2))$.

Therefore, $\varphi(t)$ satisfies the SG, and from Theorem 2.19 it follows that $\varphi(t)$ is an almost periodic function.

Let now $\varphi(t)$ be a solution of (2.80) in K which by the above is an almost periodic function, and let we choice $\alpha'_n = n$. Then, there exists $\alpha \subset \alpha'$ such that the limits $S_\alpha(g, G_k, \sigma_k) = (h, L_k, l_k)$, $S_{-\alpha}(h, L_k, l_k) = (g, G_k, \sigma_k)$ exist uniformly and $S_\alpha(\varphi(t), T) = (r(t), P)$, $S_{-\alpha}(r(t), P)$ exist uniformly on K, where $S_{-\alpha}(r(t), P)$ is the solution of (2.80).

From condition 2 of Theorem 2.21 it is easy to see that $(r(t), P) = S_\alpha(\varphi(t), T)$ and thus $S_{-\alpha}(r(t), P)$ exists uniformly and $\varphi(t)$ is almost periodic.   □

2.8   Impulsive Differential Equations in Banach Space

The abstract differential equations arise in many areas of applied mathematics, and for this reason these equations have received much attention in the recent years. Natural generalizations of the abstract differential equations are impulsive differential equations in Banach space.

In this paragraph, we shall investigate the existence of almost periodic solutions of these equations.

Let $(X, ||.||_X)$ be an abstract Banach space.
Consider the impulsive differential equation

$$\dot{x}(t) = Ax + F(t, x) + \sum_{k=\pm 1, \pm 2, \dots} \left[Bx + H_k(x) \right] \delta(t - t_k), \qquad (2.81)$$

where $A : \mathcal{D}(A) \subset X \to X$, $B : \mathcal{D}(B) \subset X \to X$ are linear bounded operators with domains $\mathcal{D}(A)$ and $\mathcal{D}(B)$, respectively. The function $F : \mathcal{D}(\mathbb{R} \times X) \to X$ is continuous with respect to $t \in \mathbb{R}$ and with respect to $x \in X$, $H_k : \mathcal{D}(H_k) \subset X \to X$ are continuous impulse operators, $\delta(.)$ is the Dirac's delta-function, $\{t_k\} \in \mathcal{B}$.

Denote by $x(t) = x(t; t_0, x_0)$, the solution of (2.81) with the initial condition $x(t_0^+) = x_0$, $t_0 \in \mathbb{R}$, $x_0 \in X$.

The solutions of (2.81) are piecewise continuous functions [16], with points of discontinuity at the moments t_k, $k = \pm 1, \pm 2, \dots$ at which they are continuous from the left, i.e. the following relations are valid:

$$x(t_k^-) = x(t_k), \ \ x(t_k^+) = x(t_k) + Bx(t_k) + H_k(x(t_k)), \ \ k = \pm 1, \pm 2, \dots .$$

Let $PC[\mathbb{R}, X] = \{\varphi : \mathbb{R} \to X, \ \varphi$ is a piecewise continuous function with points of discontinuity of the first kind at the moments t_k, $\{t_k\} \in \mathcal{B}$ at which $\varphi(t_k^-)$ and $\varphi(t_k^+)$ exist, and $\varphi(t_k^-) = \varphi(t_k)\}$.

With respect to the norm $||\varphi||_{PC} = \sup_{t \in \mathbb{R}} ||\varphi(t)||_X$, $PC[\mathbb{R}, X]$ is a Banach space [16].

Denote by $PCB[\mathbb{R}, X]$ the subspace of $PC[\mathbb{R}, X]$ of all bounded piecewise continuous functions, and together with (2.81) we consider the respective linear non-homogeneous impulsive differential equation

$$\dot{x} = Ax + f(t) + \sum_{k=\pm 1, \pm 2, \dots} \left[Bx + b_k \right] \delta(t - t_k), \qquad (2.82)$$

where $f \in PCB[\mathbb{R}, X]$, $b_k : \mathcal{D}(b_k) \subset X \to X$, and the homogeneous impulsive differential equation

$$\dot{x}(t) = Ax + \sum_{k=\pm 1, \pm 2, \dots} Bx \delta(t - t_k). \qquad (2.83)$$

Introduce the following conditions:

H2.55. The operators A and B commute with each other, and for the operator $I + B$ there exists a logarithm operator $Ln(I + B)$, I is the identity operator on the space X.

H2.56. The set of sequences $\{t_k^j\}$, $t_k^j = t_{k+j} - t_k$, $k = \pm 1, \pm 2, \dots$, $j = \pm 1, \pm 2, \dots$, is uniformly almost periodic, and $inf_k t_k^1 = \theta > 0$.

Following [16], we denote by $\Phi(t, s)$, the Cauchy evolutionary operator for (2.83),

$$\Phi(t,s) = e^{\Lambda(t-s)}(I+B)^{-p(t-s)+i(t,s)},$$

where $\Lambda = A + pL_n(I + B)$, $i(t, s)$ is the number of points t_k in the interval (t, s), and $p > 0$ is defined in Lemma 1.1.

Lemma 2.24. *Let conditions H2.55–H2.56 hold, and the spectrum $\sigma(\Lambda)$ of the operator Λ does not intersect the imaginary axis, and lying in the left half-planes.*

Then for the Cauchy evolutionary operator $\Phi(t, s)$ of (2.83) there exist positive constants K_1 and α such that

$$||\Phi(t,s)||_X \le K_1 e^{-\alpha(t-s)}, \tag{2.84}$$

where $t \ge s$, $t, s \in \mathbb{R}$.

Proof. Let $\varepsilon > 0$ be arbitrary. Then

$$||(I+B)^{-p(t-s)+i(s,t)}||_X \le \delta(\varepsilon)exp\{\varepsilon||Ln(I+B)||_X(t-s)\},$$

where $\delta(\varepsilon) > 0$ is a constant.

On the other hand [50], if $\alpha_1 > 0$ and

$$\delta_1 \in (\alpha_1, \lambda^*(\alpha_1)), \ \lambda^*(\alpha_1) = inf\{|Re\lambda|, \ \lambda \in \sigma(\Lambda)\},$$

then,

$$||e^{\Lambda(t-s)}||_X \le K_1 e^{-\alpha_1(t-s)}, \ t > s$$

and (2.84) follows immediately.   □

The next definition is for almost periodic functions in a Banach space of the form $PC[\mathbb{R}, X]$.   □

Definition 2.18. The function $\varphi \in PC[\mathbb{R}, X]$ is said to be *almost periodic*, if:

(a) The set of sequences $\{t_k^j\}$, $t_k^j = t_{k+j} - t_k$, $k = \pm 1, \pm 2, \ldots$, $j = \pm 1, \pm 2, \ldots$, $\{t_k\} \in \mathcal{B}$ is uniformly almost periodic.
(b) For any $\varepsilon > 0$ there exists a real number $\delta(\varepsilon) > 0$ such that, if the points t' and t'' belong to one and the same interval of continuity of $\varphi(t)$ and satisfy the inequality $|t' - t''| < \delta$, then $||\varphi(t') - \varphi(t'')||_X < \varepsilon$.
(c) For any $\varepsilon > 0$ there exists a relatively dense set $\overline{T}$ such that, if $\tau \in \overline{T}$, then $||\varphi(t+\tau) - \varphi(t)||_X < \varepsilon$ for all $t \in \mathbb{R}$ satisfying the condition $|t - t_k| > \varepsilon$, $k = \pm 1, \pm 2, \ldots$.

The elements of $\overline{T}$ are called $\varepsilon - almost \ periods$.

Introduce the following conditions:

H2.57. The function $f(t)$ is almost periodic.
H2.58. The sequence $\{b_k\}$, $k = \pm 1, \pm 2, \ldots$ is almost periodic.

We shall use the next lemma, similar to Lemma 1.7.

Lemma 2.25. *Let conditions H2.56–H2.58 hold.*
Then for each $\varepsilon > 0$ there exist ε_1, $0 < \varepsilon_1 < \varepsilon$, a relatively dense set $\overline{T}$ of real numbers, and a set P of integer numbers such that the following relations are fulfilled:

(a) $\|f(t + \tau) - f(t)\|_X < \varepsilon$, $t \in \mathbb{R}$, $\tau \in \overline{T}$, $|t - t_k| > \varepsilon$, $k = \pm 1, \pm 2, \ldots$.
(b) $\|b_{k+q} - b_k\|_X < \varepsilon$, $q \in P$, $k = \pm 1, \pm 2, \ldots$.
(c) $|\tau_k^q - \tau| < \varepsilon_1$, $q \in P$, $\tau \in \overline{T}$, $k = \pm 1, \pm 2, \ldots$.

We shall prove the next theorem.

Theorem 2.22. *Let the following conditions hold:*

1. *Conditions H2.55–H2.58 are met.*
2. *The spectrum $\sigma(\Lambda)$ of the operator Λ does not intersect the imaginary axis, and lying in the left half-planes.*

Then:

1. *There exists a unique almost periodic solution $x(t) \in PCB[\mathbb{R}, X]$ of (2.82).*
2. *The almost periodic solution $x(t)$ is asymptotically stable.*

Proof. We consider the function

$$x(t) = \int_{-\infty}^{t} \Phi(t, s) f(s) ds + \sum_{t_k < t} \Phi(t, t_k) b_k. \qquad (2.85)$$

It is immediately verified, that the function $x(t)$ is a solution of (2.82). From conditions H2.57 and H2.58, it follows that $f(t)$ and $\{b_k\}$ are bounded and let

$$max\{\|f(t)\|_{PC}, \|b_k\|_X\} \le C_0, \ C_0 > 0.$$

Using Lemmas 1.1 and 2.24, we obtain

$$\|x(t)\|_{PC} = \int_{-\infty}^{t} \|\Phi(t, s)\|_{PC} \|f(s)\|_{PC} ds + \sum_{t_k < t} \|\Phi(t, t_k)\|_{PC} \|b_k\|_X$$

$$\le \int_{-\infty}^{t} K_1 e^{-\alpha(t-s)} \|f(s)\|_{PC} ds + \sum_{t_k < t} K e^{-\alpha(t-t_k)} \|b_k\|_X$$

$$\le K_1 \left(\frac{C_0}{\alpha} + \frac{C_0 N}{1 - e^{-\alpha}} \right) = \overline{K}. \qquad (2.86)$$

From (2.86) it follows that $x(t) \in PCB[\mathbb{R}, X]$.

Let $\varepsilon > 0$, $\tau \in T$, $q \in Q$, where the sets T and P are from Lemma 2.25. Then,

$$||x(t + \tau) - x(t)||_{PC}$$

$$\leq \int_{-\infty}^{t} ||\Phi(t, s)||_{PC} ||f(s + \tau) - f(s)||_{PC} ds$$

$$+ \sum_{t_k < t} ||\Phi(t, t_k)||_{PC} ||b_{k+q} - b_k||_X \leq M\varepsilon,$$

where $|t - t_k| > \varepsilon$, $M > 0$.

The last inequality implies that the function $x(t)$ is almost periodic. The uniqueness of this solution follows from the fact that the (2.83) has only the zero bounded solution under conditions H2.55 and H2.56.

Let $\tilde{x} \in PCB[\mathbb{R}, X]$ be an arbitrary solution of (2.82), and $y = \tilde{x} - x$. Then $y \in PCB[\mathbb{R}, X]$ and

$$y = \Phi(t, t_0) y(t_0). \tag{2.87}$$

The proof that $x(t)$ is asymptotically stable follows from (2.87), the estimates from Lemma 2.24, and the fact that $i(t_0, t) - p(t - t_0) = o(t)$ for $t \to \infty$. $\square$

Now, we shall investigate almost periodic solutions of (2.81).

Theorem 2.23. *Let the following conditions hold:*

1. *Conditions H2.55–H2.58 are met.*
2. *The spectrum $\sigma(\Lambda)$ of the operator Λ does not intersect the imaginary axis, and lying in the left half-planes.*
3. *The function $F(t, x)$ is almost periodic with respect to $t \in \mathbb{R}$ uniformly at $x \in \Omega$ and the sequence $\{H_k(x)\}$ is almost periodic uniformly at $x \in \Omega$, Ω is every compact from X, and*

$$||x||_X < h, \ h > 0.$$

4. *The functions $F(t, x)$ and $H_k(x)$ are Lipschitz continuous with respect to $x \in \Omega$ uniformly for $t \in \mathbb{R}$ with a Lipschitz constant $L > 0$,*

$$||F(t, x) - F(t, y)||_X \leq L||x - y||_X, \ ||H_k(x) - H_k(y)||_X \leq L||x - y||_X.$$

5. *The functions $F(t, x)$ and $H_k(x)$ are bounded,*

$$max\Big\{ ||F(t, x)||_X, \ ||H_k(x)||_X \Big\} \leq C,$$

where $C > 0$, $x \in \Omega$.

Then, if:

$$\overline{K}C < h \text{ and } \overline{K}L < 1,$$

where $\overline{K}$ was defined by (2.86), it follows:

1. *There exists a unique almost periodic solution $x(t) \in PCB[\mathbb{R}, X]$ of (2.81).*
2. *The almost periodic solution $x(t)$ is asymptotically stable.*

Proof. We denote by $D^* \subset PCB[\mathbb{R}, X]$ the set of all almost periodic functions with points of discontinuity of the first kind t_k, $k = \pm 1, \pm 2, \ldots$, satisfying the inequality $\|\varphi\|_{PC} < h$.

In D^*, we define an operator S in the following way. If $\varphi \in D^*$, then $y = S\varphi(t)$ is the almost periodic solution of the system

$$\dot{y}(t) = Ay + F(t, \varphi(t)) + \sum_{k=\pm 1, \pm 2, \ldots} \big[By + H_k(\varphi(t_k)) \big] \delta(t - t_k), \qquad (2.88)$$

determined by Theorem 2.22. Then, from (2.86) and the conditions of Theorem 2.23, it follows that $\mathcal{D}(S) \subset D^*$.

Let φ, $\psi \in D^*$. Then, we obtain

$$\|S\varphi(t) - S\psi(t)\|_{PC} \leq \overline{K}L.$$

From the last inequality, and the conditions of the theorem, it follows that the operator S is a contracting operator in D^*. $\qquad \qquad \square$

Example 2.4. In this example, we shall investigate materials with fading memory with impulsive perturbations at fixed moments of time.

We shall investigate the existence of almost periodic solutions of the following impulsive differential equation

$$\begin{cases} \ddot{x}(t) + \beta(0)\dot{x}(t) = \gamma(0)\Delta x(t) + f_1(t)f_2(x(t)), \ t \neq t_k, \\ x(t_k^+) = x(t_k) + b_k^1, \\ \dot{x}(t_k^+) = \dot{x}(t_k) + b_k^2, \ k = \pm 1, \pm 2, \ldots, \end{cases} \qquad (2.89)$$

where $t_k = k + l_k$, $l_k = \frac{1}{4}|\cos k - \cos k\sqrt{2}|$, $k = \pm 1, \pm 2, \ldots$.

If $y(t) = \dot{x}(t)$ and

$$z(t) = \begin{bmatrix} x(t) \\ y(t) \end{bmatrix}, \quad A = \begin{bmatrix} 0 & 1 \\ \gamma(0)\Delta & -\beta(0) \end{bmatrix}, \quad \dot{z}(t) = \begin{bmatrix} \dot{x}(t) \\ \dot{y}(t) \end{bmatrix},$$

$$F(t, z) = \begin{bmatrix} 0 \\ f_1(t)f_2(x) \end{bmatrix}, \quad B = \begin{bmatrix} 0 & 1 \\ 1 & 0 \end{bmatrix}, \quad b_k = \begin{bmatrix} b_k^1 \\ b_k^2 \end{bmatrix}, \quad k = \pm 1, \pm 2, \ldots,$$

then the (2.89) rewrites in the form

$$\dot{z}(t) = Az + F(t, z) + \sum_{k=\pm 1, \pm 2, \ldots}^{\infty} \left[Bz + b_k \right] \delta(t - t_k). \qquad (2.90)$$

From [138], it follows that the set of sequences $\{t_k^j\}$, $k = \pm 1, \pm 2, \ldots$, $j = \pm 1, \pm 2, \ldots$, is uniformly almost periodic and for the (2.90) the conditions of Lemma 1.2 hold.

Let $X = H_0^1(\omega) \times L^2(\omega)$, where $\omega \subset R^3$ is an open set with smooth boundary of the class C^∞, $\beta(t)$, $\gamma(t)$ are bounded and uniformly continuous $\mathbb{R}$ valued functions of the class C^2 on $[0, \infty)$, $\beta(0) > 0$, $\gamma(0) > 0$.

If $A : \mathcal{D}(A) = H^2(\omega) \cap H_0^1(\omega) \times H_0^1(\omega) \to X$ is the operator from (2.90) and Δ is Laplacian on ω with boundary condition $y|_{\partial \omega} = 0$, then it follows that A is the infinitesimal generator of a C_0-semigroup and the conditions of Lemma 2.24 hold.

By Theorem 2.23 and similar arguments, we conclude with the following theorem.

Theorem 2.24. *Let for (2.89) the following conditions hold:*

1. *The sequences $\{b_k^i\}$, $k = \pm 1, \pm 2, \ldots$, $i = 1, 2$, are almost periodic.*
2. *The function $f_1(t)$ is almost periodic in the sense of Bohr.*
3. *The function $f_2(x)$ is Lipschitz continuous with respect to $||x||_X < h$ with a Lipschitz constant $L > 0$,*

$$||f_2(x_1) - f_2(x_2)||_X \le L||x_1 - x_2||_X, \ ||x_i||_X < h, \ i = 1, 2.$$

4. *The function $f_2(x)$ is bounded, $||f_2(x)||_X \le C$, where $C > 0$ and $x \in \omega$.*

   Then, if
   $$\overline{K}C < h \text{ and } \overline{K}L < 1,$$

where $\overline{K}$ was defined by (2.86), it follows:

1. *There exists a unique almost periodic solution $x \in PCB[\mathbb{R}, X]$ of (2.89).*
2. *The almost periodic solution $x(t)$ is asymptotically stable.*

Now, we shall study the existence and uniqueness of almost periodic solutions of impulsive abstract differential equations out by means of the infinitesimal generator of an analytic semigroup and fractional powers of this generator.

Let the operator A in (2.81)–(2.83) be the infinitesimal operator of analytic semigroup $S(t)$ in Banach space X. For any $\alpha > 0$, we define the fractional power $A^{-\alpha}$ of the operator A by

$$A^{-\alpha} = \frac{1}{\Gamma(\alpha)} \int_0^\infty t^{\alpha - 1} S(t) dt,$$

where $\Gamma(\alpha)$ is the Gamma function. The operators $A^{-\alpha}$ are bounded, bijective and $A^{\alpha} = (A^{-\alpha})^{-1}$, is a closed linear operator such that $\mathcal{D}(A^{\alpha}) = \mathcal{R}(A^{-\alpha})$, where $\mathcal{R}(A^{-\alpha})$ is the range of $A^{-\alpha}$. The operator A^0 is the identity operator in X and for $0 \leq \alpha \leq 1$, the space $X_{\alpha} = \mathcal{D}(A^{\alpha})$ with norm $||x||_{\alpha} = ||A^{\alpha}x||_X$ is a Banach space [50, 58, 68, 115, 126].

We shall use the next lemmas.

Lemma 2.26 ([115, 126]). *Let A be the infinitesimal operator of an analytic semigroup $S(t)$.*

Then:

1. $S(t) : X \to \mathcal{D}(A^{\alpha})$ *for every* $t > 0$ *and* $\alpha \geq 0$.
2. *For every* $x \in \mathcal{D}(A^{\alpha})$ *it follows that* $S(t)A^{\alpha}x = A^{\alpha}S(t)x$.
3. *For every* $t > 0$ *the operator* $A^{\alpha}S(t)$ *is bounded, and*

$$||A^{\alpha}S(t)||_X \leq K_{\alpha}t^{-\alpha}e^{-\lambda t}, \ K_{\alpha} > 0, \ \lambda > 0.$$

4. *For* $0 < \alpha \leq 1$ *and* $x \in \mathcal{D}(A^{\alpha})$, *we have*

$$||S(t)x - x||_X \leq C_{\alpha}t^{\alpha}||A^{\alpha}x||_X, \ C_{\alpha} > 0.$$

Lemma 2.27. *Let conditions H2.56–H2.58 hold, and A be the infinitesimal operator of an analytic semigroup $S(t)$.*

Then:

1. *There exists a unique almost periodic solution* $x(t) \in PCB[\mathbb{R}, X]$ *of* (2.82).
2. *The almost periodic solution* $x(t)$ *is asymptotically stable.*

Proof. We consider the function

$$x(t) = \int_{-\infty}^{t} S(t-s)f(s)ds + \sum_{t_k < t} S(t-t_k)b_k. \qquad (2.91)$$

First, we shall show that the right hand of (2.91) is well defined.
From H2.57 and H2.58, it follows that $f(t)$ and $\{b_k\}$ are bounded, and let

$$max\{||f(t)||_{PC}, ||b_k||_X\} \leq M_0, \ M_0 > 0.$$

Using Lemma 2.26 and the definition for the norm in X^{α}, from (2.91), we obtain

$$||x(t)||_{\alpha} = \int_{-\infty}^{t} ||A^{\alpha}S(t-s)||_X ||f(s)||_{PC}ds$$

$$+ \sum_{t_k < t} ||A^{\alpha}S(t-t_k)||_X ||b_k||_X$$

$$\leq \int_{-\infty}^{t} K_\alpha(t-s)^{-\alpha} e^{-\lambda(t-s)} \|f(s)\|_{PC} ds$$

$$+ \sum_{t_k < t} K_\alpha(t-t_k)^{-\alpha} e^{-\lambda(t-t_k)} \|b_k\|_X. \qquad (2.92)$$

We can easy to verify, that

$$\int_{-\infty}^{t} K_\alpha(t-s)^{-\alpha} e^{-\lambda(t-s)} \|f(s)\|_{PC} ds$$

$$\leq K_\alpha M_0 \int_{-\infty}^{t} (t-s)^{-\alpha} e^{-\lambda(t-s)} ds$$

$$\leq K_\alpha M_0 \frac{\Gamma(1-\alpha)}{\lambda^{1-\alpha}}. \qquad (2.93)$$

Let $m = min\{t - t_k, \ 0 < t - t_k \leq 1\}$. Then from H2.58 and Lemma 1.2, the sum of (2.92) can be estimated as follows

$$\sum_{t_k < t} K_\alpha(t-t_k)^{-\alpha} e^{-\lambda(t-t_k)} \|b_k\|_X$$

$$\leq K_\alpha M_0 \sum_{t_k < t} (t-t_k)^{-\alpha} e^{-\lambda(t-t_k)}$$

$$= K_\alpha M_0 \Big[\sum_{0 < t-t_k \leq 1} (t-t_k)^{-\alpha} e^{-\lambda(t-t_k)}$$

$$+ \sum_{j=1}^{\infty} \sum_{j < t-t_k \leq j+1} (t-t_k)^{-\alpha} e^{-\lambda(t-t_k)} \Big]$$

$$\leq 2 K_\alpha M_0 N \Big(\frac{m^{-\alpha}}{e^{-\lambda}} + \frac{1}{e^\lambda - 1} \Big). \qquad (2.94)$$

From (2.93), (2.94), and equality

$$\Gamma(\alpha)\Gamma(1-\alpha) = \frac{\pi}{sin\pi\alpha}, \ 0 < \alpha < 1,$$

we have

$$\|x(t)\|_\alpha \leq K_\alpha M_0 \Big[\frac{\pi}{\Gamma(\alpha) sin\pi\alpha \lambda^{1-\alpha}} + 2N \Big(\frac{m^{-\alpha}}{e^{-\lambda}} + \frac{1}{e^\lambda - 1} \Big) \Big],$$

and $x \in PCB[\mathbb{R}, X]$.

On the other hand, it is easy to see that the function $x(t)$ is a solution of (2.82).

Let $\varepsilon > 0$, $\tau \in T$, $q \in P$, where the sets T and P are from Lemma 2.25.

Then,

$$||\varphi(t+\tau) - \varphi(t)||_\alpha = ||A^\alpha(x(t+\tau) - x(t))||_{PC}$$

$$\leq \int_{-\infty}^t ||A^\alpha S(t-s)||_X ||f(s+\tau) - f(s)||_{PC} ds$$

$$+ \sum_{t_k < t} ||A^\alpha S(t-t_k)||_X ||b_{k+q} - b_k||_X \leq M_\alpha \varepsilon,$$

where $|t - t_k| > \varepsilon$, $M_\alpha > 0$.

The last inequality implies, that the function $x(t)$ is almost periodic. The uniqueness of this solution follows from conditions H2.56–H2.58 [126].

Let now, $\tilde{x} \in PCB[\mathbb{R}, X]$ be an arbitrary solution of (2.82), and $y = \tilde{x} - x$. Then, $y \in PCB[\mathbb{R}, X]$ and

$$y = S(t - t_0)y(t_0). \tag{2.95}$$

The proof that $x(t)$ is asymptotically stable follows from (2.95), the estimates from Lemma 2.26 and the fact that $i(t_0, t) - p(t - t_0) = o(t)$ for $t \to \infty$. □

Now, we shall investigate the almost periodic solutions of (2.81). Introduce the following conditions:

H2.59. The function $F(t, x)$ is almost periodic with respect to $t \in \mathbb{R}$ uniformly at $x \in \Omega$, Ω is compact from X, and there exist constants $L_1 > 0$, $1 > \kappa > 0$, $1 > \alpha > 0$ such that

$$||F(t_1, x_1) - F(t_2, x_2)||_X \leq L_1(|t_1 - t_2|^\kappa + ||x_1 - x_2||_\alpha),$$

where $(t_i, x_i) \in \mathbb{R} \times \Omega$, $i = 1, 2$.

H2.60. The sequence of functions $\{H_k(x)\}$, $k = \pm 1, \pm 2, \ldots$ is almost periodic uniformly at $x \in \Omega$, Ω is every compact from X, and there exist constants $L_2 > 0$, $1 > \alpha > 0$ such that

$$||H_k(x_1) - H_k(x_2)||_X \leq L_2||x_1 - x_2||_\alpha ,$$

where x_1, $x_2 \in \Omega$.

Theorem 2.25. *Let the following conditions hold:*

1. *Conditions H2.58–H2.60 hold.*
2. *A is the infinitesimal generator of the analytic semigroup $S(t)$.*
3. *The functions $F(t, x)$ and $H_k(x)$ are bounded:*

$$max\Big\{||F(t, x)||_X, \ ||H_k(x)||_X\Big\} \leq M,$$

where $t \in \mathbb{R}$, $k = \pm 1, \pm 2, \ldots$, $x \in \Omega$, $M > 0$.

Then if $L = max\{L_1, L_2\}$, $L > 0$ is sufficiently small it follows that:

1. *There exists a unique almost periodic solution $x \in PCB[\mathbb{R}, X]$ of (2.81).*

2. *The almost periodic solution $x(t)$ is asymptotically stable.*

Proof. We denote by $D^* \subset PCB[\mathbb{R}, X]$ the set of all almost periodic functions with points of discontinuity of the first kind t_k, $k = \pm 1, \pm 2, \ldots$, satisfying the inequality $||\varphi||_{PC} < h$, $h > 0$.

In D^*, we define the operator S^* in the following way

$$S^* \varphi(t) = \int_{-\infty}^{t} A^\alpha S(t - s) F(t, A^{-\alpha} \varphi(s)) ds$$

$$+ \sum_{t_k < t} A^\alpha S(t - t_k) H_k(A^{-\alpha} \varphi(t_k)). \tag{2.96}$$

The facts that S^* is well defined, and $S^* \varphi(t)$ is almost periodic function follow in the same way as in the proof of Lemma 2.27. Now, we shall show, that S^* is a contracting operator in D^*.

Let φ, $\psi \in D^*$. Then, we obtain

$$||S^* \varphi(t) - S^* \psi(t)||_X$$

$$\leq \int_{-\infty}^{t} ||A^\alpha S(t - s)||_X ||F(t, A^{-\alpha} \varphi(t)) - F(t, A^{-\alpha} \psi(t))||_X ds$$

$$+ \sum_{t_k < t} ||A^\alpha S(t - t_k)||_X ||H_k(A^{-\alpha} \varphi(t_k)) - H_k(A^{-\alpha} \psi(t_k))||_X$$

$$\leq L K_\alpha ||\varphi(t) - \psi(t)||_X \left[\int_{-\infty}^{t} (t - s)^{-\alpha} e^{-\lambda(t-s)} ds \right.$$

$$+ \left. \sum_{t_k < t} (t - t_k)^{-\alpha} e^{-\lambda(t-t_k)} \right].$$

With similar arguments like in (2.94), for the last inequality, we have

$$||S^* \varphi(t) - S^* \psi(t)||_X \leq L K_\alpha \left[\frac{\Gamma(1 - \alpha)}{\lambda^{1-\alpha}} \right.$$

$$+ \left. 2N \left(\frac{m^{-\alpha}}{e^{-\lambda}} + \frac{1}{e^\lambda - 1} \right) \right] ||\varphi(t) - \psi(t)||_X.$$

Then, if L is sufficiently small and

$$L \leq \left(K_\alpha \left[\frac{\pi}{\Gamma(\alpha) \sin \pi \alpha \lambda^{1-\alpha}} + 2N \left(\frac{m^{-\alpha}}{e^{-\lambda}} + \frac{1}{e^\lambda - 1} \right) \right] \right)^{-1},$$

it follows that the operator S^* is a contracting operator in D^*.

Consequently, there exists $\varphi \in D^*$ such that

$$\varphi(t) = \int_{-\infty}^{t} A^\alpha S(t-s) F(t, A^{-\alpha}\varphi(s)) ds$$

$$+ \sum_{t_k < t} A^\alpha S(t - t_k) H_k(A^{-\alpha}\varphi(t_k)). \qquad (2.97)$$

On the other hand, since A^α is closed, we get

$$A^{-\alpha}\varphi(t) = \int_{-\infty}^{t} S(t-s) F(t, A^{-\alpha}\varphi(s)) ds$$

$$+ \sum_{t_k < t} S(t - t_k) H_k(A^{-\alpha}\varphi(t_k)). \qquad (2.98)$$

Now, let $h \in (0, \theta)$, where θ is the constant from H2.56, and $t \in (t_k, t_{k+1} - h]$.

Then,

$$||\varphi(t+h) - \varphi(t)||_\alpha$$

$$\leq ||\int_{-\infty}^{t} (S(h) - I) A^\alpha S(t-s) F(t, A^{-\alpha}\varphi(s)) ds||_\alpha$$

$$+ ||\int_{t}^{t+h} A^\alpha S(t+h-s) F(t, A^{-\alpha}\varphi(s)) ds||_\alpha. \qquad (2.99)$$

From Lemma 2.26 for (2.99), it follows that

$$||\varphi(t+h) - \varphi(t)||_\alpha \leq K_{\alpha+\beta} M C_\beta h^\beta + K_\alpha M \frac{h^{1-\alpha}}{1-\alpha}.$$

Then, there exists a constant $C > 0$ such that

$$||\varphi(t+h) - \varphi(t)||_\alpha \leq C h^\beta.$$

On the other hand, from H2.59 it follows that $F(t, A^{-\alpha}\varphi(t))$ is locally Hölder continuous. From H2.60 and the conditions of the theorem, $H_k(A^{-\alpha}\varphi(t_k))$ is a bounded almost periodic sequence.

Let $\varphi(t)$ be a solution of (2.97), and let consider the equation

$$\dot{x}(t) = Ax + F(t, A^{-\alpha}\varphi(t)) + \sum_{k=-\infty}^{\infty} H_k(A^{-\alpha}\varphi(t_k))\delta(t - t_k). \qquad (2.100)$$

Using the condition H2.60 and Lemma 2.27, it follows that for (2.100) there exists a unique asymptotically stable solution in the form

$$\psi(t) = \int_{-\infty}^{t} S(t-s)F(s, A^{-\alpha}\varphi(s))ds + \sum_{t_k<t} S(t-t_k)H_k(A^{-\alpha}\varphi(t_k)),$$

where $\psi \in \mathcal{D}(A^{\alpha})$.

Then,

$$A^{\alpha}\psi(t) = \int_{-\infty}^{t} A^{\alpha} S(t-s)F(s, A^{-a}\varphi(s))ds$$

$$+ \sum_{t_k<t} A^{\alpha} H_k(A^{-\alpha}\varphi(t_k)) = \varphi(t).$$

The last equality shows that $\psi(t) = A^{-\alpha}\varphi(t)$ is a solution of (2.81), and the uniqueness follows from the uniqueness of the solution of (2.97), (2.100) and Lemma 2.27. □

Example 2.5. Here, we shall consider a two-dimensional impulsive predator–prey system with diffusion, when biological parameters assumed to change in almost periodical manner. The system is affected by impulses, which can be considered as a control.

Assuming that the system is confined to a fixed bounded space domain $\Omega \subset \mathbb{R}^n$ with smooth boundary $\partial\Omega$, non-uniformly distributed in the domain $\overline{\Omega} = \Omega \times \partial\Omega$ and subjected to short-term external influence at fixed moment of time. The functions $u(t,x)$ and $v(t,x)$ determine the densities of predator and pray, respectively, $\Delta = \frac{\partial^2}{\partial x_1^2} + \frac{\partial^2}{\partial x_2^2} + \ldots + \frac{\partial^2}{\partial x_n^2}$ is the Laplace operator and $\frac{\partial}{\partial n}$ is the outward normal derivative.

The system is written in the form

$$\begin{cases} \dfrac{\partial u}{\partial t} = \mu_1 \Delta u + u\Big[a_1(t,x) - b(t,x)u - \dfrac{c_1(t,x)v}{r(t,x)v+u}\Big], \ t \neq t_k, \\[2mm] \dfrac{\partial v}{\partial t} = \mu_2 \Delta v + v\Big[-a_2(t,x) + \dfrac{c_2(t,x)u}{r(t,x)u+v}\Big], \ t \neq t_k, \\[2mm] u(t_k^+, x) = u(t_k^-, x)I_k(x, u(t_k, x), v(t_k, x)), \ k = \pm 1, \pm 2, \ldots, \\[2mm] v(t_k^+, x) = v(t_k^-, x)J_k(x, u(t_k, x), v(t_k, x)), \ k = \pm 1, \pm 2, \ldots, \\[2mm] \dfrac{\partial u}{\partial n}\Big|_{\partial\Omega} = 0, \ \dfrac{\partial v}{\partial n}\Big|_{\partial\Omega} = 0. \end{cases} \qquad (2.101)$$

The boundary condition characterize the absence of migration, $\mu_1 > 0$, $\mu_2 > 0$ are diffusion coefficients. We assume that, the predator functional response has the form of the ratio function $\dfrac{c_1 v}{rv + u}$. The ratio function $\dfrac{c_2 u}{rv + u}$ represents the conversion of prey to predator, a_1, a_2, c_1 and c_2 are positive functions that stand for prey intrinsic growth rate, capturing rate of the predator, death rate of the predator and conversion rate, respectively, $\dfrac{a_1(t, x)}{b(t, x)}$ gives the carrying capacity of the prey, and $r(t, x)$ is the half saturation function.

We note that the problems of existence, uniqueness, and continuability of solutions of impulsive differential equations (2.101) have been investigated in [7].

Introduce the following conditions:

H2.61. The functions $a_i(t, x)$, $c_i(t, x)$, $i = 1, 2$, $b(t, x)$ and $r(t, x)$ are almost periodic with respect to t, uniformly at $x \in \overline{\Omega}$, positive-valued on $\mathbb{R} \times \overline{\Omega}$ and locally Hölder continuous with points of discontinuity at the moments t_k, $k = \pm 1, \pm 2, \ldots$, at which they are continuous from the left.

H2.62. The sequences of functions $\{I_k(x, u, v)\}$, $\{J_k(x, u, v)\}$, $k = \pm 1, \pm 2, \ldots$ are almost periodic with respect to k, uniformly at $x, u, v \in \overline{\Omega}$.

Set $w = (u, v)$, and

$$
A = \begin{bmatrix} \lambda - \mu_1 \Delta & 0 \\ 0 & \lambda - \mu_2 \Delta \end{bmatrix},
$$

$$
F(t, w) - \begin{bmatrix} u\left[a_1(t, x) - b(t, x)u - \dfrac{c_1(t, x)v}{r(t, x)v + u}\right] + \lambda u \\ v\left[-a_2(t, x) + \dfrac{c_2(t, x)u}{r(t, x)u + v}\right] + \lambda v \end{bmatrix},
$$

$$
H_k(w(t_k)) = \begin{bmatrix} u(t_k, x)I_k(x, u(t_k, x), v(t_k, x)) - u(t_k, x) \\ v(t_k, x)J_k(x, u(t_k, x), v(t_k, x)) - v(t_k, x) \end{bmatrix},
$$

where $\lambda > 0$.

Then, the system (2.101) moves to the equation

$$
\dot{w}(t) = Aw + F(t, w) + \sum_{k=\pm 1, \pm 2, \ldots} G_k(w)\delta(t - t_k). \tag{2.102}
$$

It is well-known [68], that the operator A is sectorial, and $\mathrm{Re}\,\sigma(A) \leq -\lambda$, where $\sigma(A)$ is the spectrum of A. Now, the analytic semigroup of the operator A is e^{-At}, and

$$
A^{-\alpha} = \frac{1}{\Gamma(\alpha)} \int_0^\infty t^{\alpha - 1} e^{-At} dt.
$$

Theorem 2.26. *Let for the equation (2.102) the following conditions hold:*

1. *Conditions H2.56, H2.61 and H2.62 are met.*
2. *For the functions $F(t, w)$ there exist constants $L_1 > 0$, $1 > \kappa > 0$, $1 > \alpha > 0$ such that*

$$||F(t_1, w_1) - F(t_2, w_2)||_X \leq L_1 \big(|t_1 - t_2|^\kappa + ||w_1 - w_2||_\alpha \big),$$

   where $(t_i, w_i) \in \mathbb{R} \times X_\alpha$, $i = 1, 2$.
3. *For the set of functions $\{H_k(w)\}$, $k = \pm 1, \pm 2, \ldots$ there exist constants $L_2 > 0$, $1 > \alpha > 0$ such that*

$$||H_k(w_1) - H_k(w_2)||_X \leq L_2 ||w_1 - w_2||_\alpha.$$

   where w_1, $w_2 \in X_\alpha$
4. *The functions $F(t, w)$ and $H_k(w)$ are bounded for $t \in \mathbb{R}, w \in X_\alpha$ and $k = \pm 1, \pm 2, \ldots$.*

Then, if $L = max\{L_1, L_2\}$ is sufficiently small, it follows:

1. *There exists a unique almost periodic solution $x \in PCB[\mathbb{R}, X]$ of (2.101).*
2. *The almost periodic solution $x(t)$ is asymptotically stable.*

Proof. From conditions H2.61, H2.62 and conditions of the theorem, it follows that all conditions of Theorem 2.25 hold. Then, for (2.102) and consequently for (2.101) there exists a unique almost periodic solution of (2.101), which is asymptotically stable.   □

Chapter 3
Lyapunov Method and Almost Periodicity

The present chapter will deal with the existence and uniqueness of almost periodic solutions of impulsive differential equations by Lyapunov method.

Section 3.1 will offer almost periodic Lyapunov functions. The existence results of almost periodic solutions for different kinds of impulsive differential equations will be given.

In Sect. 3.2, we shall use the comparison principle for the existence theorems of almost periodic solutions of impulsive integro-differential equations.

Section 3.3 will deal with the existence of almost periodic solutions of impulsive differential equations with time-varying delays. The investigations are carried out by using minimal subsets of a suitable space of piecewise continuous Lyapunov functions.

In Sect. 3.4, we shall continue to use Lyapunov method, and we shall investigate the existence and stability of almost periodic solutions of nonlinear impulsive functional differential equations.

Finally, in Sect. 3.5, by using the concepts of uniformly positive definite matrix functions and Hamilton–Jacobi–Riccati inequalities, we shall prove the existence theorems for almost periodic solutions of uncertain impulsive dynamical equations.

3.1 Lyapunov Method and Almost Periodic Solutions

In this part, we shall consider the system of impulsive differential equations

$$\begin{cases} \dot{x} = f(t, x), \ t \neq t_k, \\ \Delta x(t_k) = I_k(x(t_k)), \ k = \pm 1, \pm 2, \ldots, \end{cases} \tag{3.1}$$

where $t \in \mathbb{R}$, $\{t_k\} \in \mathcal{B}$, $f : \mathbb{R} \times \mathbb{R}^n \to \mathbb{R}^n$, $I_k : \mathbb{R}^n \to \mathbb{R}^n$, $k = \pm 1, \pm 2, \ldots$.

G.T. Stamov, *Almost Periodic Solutions of Impulsive Differential Equations*,
Lecture Notes in Mathematics 2047, DOI 10.1007/978-3-642-27546-3_3,
© Springer-Verlag Berlin Heidelberg 2012

We shall introduce the following conditions:

H3.1.   The function $f \in C[\mathbb{R} \times \mathbb{R}^n, \mathbb{R}^n]$ and $f(t,0) = 0$ for $t \in \mathbb{R}$.

H3.2.   The function f is Lipschitz continuous with respect to $x \in \mathbb{R}^n$ with a Lipschitz constant $L_1 > 0$ uniformly on $t \in \mathbb{R}$, i.e.

$$||f(t,\overline{x}) - f(t,x)|| \leq L_1||\overline{x} - x||,$$

for $\overline{x}, x \in \mathbb{R}^n$.

H3.3.   The functions $I_k \in C[\mathbb{R}^n, \mathbb{R}^n]$, $k = \pm 1, \pm 2, \ldots$.

H3.4. $I_k(0) = 0$, $k = \pm 1, \pm 2, \ldots$.

H3.5.   The functions I_k, $k = \pm 1, \pm 2, \ldots$ are Lipschitz continuous with respect to $x \in \mathbb{R}^n$ with a Lipschitz constant $L_2 > 0$, i.e.

$$||I_k(\overline{x}) - I_k(x)|| \leq L_2||\overline{x} - x||,$$

for $\overline{x}, x \in \mathbb{R}^n$.

We note that [94], if conditions H3.1–H3.5 are satisfied, then for the system (3.1) with initial condition $x(t_0^+) = x_0$, there exists a unique solution $x(t) = x(t; t_0, x_0)$.

We shall use the sets G_k, the set

$$G = \bigcup_{k=\pm 1, \pm 2, \ldots} G_k,$$

the class of piecewise continuous functions V_0, introduced in Chap. 1 for $\Omega \equiv \mathbb{R}^n$, and the class of function K,

$$K = \left\{ a \in C[\mathbb{R}^+, \mathbb{R}^+]; \ a \text{ is increasing and } a(0) = 0 \right\}.$$

Introduce the following conditions:

H3.6.   The function $f(t,x)$ is almost periodic in t uniformly with respect to $x \in \mathbb{R}^n$.

H3.7.   The sequence $\{I_k(x)\}$, $k = \pm 1, \pm 2, \ldots$ is almost periodic uniformly with respect to $x \in \mathbb{R}^n$.

H3.8.   The set of sequences $\{t_k^j\}$, $t_k^j = t_{k+j} - t_k$, $k = \pm 1, \pm 2, \ldots$, $j = \pm 1, \pm 2, \ldots$ is uniformly almost periodic, and $inf_k t_k^1 = \theta > 0$.

Let the assumptions H3.1–H3.8 are satisfied, and let $\{s_n\}$ be an arbitrary sequence of real numbers. Like in Chap. 1, the system (3.1) moves to the system

$$\begin{cases} \dot{x} = f^s(t, x), \ t \neq t_k^s, \\ \Delta x(t_k^s) = I_k^s(x(t_k^s)), \ k = \pm 1, \pm 2, \ldots, \end{cases} \quad (3.2)$$

and the set of systems (3.2) we shall denote by $H(f, I_k, t_k)$.

In this paragraph we shall use Definition 1.6, Definition 1.7, and the next definitions.

Definition 3.1. The zero solution $x(t) \equiv 0$ of system (3.1) is said to be:

1. *Globally asymptotically stable*, if it is stable and if every solution of (3.1) with an initial state in a neighborhood of the zero tends to zero as $t \to \infty$;
2. *Globally quasi-equi-asymptotically stable*, if

$$(\forall \alpha > 0)(\forall \varepsilon > 0)(\forall t_0 \in \mathbb{R})(\exists T > 0)(\forall x_0 \in B_\alpha)(\forall t \geq t_0 + T):$$

$$||x(t; t_0, x_0)|| < \varepsilon.$$

Definition 3.2. The solution $x(t; t_0, x_0)$ of system (3.1) is *equi-bounded*, if

$$(\forall \alpha > 0)(\forall t_0 \in \mathbb{R})(\exists \beta > 0)(\forall x_0 \in B_\alpha)(\forall t \geq t_0): \ ||x(t; t_0, x_0)|| < \beta.$$

Definition 3.3. The zero solution $x(t) \equiv 0$ of (3.1) is said to be *globally perfectly uniform-asymptotically stable*, if it is uniformly stable, the number β in Definition 3.2 and the T in the part 2 of Definition 3.2 are independent of $t_0 \in \mathbb{R}$.

3.1.1 Almost Periodic Lyapunov Functions

In the further considerations we shall use the following lemma.

Lemma 3.1 ([116]). *Given any real function $A(r, \varepsilon)$ of real variables, defined, continuous and positive in $Q = \{(r, \varepsilon) : \ r \in \mathbb{R}^+ \text{ and } \varepsilon > 0\}$, there exist two continuous functions $h = h(r), h(r) > 0$ and $g = g(\varepsilon)$, $g(\varepsilon) > 0, g(0) = 0$ such that $h(r)g(\varepsilon) \leq A(r, \varepsilon)$ in Q.*
Now, we shall prove a Massera's type theorem.

Theorem 3.1. *Let conditions H3.1–H3.8 hold, and suppose that the zero solution of system (3.1) is globally perfectly uniform-asymptotically stable.*
Then there exists a Lyapunov function V, defined on $\mathbb{R} \times \mathbb{R}^n$, $V \in V_0$, which is almost periodic in t uniformly with respect to $x \in \mathbb{R}^n$, and satisfies the following conditions

$$a(||x||) \leq V(t, x) \leq b(||x||), \ (t, x) \in \mathbb{R} \times \mathbb{R}^n, \tag{3.3}$$

where $a, b \in K$, $a(r), b(r) \to \infty$ as $r \to \infty$,

$$V(t^+, x) \leq V(t, x), \ x \in \mathbb{R}^n, \ t = t_k, \ k = \pm 1, \pm 2, ...,$$

and

$$D^+V(t, x) \leq -cV(t, x), \quad (t, x) \in G \tag{3.4}$$

for $c = const > 0$.

Proof. Fallow [15] let $\Gamma^*(\sigma, \alpha) = \{(t, x) : t \in (-\sigma, \sigma), \ x \in B_\alpha\}$, where σ and α are arbitrary positive constants. From [21] and by the global perfect uniform-asymptotic stability of the zero solution of (3.1), it follows that the solutions of system (3.1) are equi-bounded, i.e. there exists a constant $\beta = \beta(\alpha) > 0$ such that for $(t_0, x_0) \in \Gamma^*(\sigma, \alpha)$, we have $\|x(t; t_0, x_0)\| < \beta(\alpha)$, where $t \geq t_0$.

Moreover, there exists a $T(\alpha, \varepsilon) > 0$ such that from $(t_0, x_0) \in \Gamma^*(\sigma, \alpha)$, we obtain $\|x(t; t_0, x_0)\| < \varepsilon$ for $t \geq t_0 + T(\alpha, \varepsilon)$. If $\varepsilon > 1$, we set $T(\alpha, \varepsilon) = T(\alpha, 1)$.

From conditions H3.2 and H3.4, it follows that there exist $L_1(\alpha, \varepsilon) > 0$ and $L_2(\alpha, \varepsilon) > 0$ such that if $0 \leq t \leq \sigma + T(\alpha, \varepsilon)$, $x_1, x_2 \in B_{\beta(\alpha)}$, we get

$$\|f(t, x_1) - f(t, x_2)\| \leq L_1(\alpha, \varepsilon)\|x_1 - x_2\|,$$

$$\|I_k(x_1) - I_k(x_2)\| \leq L_2(\alpha, \varepsilon)\|x_1 - x_2\|, \ k = \pm 1, \pm 2, \dots.$$

Let

$$f^* = 1 + \max\|f(t, x)\|, \quad 0 \leq t \leq T(\alpha, \varepsilon), \ x \in B_{\beta(\alpha)},$$

$$I^* = \max\|I_k(x)\|, \quad x \in B_{\beta(\alpha)}, \ k = \pm 1, \pm 2, \dots,$$

and let $c = const > 0$.

We set

$$A(\alpha, \varepsilon) = e^{cT(\alpha, \varepsilon)} \times \left\{ \left(2\left(f^* + I^* \frac{1}{p} \right) \right. \right.$$

$$\left. \left. + \left(\frac{1}{p} + 1 \right) T(\alpha, \varepsilon) \right) e^{L_1(\alpha, \varepsilon) + \frac{1}{p} \ln(1 + L_2(\alpha, \varepsilon))} + \beta(\alpha) \right\}. \tag{3.5}$$

From Lemma 3.1, it follows that there exist two functions $h(\alpha) > 0$ and $g(\varepsilon) > 0$ such that $\varepsilon > 0$, $g(0) = 0$ and

$$g(\varepsilon)A(\alpha, \varepsilon) \leq h(\alpha). \tag{3.6}$$

For $i = 1, 2, \dots$, let we define $V_i(t, x)$ by

$$V_i(t, x) = g\left(\frac{1}{i} \right) \sup_{\tau \geq 0} Y_i\big(\|x(t + \tau, t, x)\| \big) e^{c\tau}, \quad t \neq t_k,$$

$$V_i(t_k, x) = V_i(t_k^-, x), \ k = \pm 1, \pm 2, \dots, \tag{3.7}$$

where

$$Y_i(z) = \begin{cases} z - \frac{1}{i}, & if \ z \geq \frac{1}{i}, \\ 0, & if \ 0 \leq z \leq \frac{1}{i}. \end{cases}$$

Clearly, $Y_i(z) \to \infty$ as $z \to \infty$, for each i, and

$$\|Y_i(z_1) - Y_i(z_2)\| \le |z_1 - z_2|, \tag{3.8}$$

where $z_1, z_2 \ge 0$.

From the definition of $V_i(t, x)$ it follows that

$$g\left(\frac{1}{i}\right) Y_i(\|x\|) \le V_i(t, x), \tag{3.9}$$

and $V_i(t, 0) \equiv 0$ as $(t, x) \in \Gamma^*(\sigma, \alpha)$.

On the other hand, from (3.5) and (3.6), we have

$$V_i(t, x) \le g\left(\frac{1}{i}\right) Y_i(\beta(\alpha)) e^{cT(\alpha, \frac{1}{i})} \le g\left(\frac{1}{i}\right) \beta(\alpha) e^{cT(\alpha, \frac{1}{i})} \le h(\alpha). \tag{3.10}$$

Then from (3.9) and (3.10) for the function $V_i(t, x)$ it follows that (3.3) holds. For $(t', x'), (t, x) \in \Gamma^*(\sigma, \alpha)$ and $t < t'$, we get

$$\left| V_i(t', x') - V_i(t, x) \right| \le g\left(\frac{1}{i}\right) \sup_{\tau \ge 0} \left| Y_i(\|x(t' + \tau; t', x')\|) \right.$$

$$\left. - Y_i(\|x(t + \tau; t, x)\|) \right| e^{c\tau}$$

$$\le g\left(\frac{1}{i}\right) \sup_{0 \le \tau \le T(\alpha, \frac{1}{i})} e^{c\tau} \|x(t' + \tau; t', x') - x(t + \tau; t, x)\|$$

$$\le g\left(\frac{1}{i}\right) \sup_{\tau \ge 0} e^{c\tau} \Big\{ \|x(t' + \tau; t', x') - x(t' + \tau; t, x)\|$$

$$+ \|x(t' + \tau; t, x) - x(t + \tau; t, x)\| \Big\}. \tag{3.11}$$

Then

$$\|x(t' + \tau; t, x) - x(t + \tau; t, x)\|$$

$$\le \int_{t+\tau}^{t'+\tau} \|f(s, x(s))\| \, ds + \sum_{t+\tau < t_k < t'+\tau} \|I_k(x(\tau_k))\|$$

$$\le \max_{\substack{t+\tau \le s \le t'+\tau \\ x \in B_{\beta(\alpha)}}} \|f(s, x(s))\| (t' - t) + \max_{t+\tau \le t_k \le t'+\tau} \|I_k(x(t_k))\| \, i(t + \tau, t' + \tau)$$

$$\le \left(f^* + I^* \frac{1}{p} \right) (t' - t), \tag{3.12}$$

where $i(t + \tau, t' + \tau)$ is the number of points on the interval $(t + \tau, t' + \tau)$.

Let $X = x(t'; t, x)$. From Theorem 1.9, we obtain

$$\|x(t' + \tau; t, x) - x(t' + \tau; t', x')\|$$

$$\leq \|X - x'\| \exp\left\{L_1\left(\alpha, \frac{1}{i}\right) + \frac{1}{p}\ln\left(1 + L_2\left(\alpha, \frac{1}{i}\right)\right)T\left(\alpha, \frac{1}{i}\right)\right\}$$

$$\leq (\|X - x\| + \|x - x'\|) \exp\left\{L_1\left(\alpha, \frac{1}{i}\right) + \frac{1}{p}\ln\left(1 + L_2\left(\alpha, \frac{1}{i}\right)\right)T\left(\alpha, \frac{1}{i}\right)\right\}$$

$$\leq \left(\left(f^* + I^*\frac{1}{p}\right)(t' - t) + \|x - x'\|\right)$$

$$\times \exp\left\{L_1\left(\alpha, \frac{1}{i}\right) + \frac{1}{p}\ln\left(1 + L_2\left(\alpha, \frac{1}{i}\right)\right)T\left(\alpha, \frac{1}{i}\right)\right\}. \qquad (3.13)$$

Then from (3.12), (3.13) for (3.11), it follows

$$|V_i(t, x) - V_i(t', x')| \leq g\left(\frac{1}{i}\right)\sup_{0 \leq \tau \leq T(\alpha, \frac{1}{i})} e^{c\tau}\left(\left(f^* + I^*\frac{1}{p}\right) + \left(f^* + I^*\frac{1}{p}\right)(t' - t)\right.$$

$$\left. + \|x - x'\|\right) \times \exp\left\{L_1\left(\alpha, \frac{1}{i}\right) + \frac{1}{p}\ln\left(1 + L_2\left(\alpha, \frac{1}{i}\right)\right)T\left(\alpha, \frac{1}{i}\right)\right\}$$

$$\leq g\left(\frac{1}{i}\right)2\left(f^* + I^*\frac{1}{p}\right)\exp\left\{L_1\left(\alpha, \frac{1}{i}\right) + \frac{1}{p}\ln\left(1 + L_2\left(\alpha, \frac{1}{i}\right)\right)T\left(\alpha, \frac{1}{i}\right)\right\}$$

$$\times (|t' - t| + \|x - x'\|) \leq h(\alpha)(|t' - t| + \|x - x'\|). \qquad (3.14)$$

On the other hand, as $x \in B_{\beta(\alpha)}$ and $t = t_k$, from (3.14) it follows that $V_i(t, x)$ is continuous, and for $t = t'$ we obtain that the function $V_i(t, x)$ is locally Lipschitz continuous.

Let t_k are fixed, $t', t'' \in (t_k, t_{k+1}]$, $x', x'' \in B_{\beta(\alpha)}$ and $u' = x(t'; t_k, x')$, $u'' = x(t''; t_k, x'')$.

Then

$$|V_i(t', x') - V_i(t'', x'')| \leq |V_i(t', x') - V_i(t', u')|$$

$$+ |V_i(t'', x'') - V_i(t'', u'')| + |V_i(t', u') - V_i(t'', u'')|. \qquad (3.15)$$

By the fact that the functions $V_i(t, x)$ and $f(t, x)$ are Lipschitz continuous, we obtain the estimates

$$|V_i(t', x') - V_i(t', u')| \leq h(\alpha)\|x' - u'\|,$$

$$\|r' - u'\| \leq \|x' - x\| + \|u' - x\|,$$

$$\|u' - x\| \leq \int_{t_k}^{t'} L_1\left(\alpha, \frac{1}{i}\right)\exp\left\{\int_{t_k}^{s} L_1\left(\alpha, \frac{1}{i}\right)d\tau\right\}ds\|x\| \equiv N(t')\|x\|.$$

Then

$$\left|V_i(t',x') - V_i(t',u')\right| \le h(\alpha)\|x' - x\| + h(\alpha)N(t')\|x\|. \tag{3.16}$$

By analogy,

$$\left|V_i(t'',x'') - V_i(t'',u'')\right| \le h(\alpha)\|x'' - x\| + h(\alpha)N(t'')\|x\|. \tag{3.17}$$

Since $a_i(\delta) = \sup_{\tau > \delta} Y_i(\|x(t_k+\tau, t_k, x)\|)e^{c\tau}$ is non-increasing and $\lim_{\delta \to 0^+} a_i(\delta) = a_i(0)$, it follows that

$$\left|V_i(t',u') - V_i(t'',u'')\right| \le g\!\left(\frac{1}{i}\right)\left| \sup_{s>0} Y_i\big(\|x(t'+s;t',u')\|\big)e^{cs}\right.$$

$$\left. - \sup_{s>0} Y_i\big(\|x(t''+s;t'',u'')\|\big)e^{cs}\right|$$

$$\le g\!\left(\frac{1}{i}\right)\left|a(t'-t_k)e^{-c(t'-t_k)} - a(t''-t_k)e^{-c(t''-t_k)}\right| \to 0$$

as $t' \to t_k^+$ and $t'' \to t_k^+$. From (3.15)–(3.17), we obtain that there exists the limit $V_i(t_k^+, x)$.

The proof of the existence of the limit $V_i(t_k^-, x)$ follows by analogy.

Let $\eta(t; t_0, x_0)$ be the solution of the initial value problem

$$\begin{cases} \dot{\eta} = f(t,\eta), \\ \eta(t_0) = x_0. \end{cases}$$

Since $t_{k-1} < \lambda < t_k < \mu < t_{k+1}$ and $s > \mu$ it follows that

$$x\big(s; \mu, \eta(\mu; t_k, x + I_k(x))\big) = x\big(s; \lambda, \eta(\lambda, t_k, x)\big).$$

Then

$$V_i\big(\mu, \eta(\mu; t_k, x + I_k(x))\big) \le V_i\big(\lambda, \eta(\lambda; t_k, x)\big)$$

and passing to the limits as $\mu \to t_k^+$ and $\lambda \to t_k^-$, we obtain

$$V_i\big(t_k^+, x + I_k(x)\big) \le V_i(t_k^-, x) = V_i(t_k, x). \tag{3.18}$$

Let $x \in B_{\beta(\alpha)}$, $t \ne t_k$, $h > 0$, and $x' = x(t + h; t, x)$.
Then

$$V_i(t + h, x') = g\!\left(\frac{1}{i}\right) \sup_{s \ge 0} Y_i\big(\|x(t + h + s, t + h, x')\|\big)e^{cs}$$

$$= g\!\left(\frac{1}{i}\right) \sup_{\tau > h} Y_i\big(\|x(t + \tau, t + h, x')\|\big)e^{c\tau}e^{-ch} \le V_i(t, x)e^{-ch}$$

or

$$\frac{1}{h}\big(V_i(t+h,x') - V_i(t,x)\big) \le \frac{1}{h}(e^{-ch}-1)V_i(t,x).$$

Consequently, $D^+V_i(t,x) \le -cV_i(t,x)$. From this inequality, we obtain (3.4) for the function $V_i(t,x)$.

Now, we define the desired function $V(t,x)$ by setting

$$\begin{cases} V(t,x) = \displaystyle\sum_{i=1}^{\infty} \frac{1}{2^i} V_i(t,x), \ t \ne t_k, \\ V(t_k,x) = V(t_k^-,x), \ k = \pm 1, \pm 2, ... \end{cases} \tag{3.19}$$

Since (3.11) implies the uniform convergence of the series (3.19) in $\Gamma^*(\sigma,\alpha)$, $V(t,x)$ is defined on $\mathbb{R} \times \mathbb{R}^n$, piecewise continuous along t, with points of discontinuity at the moments t_k, $k = \pm 1, \pm 2, ...$ and it is continuous along x.

From (3.9) it follows that $V(t,0) \equiv 0$. For x such that $||x|| \ge 1$, from (3.9) and (3.19), we obtain

$$V(t,x) > \frac{1}{2}V_1(t,x) \ge \frac{1}{2}g(1)Y_1(||x||) \ge \frac{1}{2}(||x||-1). \tag{3.20}$$

For x such that $\dfrac{1}{i} \le ||x|| \le \dfrac{1}{i-1}$, we obtain

$$V(t,x) \ge \frac{1}{2^{i+1}}V_{i+1}(t,x)$$

$$\ge \frac{1}{2^{i+1}}g\Big(\frac{1}{i+1}\Big)Y_{i+1}(||x||)$$

$$\ge \frac{1}{2^{i+1}}g\Big(\frac{1}{i+1}\Big)Y_{i+1}\Big(||x|| - \frac{1}{i+1}\Big)$$

$$\ge \frac{1}{2^{i+1}}g\Big(\frac{1}{i+1}\Big)\frac{1}{i(i+1)}. \tag{3.21}$$

From (3.20) and (3.21) we can find $a \in K$ such that $a(r) \to \infty$ as $r \to \infty$ and $a(||x||) \le V(t,x)$.

Let $(t,x),(t',x') \in \Gamma^*(\sigma,\alpha)$ with $t < t'$, and then

$$\big|V(t,x) - V(t',x')\big| \le \sum_{i=1}^{\infty} \frac{1}{2^i}\big|V_i(t,x) - V_i(t',x')\big|$$

$$\le \sum_{i=1}^{\infty} \frac{1}{2^i}h(\alpha)\big(|t-t'| + ||x-x'||\big)$$

$$\le h(\alpha)\big(|t-t'| + ||x-x'||\big). \tag{3.22}$$

From (3.22) it follows that for $x \in B_{\beta(\alpha)}$ and $t \neq t_k$ the function $V(t, x)$ is continuous, and for $t = t'$, we obtain $V(t^+, x + I_k(x)) \leq V(t, x)$.

Let $t_k \in \mathbb{R}, x \in B_{\beta(\alpha)}$ be fixed and $\xi_j \in (t_k, t_{k+1}], \ x_j \in B_{\beta(\alpha)}$, where $u_j = x(\xi_j; t_k, x_j), \ (j = 1, 2)$.

Then

$$\left| V(\xi_j, x_j) - V(\xi_j, u_j) \right| = \sum_{i=1}^{\infty} \frac{1}{2^i} \left| V_i(\xi_j, x_j) - V_i(\xi_j, u_j) \right|$$

$$\leq \sum_{i=1}^{\infty} \frac{1}{2^i} g\left(\frac{1}{2^i} \right) \left| a(\xi_1 - t_k) e^{-c(\xi_1 - t_k)} - a(\xi_2 - t_k) e^{-c(\xi_2 - t_k)} \right| \to 0$$

for $\xi_j \to t_k^+ \ (j = 1, 2)$, i.e. there exists the limit $V(t_k^+, x)$. The proof of the existence of the limit $V(t_k^-, x)$ follows by analogy.

Let now $t_{k-1} < \lambda < t_k < \mu < t_{k+1}, s > \mu$ and from (3.18), we get

$$V\left(t_k^+, x + I_k(x)\right) = \sum_{i=1}^{\infty} \frac{1}{2^i} V_i\left(t_k^+, x + I_k(x)\right) \leq \sum_{i=1}^{\infty} V\left(t_k^-, x\right)$$

$$\leq \sum_{i=1}^{\infty} \frac{1}{2^i} V_i(t_k, x) = V(t_k, x). \tag{3.23}$$

Let $x \in B_{\beta(\alpha)}, \ t \neq t_k$ and $h > 0$. Then from (3.19), we obtain

$$D^+ V(t, x) = \sum_{i-1}^{\infty} \lim_{h \to 0^+} \sup \frac{1}{h} \left[V_i(t + h, x(t + h; t, x)) - V_i(t, x) \right]$$

and

$$\sum_{i=1}^{\infty} \frac{1}{2^i} \left(-cV_i(t, x) \right) \leq -cV(t, x).$$

Then

$$D^+ V(t, x) \leq -cV(t, x). \tag{3.24}$$

Consequently, from (3.24) it follows that there exists $V(t, x)$ from V_0 such that (3.3) and (3.4) are fulfilled.

Here, we shall show that the function $V(t, x)$ is almost periodic in t uniformly with respect to $x \in B_{\beta(\alpha)}$.

From conditions of the theorem it follows that if $x \in B_{\beta(\alpha)}$, then there exists $\beta(\alpha) > 0$ such that $||x(t; \tau, x)|| \leq \beta(\alpha)$ for any $t \geq \tau, \ \tau \in \mathbb{R}$. From conditions H3.6–H3.8, we get that for an arbitrary sequence $\{s_m'\}$ of real numbers there exists a subsequence $\{s_n\}, \ s_n = s_{m_n}'$ moving (3.1) in $H(f, I_k, t_k)$.

Then, as $x \in B_{\beta(\alpha)}$, we obtain

$$\left| V_i(t + s_n, x) - V_i(t + s_p, x) \right| \le g\left(\frac{1}{i}\right) \sup_{\tau \ge 0} e^{c\tau} \left| Y_i\left(\left\| x(t + s_n + \tau; t + s_n, x) \right\| \right) \right.$$

$$\left. - Y_i\left(\left\| x(t + s_p + \tau; t + s_p, x) \right\| \right) \right|$$

$$\le g\left(\frac{1}{i}\right) \sup_{0 \le \tau \le T(\alpha, \frac{1}{i})} e^{c\tau} \left\| x\left(t + s_n + \tau; t + s_n, x\right) \right.$$

$$\left. - x\left(t + s_p + \tau; t + s_n, x\right) \right\|. \tag{3.25}$$

On the other hand,

$$x\left(t + s_n + \tau; t + s_n, x\right)$$

$$= x + \int_t^{t+\tau} f\left(\sigma + s_n, x(\sigma + s_n; t + s_n, x)\right) d\sigma$$

$$+ \sum_{t < \sigma_i(s_n) < t+\tau} I_{i+i(s_n)}\left(x\left(\sigma_i(s_n) + s_n; t + s_n, x\right)\right) \tag{3.26}$$

and

$$x\left(t + s_p + \tau; t + s_p, x\right)$$

$$= x + \int_t^{t+\tau} f\left(\sigma + s_p, x(\sigma + s_p; t + s_p, x)\right) d\sigma$$

$$+ \sum_{t < \sigma_i(s_p) < t+\tau} I_{i+i(s_p)}\left(x\left(\sigma_i(s_p) + s_p; t + s_p, x\right)\right), \tag{3.27}$$

where $\sigma_i(s_j) = t_k - s_j$, $j = n, p$, and the numbers $i(s_n)$ and $i(s_p)$ are such that $i + i(s_j) = k$.

From (3.26) and (3.27), it follows

$$\left\| x(t + s_n + \tau; t + s_n, x) - x(t + s_p + \tau, t + s_p, x) \right\|$$

$$\le \int_t^{t+\tau} \left\| f\left(\sigma + s_n, x(\sigma + s_n; t + s_n, x)\right) \right.$$

$$\left. - f\left(\sigma + s_p, x(\sigma + s_n; t + s_n, x)\right) \right\| d\sigma$$

$$+ \int_t^{t+\tau} \left\| f\left(\sigma + s_p, x(\sigma + s_n; t + s_n, x)\right) \right.$$

$$- f\big(\sigma + s_p, x(\sigma + s_p; t + s_p, x)\big)\Big\|\,d\sigma$$

$$+ \sum_{t < \sigma_i(s_n) < t+\tau} \Big\|I_{i+i(s_n)}\big(x(\sigma_i(s_n) + s_n; t + s_n, x)\big)$$

$$- I_{i+i(s_p)}\big(x(\sigma_i(s_n) + s_n; t + s_n, x)\big)\Big\|$$

$$+ \sum_{t < \sigma_i(s_p) < t+\tau} \Big\|I_{i+i(s_p)}\big(x(\sigma_i(s_p) + s_n; t + s_n, x)\big)$$

$$- I_{i+i(s_p)}\big(x(\sigma_i(s_p) + s_p; t + s_p, x)\big)\Big\|.$$

Now, from $x(\sigma + s_n; t + s_n, x) \in B_{\beta(\alpha)}$ it follows that for any $\varepsilon > 0$ there exists a number $N(\varepsilon) > 0$ such that as $n, p \geq N(\varepsilon)$, we obtain

$$\Big\|f\big(\sigma + s_n, x(\sigma + s_n; t + s_n, x)\big)$$

$$- f\big(\sigma + s_p, x(\sigma + s_n; t + s_n, x)\big)\Big\| < \varepsilon, \tag{3.28}$$

$$\Big\|I_{i+i(s_n)}\big(x(\sigma_i(s_n) + s_n; t + s_n, x)\big)$$

$$- I_{i+i(s_p)}\big(x(\sigma_i(s_n) + s_n; t + s_n, x)\big)\Big\| < \varepsilon. \tag{3.29}$$

Then from (3.28), (3.29) and conditions H3.2 and H3.4, we obtain

$$\Big\|x(t + s_n + \tau; t + s_n, x) - x(t + s_p + \tau, t + s_p, x)\Big\|$$

$$\leq \varepsilon\tau\left(1 + \frac{1}{p}\right) + \int_t^{t+\tau} L_1\left(\alpha, \frac{1}{i}\right)\Big\|x(\sigma + s_n; t + s_n, x)$$

$$- x(\sigma + s_p, t + s_p, x)\Big\|\,d\sigma$$

$$+ \sum_{t < \sigma_i < t+\tau} L_2\left(\alpha, \frac{1}{i}\right)\Big\|x(\sigma_i(s_n) + s_n; t + s_n, x)$$

$$- x(\sigma_i(s_p) + s_p; t + s_p, x)\Big\|. \tag{3.30}$$

On the other hand, from Theorem 1.9 and (3.30), we obtain

$$\Big\|x(t + s_n + \tau; t + s_n, x) - x(t + s_p + \tau, t + s_p, x)\Big\|$$

$$\leq \varepsilon\tau\left(1 + \frac{1}{p}\right)\exp\left\{\left(L_1\left(\alpha, \frac{1}{i}\right) + \frac{1}{p}\ln(1 + L_2\left(\alpha, \frac{1}{i}\right))\tau\right\}. \tag{3.31}$$

From (3.31) and (3.25) it follows

$$\left|V_i(t+s_n,x) - V_i(t+s_p,x)\right|$$

$$\leq g\left(\frac{1}{i}\right)\left(1+\frac{1}{p}\right)T\left(\alpha,\frac{1}{i}\right)\exp\left\{(c+L_1(\alpha,\frac{1}{i})\right.$$

$$\left. +\frac{1}{p}\ln(1+L_2(\alpha,\frac{1}{i}))T(\alpha,\frac{1}{i}))\right\}\varepsilon$$

$$\leq h(\alpha)\varepsilon. \tag{3.32}$$

From (3.32) we get that $V_i(t+s_n,x)$ is uniformly convergent with respect to $t \in \mathbb{R}$ and $x \in B_{\beta(\alpha)}$. Then, $V_i(t,x)$ is almost periodic on t uniformly with respect to $x \in B_{\beta(\alpha)}$.

Inequality (3.19) implies that for $n,p \in N(\varepsilon)$ and $x \in B_{\beta(\alpha)}$ we obtain

$$\left|V(t+s_n,x) - V(t+s_p,x)\right| \leq h(\alpha)\varepsilon,$$

i.e. $V(t,x)$ is almost periodic in t uniformly with respect to $x \in B_{\beta(\alpha)}$.   □

3.1.2 Almost Periodic Solutions of Impulsive Differential Equations

In this part of the paragraph, together with system (3.1), we shall consider the comparison equation (1.16).

The proof of the next lemma is analogous to the proof of Theorem 1.8.

Lemma 3.2. *Let the following conditions hold:*

1. *Conditions H3.1–H3.3 are met.*
2. *The function* $V \in \mathcal{V}_1$,

$$V\left(t^+, x+I_k(x), y+I_k(y)\right) \leq \psi_k\left(V(t,x,y)\right),\ x,y \in \mathbb{R}^n,\ t=t_k,$$

$$k=\pm 1,\pm 2,...,$$

and the inequality

$$D^+V(t,x(t),y(t)) \leq g(t,V(t,x(t),y(t))),\ t \neq t_k,\ k=\pm 1,\pm 2,...$$

holds for any $t \geq t_0$ *and for any* $x,y \in PC[\mathbb{R},\mathbb{R}^n]$.
3. *The maximal solution* $r(t;t_0,u_0)$ *of* (3.16), *for which* $u_0 \geq V(t_0^+,x_0,y_0)$, *is defined on the interval* $[t_0,\infty)$.

Then
$$V(t, x(t; t_0, x_0), y(t; t_0, y_0)) \leq r(t; t_0, u_0),$$

as $t \geq t_0$, where $x(t) = x(t; t_0, x_0)$, $y(t) = y(t; t_0, y_0)$ are solutions of (3.1).

Lemma 3.3 ([94]). *Let the following conditions hold:*

1. *Conditions H3.1–H3.8 are met.*
2. *For any system from $H(f, I_k, t_k)$ there exist functions $W \in W_0$, $a, b \in K$ such that*

$$a(||x||) \leq W(t, x) \leq b(||x||), \ (t, x) \in \mathbb{R} \times \mathbb{R}^n,$$

$$W(t^+, x + I_k^s(x)) \leq W(t, x), \ x \in \mathbb{R}^n, \ t = t_k^s, \ k = \pm 1, \pm 2, ...,$$

and the inequality

$$D^+ W(t, x(t)) \leq -cW(t, x(t)), \ t \neq t_k^s, \ k = \pm 1, \pm 2, ...$$

is valid for $t \geq t_0$, $x \in PC[[t_0, \infty), \mathbb{R}^n]$.

Then the zero solution of (3.2) is uniformly asymptotically stable.

We shall proof the main theorem in this part.

Theorem 3.2. *Let the following conditions hold:*

1. *Conditions H3.1–H3.8 are met.*
2. *There exists a function $V(t, x, y)$, $V \in V_2$ and $a, b \in K$ such that*

$$a(||x - y||) \leq V(t, x, y) \leq b(||x - y||), \ (t, x, y) \in \mathbb{R} \times B_\alpha \times B_\alpha,$$

$$V(t^+, x + I_k(x), y + I_k(y)) \leq V(t, x, y), \ x, y \in B_\alpha, \ t = t_k,$$

$$k = \pm 1, \pm 2, ...,$$

$$D^+ V(t, x, y) \leq -cV(t, x, y), \ x, y \in B_\alpha, \ t \neq t_k, \ k = \pm 1, \pm 2, ...,$$

where $c > 0$, $\alpha > 0$.
3. *There exists a solution $x(t; t_0, x_0)$ of (3.1) such that $||x(t; t_0, x_0)|| \leq \alpha_1$, where $t \geq t_0$, $\alpha_1 < \alpha$.*

Then in B_α, there exists a unique almost periodic solution $\omega(t)$ of (3.1) such that:

(a) $||\omega(t)|| \leq \alpha_1$.
(b) $\omega(t)$ is uniformly asymptotically stable.
(c) $H(\omega, t_k) \subseteq H(f, I_k, t_k)$.

Proof. Let $\{s_i\}$ be an arbitrary sequence of real numbers, such that $s_i \to \infty$ as $i \to \infty$, and $\{s_i\}$ moves the (3.1) into $H(f, I_k, t_k)$.

For the real number β, let $i_0 = i_0(\beta)$ be the smallest value of i such that $s_{i_0} + \beta \geq t_0$. Since $\|x(t; t_0, x_0)\| < \alpha_1$, $\alpha_1 < \alpha$ for all $t \geq t_0$, then $x(t + s_i; t_0, x_0) \in B_{\alpha_1}$ for $t \geq \beta$, $i \geq i_0$.

Then for any compact subset U, $U \subset [\beta, \infty)$ and any $\varepsilon > 0$, choose an integer $n_0(\varepsilon, \beta) \geq i_0(\beta)$, so large that for $l \geq i \geq n_0(\varepsilon, \beta)$ and $t \in \mathbb{R}$, it follows

$$b(2\alpha_1)e^{-c(\beta + s_i - t_0)} < \frac{a(\varepsilon)}{2}, \tag{3.33}$$

$$\|f(t + s_l, x) - f(t + s_i, x)\| < \frac{a(\varepsilon)c}{2H_1}, \tag{3.34}$$

where $a, b \in K$ and $H_1 > 0$ is the Lipschitz constant of the function $V(t, x, y)$. From condition 2 of Theorem, we obtain

$$D^+V(\sigma, x(\sigma), x(\sigma + s_l - s_i)) \leq -cV(\sigma, x(\sigma), x(\sigma + s_l - s_i))$$
$$+ H_1\|f(\sigma + s_l - s_i, x(\sigma + s_l - s_i)) - f(\sigma, x(\sigma + s_l - s_i))\|, \tag{3.35}$$

for $\sigma \neq t_k - (s_l - s_i)$.

On the other hand, from $\sigma = t_k - (s_l - s_i)$ and the system (3.5) it follows

$$V(\sigma, x(\sigma) + I_k(x(\sigma)), x(\sigma + s_l - s_i) + I_k(x(\sigma + s_l - s_i)))$$
$$\leq V(\sigma, x(\sigma), x(\sigma + s_l - s_i)). \tag{3.36}$$

Set $\zeta = \sigma - s_i$. Then,

$$f(\sigma + s_l - s_i, x(\sigma + s_l - s_i)) - f(\sigma, x(\sigma + s_l - s_i))$$
$$= f(\zeta + s_l, x(\zeta + s_l)) - f(\zeta + s_i, x(\zeta + s_i)).$$

Then from (3.34) and (3.36), we get

$$D^+V(\sigma, x(\sigma), x(\sigma + s_l - s_i)) \leq -cV(\sigma, x(\sigma), x(\sigma + s_l - s_i)) + \frac{a(\varepsilon)c}{2}. \tag{3.37}$$

From (3.35), (3.36) and (3.37) it follows that the conditions of Lemma 3.2 are fulfilled.

Consequently, for $l \geq n_0(\varepsilon, \beta)$ and any $t \in U$,

$$V(t, x(t), x(t + s_l - s_i))$$

$$\leq exp\{-c(t + s_i - t_0)\}V(t_0, x(t_0), x(t_0 + s_l - s_i)) + \frac{a(\varepsilon)}{2} \leq a(\varepsilon).$$

From the condition 2 of theorem it follows that for any $t \in U$, we get

$$||x(t + s_i) - x(t + s_l)|| < \varepsilon.$$

Consequently, there exists a function $\omega(t)$, $t \in (\beta, \infty)$ such that $x(t+s_i) - \omega(t) \to 0$, $i \to \infty$, which is bounded from α_1. Since β is arbitrary, it follows that $\omega(t)$ is defined uniformly on $t \in \mathbb{R}$.

Next, we shall show that $\omega(t)$ is a solution of (3.2).

Since $x(t; t_0, x_0)$ is a solution of (3.1), we have

$$||\dot{x}(t + s_i) - \dot{x}(t + s_l)|| \leq ||f(t + s_i, x(t + s_i)) - f(t + s_l, x(t + s_i))||$$
$$+ ||f(t + s_l, x(t + s_i)) - f(t + s_l, x(t + s_l))|| \qquad (3.38)$$

for $t + s_j \neq t_k$, $j = i, l$ and $k = \pm 1, \pm 2, ...$

As $x(t + s_i) \in B_{\beta(\alpha)}$ for large i, then for each compact subset of $\mathbb{R}$ there exists an $n_1(\varepsilon) > 0$ such that, if $l \geq i \geq n_1(\varepsilon)$ then,

$$||f(t + s_i, x(t + s_i)) - f(t + s_l, x(t + s_i))|| < \frac{\varepsilon}{2}.$$

Since $x(t + s_j) \in B_{\beta(\alpha)}$, $j = i, l$, and from Lemma 1.2 it follows that there exists $n_2(\varepsilon) > 0$ such that, if $l \geq i \geq n_2(\varepsilon)$, then

$$||f(t + s_l, x(t + s_i) - f(t + s_l, x(t + s_l))|| < \frac{\varepsilon}{2}.$$

Then for $l \geq i \geq n(\varepsilon)$, $n(\varepsilon) = max\{n_1(\varepsilon), n_2(\varepsilon)\}$, we obtain

$$||\dot{x}(t + s_i) - \dot{x}(t + s_l)|| < \varepsilon,$$

where $t + s_j \neq t_k^s$, $j = i, l$, $k = \pm 1, \pm 2, ...$, which shows that $\lim_{i \to \infty} \dot{x}(t + s_i)$ exists uniformly on all compact subsets of $\mathbb{R}$.

Now, we have $\lim_{i \to \infty} \dot{x}(t + s_i) = \dot{\omega}(t)$, and

$$\dot{\omega}(t) = \lim_{i \to \infty} \Big(f(t + s_i, x(t + s_i)) - \big(f(t + s_i, \omega(t)) + f(t + s_i, \omega(t)) \big) \Big)$$
$$= f^s(t, \omega(t)), \qquad (3.39)$$

for $t \neq t_k^s$, where $t_k^s = \lim_{i \to \infty} t_{k+i(s)}$.

On the other hand, for $t + s_i = t_k^s$, we get

$$\omega(t_k^{s+}) - \omega(t_k^{s-}) = \lim_{i \to \infty} (x(t_k^s + s_i + 0) - x(t_k^s + s_i - 0))$$
$$= \lim_{i \to \infty} I_k^s(x(t_k^s + s_i)) = I_k^s(\omega(t_k^s)). \qquad (3.40)$$

From (3.39) and (3.40) it follows that $\omega(t)$ is a solution of (3.2).

Now, we shall show that $\omega(t)$ is an almost periodic function.

Let the sequence of real number $\{s_n\}$ moves the system (3.1) to $H(f, I_k, t_k)$. Then, for any $\varepsilon > 0$ there exists $m_0(\varepsilon) > 0$ such that if $l \geq i \geq m_0(\varepsilon)$, then

$$e^{-cs_i}b(2\alpha_1) < \frac{a(\varepsilon)}{4},$$

$$\|f(\sigma + s_i, x) - f(\sigma + s_l, x)\| < \frac{a(\varepsilon)}{4H_1}.$$

For each fixed $t \in \mathbb{R}$, let $\tau_\varepsilon = \frac{a(\varepsilon)}{4H_1}$ be a translation number of f such that $t + \tau_\varepsilon \geq 0$.

Consider the function

$$V(\tau_\varepsilon + \sigma, \omega(\sigma), \omega(\sigma + s_l - s_i)),$$

where $t \leq \sigma \leq t + s_i$.

Then

$$D^+V(\tau_\varepsilon + \sigma, \omega(\sigma), \omega(\sigma + s_l - s_i))$$
$$\leq -cV(\tau_\varepsilon + \sigma, \omega(\sigma), \omega(\sigma + s_l - s_i)) + H_1\|f^s(\sigma, \omega(\sigma)) - f^s(\tau_\varepsilon + \sigma, \omega(\sigma))\|$$
$$+ H_1\|f^s(\sigma + s_l - s_i, \omega(\sigma + s_l - s_i)) - f^s(\tau_\varepsilon + \sigma, \omega(\sigma + s_l - s_i))\|$$
$$\leq -cV(\tau_\varepsilon + \sigma, \omega(\sigma), \omega(\sigma + s_l - s_i)) + \frac{3a(\varepsilon)}{4}. \qquad (3.41)$$

On the other hand,

$$V(\tau_\varepsilon + t_k^s, \omega(t_k^s) + I_k^s(\omega(t_k^s)), \omega(t_k^s + s_l - s_i) + I_k^s(\omega(t_k^s + s_l - s_i)))$$
$$\leq V(\tau_\varepsilon + t_k^s, \omega(t_k^s), \omega(t_k^s + s_l - s_i)). \qquad (3.42)$$

From (3.41), (3.42) and Lemma 3.2 it follows

$$V(\tau_\varepsilon + t + s_i, \omega(t + s_i), \omega(t + s_l))$$
$$\leq e^{-cs_i}V(\tau_\varepsilon + t, \omega(t), \omega(t + s_i - s_l)) + \frac{3a(\varepsilon)}{4} < a(\varepsilon). \qquad (3.43)$$

Then by (3.43) for $l \geq i \geq m_0(\varepsilon)$, we have

$$\|\omega(t + s_i) - \omega(t + s_l)\| < \varepsilon. \qquad (3.44)$$

From the definition of the sequence $\{s_n\}$ and (3.2) for $l \geq i \geq m_0(\varepsilon)$ it follows that $\rho(t_k + s_i, t_k + s_l) < \varepsilon$, where $\rho(.,.)$ is an arbitrary distance in the set $\mathcal{B}$.

Now, by (3.44) and the last inequality, we obtain that the sequence $\omega(t+s_i)$ converges uniformly to the function $\omega(t)$.

The assertions (a) and (c) of the theorem follow immediately. We shall proof the assertion (b).

Let $\overline{\omega}(t)$ be an arbitrary solution of (3.2).

Set

$$u(t) = \overline{\omega}(t) - \omega(t),$$

$$g^s(t, u) = f^s(t, u + \omega(t)) - f^s(t, \omega(t)),$$

$$B_k^s(u) = I_k^s(u + \omega) - I_k^s(u).$$

Let we consider the system

$$\begin{cases} \dot{u} = g^s(t, u), \ t \neq t_k^s, \\ \Delta u(t_k^s) = B_k^s(u(t_k^s)), \ k = \pm 1, \pm 2, ..., \end{cases} \tag{3.45}$$

and let $W(t, u(t)) = V(t, \omega(t), \omega(t) + u(t))$.

Then from Lemma 3.3 it follows that the zero solution $u(t) \equiv 0$ of (3.45) is uniformly asymptotically stable for $t_0 \geq 0$, and consequently, $\omega(t)$ is uniformly asymptotically stable. $\qquad \square$

3.1.3 Weakly Nonlinear Impulsive Differential Equations

We shall consider the system of impulsive differential equations

$$\begin{cases} \dot{x} = f(t, x) + \gamma(t), \ t \neq t_k, \\ \Delta x(t_k) = I_k(x(t_k)) + \gamma_k, \ k = \pm 1, \pm 2, ..., \end{cases} \tag{3.46}$$

where $\gamma \in C[\mathbb{R}, \mathbb{R}^n], \ \gamma_k \in \mathbb{R}^n$.

Introduce the following conditions:

H3.9.   The function $\gamma(t)$ is almost periodic.
H3.10.   The sequence $\{\gamma_k\}, \ k = \pm 1, \pm 2, ...$ is almost periodic.
H3.11.   The following inequality

$$||\gamma(t)|| + ||\gamma_k|| \leq \kappa,$$

where $\kappa > 0$, holds for $t \in \mathbb{R}, \ k = \pm 1, \pm 2, ...$

The next lemma is important for the proof of the main result in this part.

Lemma 3.4. *Let the following conditions hold:*

1. *Conditions H3.1–H3.11 are met.*
2. *There exist functions $V(t, x, y)$, $V \in V_2$ and $a, b \in K$ such that*

$$a(||x - y||) \leq V(t, x, y) \leq b(||x - y||), \tag{3.47}$$

$$V(t^+, x + I_k(x), y + I_k(y)) \leq V(t, x, y), \ t = t_k,$$

$$D^+ V(t, x, y) \leq -cV(t, x, y), \ t \neq t_k, \ k = \pm 1, \pm 2, ..., \tag{3.48}$$

   where $c > 0$, $x, y \in B_\alpha$, $\alpha > 0$.
3. *There exists a solution $x^0 = x^0(t)$ of (3.1) such that $||x^0|| \leq r$, where $r = const > 0$, and*

$$a^{-1}\left(\frac{H_1 \kappa}{c}\right) + r \leq \alpha_1 < \alpha,$$

   a^{-1} is the inverse function of $a \in K$.
4. *The following inequality*

$$||x_0 - x^0(t_0)|| \leq b^{-1}\left(\frac{H_1 \kappa}{c}\right),$$

   holds, where $x_0 \in B_\alpha$, $t_0 \geq 0$ and b^{-1} is the inverse function of $b \in K$.

Then, if $x(t; t_0, x_0)$ is a solution of (3.46),

$$||x(t; t_0, x_0) - x^0(t)|| \leq a^{-1}\left(\frac{H_1 \kappa}{c}\right), \tag{3.49}$$

for $t \in \mathbb{R}$.

Proof. Let $x(t; t_0, x_0)$ be an arbitrary solution of (3.46) such that

$$||x_0 - x^0(t_0)|| \leq b^{-1}\left(\frac{H_1 \kappa}{c}\right),$$

and let $||x_0 - x^0(t_0)|| \leq \alpha^*$, where $\alpha_1 < \alpha^* < \alpha$.

From (3.48) and (3.49) for $t \neq t_k$, $k \pm 1, \pm 2, ...$, we get

$$D^+ V(t, x^0(t), x(t; t_0, x_0)) \leq -cV(t, x^0(t), x(t; t_0, x_0)) + H_1 \kappa \tag{3.50}$$

and from (3.2) for $t = t_k$, $k = \pm 1, \pm 2, ...$, we obtain

$$V(t_k^+, x^0(t_k) + I_k(x^0(t_k)), x(t_k; t_0, x_0) + I_k(x(t_k; t_0, x_0)))$$

$$\leq V(t_k, x^0(t_k), x(t_k; t_0, x_0)). \tag{3.51}$$

Then from Lemma 3.3, (3.50) and (3.51) it follows that

$$V(t, x^0(t), x(t; t_0, x_0)) \leq e^{-c(t-t_0)}\left(V(t, x^0(t_0), x_0) - \frac{H_1\kappa}{c}\right) + \frac{H_1\kappa}{c}$$

$$\leq \frac{H_1\kappa}{c}.$$

Now, from (3.47), we have

$$||x(t; t_0, x_0) - x^0(t)|| \leq a^{-1}\left(\frac{H_1\kappa}{c}\right),$$

and from (3.48), it follows

$$||x(t; t_0, x_0)|| = ||x^0(t) - x(t; t_0, x_0)|| + ||x^0(t)||$$

$$\leq a^{-1}\left(\frac{H_1\kappa}{c}\right) + c \leq \alpha_1 < \alpha^*.$$

Then, (3.49) holds for every $t \geq t_0$. In particular, if $x_0 = x^0(t_0)$, then clearly $x(t; t_0, x_0)$ is a solution of (3.46), such that $||x(t; t_0, x_0)|| \leq \alpha_1$. □

Theorem 3.3. *Let conditions of Lemma 3.4 hold.*

Then for the system (3.46) *there exists a unique globally perfectly uniform-asymptotically stable almost periodic solution bounded by the constant* α_1.

Proof. Let we consider the following system associated with (3.46)

$$\begin{cases} \dot{x} = f(t, x) + \gamma(t), \ t \neq t_k, \\ \Delta x(t_k) - I_k(x(t_k)) + \gamma_k, \\ \dot{y} = f(t, y) + \gamma(t), \ t \neq t_k, \\ \Delta y(t_k) = I_k(y(t_k)) + \gamma_k, \ k = \pm 1, \pm 2, ... \end{cases} \tag{3.52}$$

Then for any α^* such that $\alpha_1 < \alpha^* < \alpha$, it follows

$$D^+V(t, x(t), y(t)) = \lim_{\delta \to 0^+} sup\frac{1}{\delta}\Big\{V\big(t + \delta, x(t) + \delta f(t, x(t))$$

$$+ \delta\gamma(t), y(t) + \delta f(t, y(t)) + \delta\gamma(t))\big) - V(t, x(t), y(t))\Big\}$$

$$\leq \lim_{\delta \to 0^+} sup\frac{1}{\delta}\Big\{V\big(t + \delta, x(t) + \delta f(t, x(t)), y(t) + \delta f(t, y(t))\big)$$

$$\leq -V(t, x(t), y(t))\Big\} \leq -cV(t, x(t), y(t)). \tag{3.53}$$

On the other hand, for $t = t_k$, we have

$$V\left(t_k^+, x(t_k) + I_k(x(t_k)) + \gamma_k, y(t_k) + I_k(y(t_k)) + \gamma_k\right)$$

$$\leq V\left(t_k^+, x(t_k) + I_k(x(t_k)), y(t_k) + I_k(y(t_k))\right). \tag{3.54}$$

Then, from Lemma 3.4, (3.53) and (3.54) it follows that for the system (3.52) conditions of Theorem 3.2 hold. Consequently, the proof follows from Theorem 3.2   □

Example 3.1. Now, we shall consider the linear systems

$$\begin{cases} \dot{x} = A(t)x, \ t \neq t_k, \\ \Delta x(t_k) = B_k x(t_k), \ k = \pm 1, \pm 2, ..., \end{cases} \tag{3.55}$$

and

$$\begin{cases} \dot{x} = A(t)x + \gamma(t), \ t \neq t_k, \\ \Delta x(t_k) = B_k x(t_k) + \gamma_k, \ k = \pm 1, \pm 2, ..., \end{cases} \tag{3.56}$$

where $A \in C[\mathbb{R}, \mathbb{R}^{n \times n}]$, and $B_k \in \mathbb{R}^{n \times n}$.

The proof of the next theorem follows immediately.

Theorem 3.4. *Let the following conditions hold:*

1. *Conditions H3.1–H3.11 are met.*
2. *The matrix $A(t)$ is almost periodic.*
3. *The sequence $\{B_k\}$, $k = \pm 1, \pm 2, ...$ is almost periodic.*
4. *The zero solution of system (3.55) is globally perfectly uniform-asymptotically stable.*

Then:

1. *All solutions of (3.56) are bounded;*
2. *For the system (3.56) there exists a unique globally perfectly uniform-asymptotically stable almost periodic solution.*

3.1.4 (h₀, h)-*Stable Impulsive Differential Equations*

We extend, in this part, Lyapunov second method to investigate almost periodic solutions for (h_0, h)-stable systems of impulsive differential equations in the form (3.1).

Let us list the following classes of functions for convenience,

$$\Gamma_h = \left\{ h \in V_1 : \inf_{x \in \mathbb{R}^n, \, y \in \mathbb{R}^n} h(t, x, y) = 0, \text{ for } t \in \mathbb{R} \right\},$$

$$\Gamma_r = \left\{ r \in C[\mathbb{R} \times \mathbb{R}^n, \mathbb{R}^+] : \inf_{x \in \mathbb{R}^n} r(t, x) = 0, \text{ for } t \in \mathbb{R} \right\}.$$

Let

$$S(r, \alpha) = \left\{ x \in \mathbb{R}^n : r(t, x) < \alpha, \; r \in \Gamma_r, \; \alpha > 0 \right\}.$$

We replace the conditions H3.1.1–H3.1.5 with the following conditions:

H'3.1. The function $f \in C[\mathbb{R} \times S(r, \alpha), \mathbb{R}^n]$, and $f(t, 0) = 0$ for $t \in \mathbb{R}$.

H'3.2. The functions $I_k \in C[S(r, \alpha), \mathbb{R}^n]$ and $I_k(0) = 0$, $k = \pm 1, \pm 2, \dots$.

H'3.3. The functions $(E + I_k) : S(r, \alpha) \to S(r, \alpha)$, $k = \pm 1, \pm 2, \dots$, where E is the identity in $\mathbb{R}^{n \times n}$.

H'3.4. The function f is Lipschitz continuous with respect to $x \in \mathbb{R}^n$ uniformly on $t \in \mathbb{R}$, with a Lipschitz constant $L_1 > 0$, i.e.

$$||f(t, x_1) - f(t, x_2)|| \le L_1 ||x_1 - x_2||,$$

for $x_1, x_2 \in S(r, \alpha)$.

H'3.5. The functions I_k, $k = \pm 1, \pm 2, \dots$ are Lipschitz continuous with respect to $x \in \mathbb{R}^n$ with a Lipschitz constant $L_2 > 0$, i.e.

$$||I_k(x_1) - I_k(x_2)|| \le L_2 ||x_1 - x_2||$$

for $x_1, x_2 \in S(r, \alpha)$.

Introduce and new conditions:

H3.12. The functions from the class Γ_h are Lipschitz continuous along its second and third arguments with a Lipschitz constant $\kappa > 0$ such that

$$|h(t, x_1, y_1) - h(t, x_2, y_2)| \le \kappa(||x_1 - x_2|| + ||y_1 - y_2||)$$

for $x_i, y_i \in \mathbb{R}^n$, $i = 1, 2$.

H3.13. For any function $h \in \Gamma_h$, it follows that

$$\left| \frac{\partial h}{\partial t}(t, x, y) \right| \le c,$$

where $c > 0$, and $x, y \in \mathbb{R}^n$.

H3.14. For any $x, y \in S(r_0, \alpha)$ and $t \in \mathbb{R}$, where $r_0 \in \Gamma_r$ it follows that

$$h(t, x, y) \le 2r_0(t, x).$$

H3.15. There exists α_0, $0 < \alpha_0 < \alpha$, such that, $h(t_k, x, y) < \alpha_0$ implies $h(t_k^+, x + I_k(x), y + I_k(y)) < \alpha, \ k = \pm 1, \pm 2, \dots.$

Definition 3.4. Let $(h_0, h) \in \Gamma_h$. The impulsive system (3.1) is said to be:

(a) $(h_0, h) - stable$, if

$$(\forall \varepsilon > 0)(\forall t_0 \in \mathbb{R})(\exists \delta > 0)(\forall x_0, y_0 \in \mathbb{R}^n, \ h_0(t_0^+, x_0, y_0) < \delta)$$
$$(\forall t \ge t_0) : h(t, x(t; t_0, x_0), y(t; t_0, y_0)) < \varepsilon;$$

(b) $(h_0, h) - uniformly\ stable$, if δ in (a) is independent of t_0;
(c) $(h_0, h) - equi - attractive$, if

$$(\forall t_0 \in \mathbb{R})(\exists \delta > 0)(\forall \varepsilon > 0)(\exists T > 0)$$
$$(\forall x_0, y_0 \in \mathbb{R}^n, \ h_0(t_0^+, x_0, y_0) < \delta)$$
$$(\forall t \ge t_0 + T) : h(t, x(t; t_0, x_0), y(t; t_0, y_0)) < \varepsilon;$$

(d) $(h_0, h) - uniformly\ attractive$, if the numbers δ and T in (c) are independent of t_0;
(e) $(h_0, h) - uniformly\ asymptotically\ stable$, if it is (h_0, h)-uniformly stable and (h_0, h) uniformly attractive.

Definition 3.5. Let $r_0, r \in \Gamma_r$. The solution $x(t; t_0, x_0)$ of system (3.1) is called $(r_0, r) - uniformly\ bounded$, if

$$(\forall \alpha > 0)(\exists \beta > 0)(\forall t_0 \in \mathbb{R})$$
$$(\forall x_0 \in \mathbb{R}^n, \ r_0(t_0, x_0) \le \alpha)(\forall t \ge t_0) : r(t, x(t; t_0, x_0)) < \beta.$$

The proofs of the next theorems are analogous to the proof of Theorems 3.1 and 3.2

Theorem 3.5. *Let the following conditions hold:*

1. *Conditions H3'.1–H'3.5, H3.6–H3.8 and H3.12–H3.15 are met.*
2. *For the system (3.1) there exists an $(r_0, r)-uniformly$ bounded solution.*
3. *The system (3.1) is (h_0, h)-uniformly asymptotically stable.*

Then there exists a function $V \in V_2$ almost periodic along t uniformly with respect to $x \in \mathbb{R}^n$ such that

$$V\big(t_k^+, x(t_k) + I_k(x(t_k)), y(t_k) + I_k(y(t_k))\big) \leq V(t_k, x(t_k), y(t_k)),$$

$$a(||x - y||) \leq V(t, x, y) \leq b(||x - y||),$$

where $h(t, x, y) < \delta$, $h_0(t, x, y) \leq \delta$, $a, b \in K$,

$$D^+V(t, x, y) \leq -cV(t, x, y)$$

for $t \neq t_k$, $(x, y) \in S(r_0, \alpha) \times S(r_0, \alpha)$, $\alpha > 0$, $c > 0$, $k = \pm 1, \pm 2, ...,$.

Theorem 3.6. *Let the following conditions hold:*

1. *Conditions H3'.1–H'3.5, H3.6–H3.8 and H3.12–H3.15 are met.*
2. *For the system* (3.1) *there exists an* (r_0, r)*-uniformly bounded solution.*
3. *The system* (3.1) *is* (h_0, h)*-uniformly asymptotically stable.*

Then, in B_α, $\alpha > 0$ there exists a unique almost periodic solution $\omega(t)$ of (3.1), *such that:*

(a) $\omega(t)$ is $(r_0, r)-$uniformly bounded.
(b) $\omega(t)$ is $(h_0, h)-$ uniformly asymptotically stable.
(c) $H(\omega(t), t_k) \subseteq H(f, I_k, t_k)$.

3.2 Impulsive Integro-Differential Equations

In this part we apply the comparison principle to the problem of existence of almost periodic solutions of the system of impulsive integro-differential equations.

We shall consider the following system of impulsive integro-differential equation

$$\begin{cases} \dot{x}(t) = f\big(t, x(t), \int_{t_0}^{t} K(t, s, x(s))ds\big), \ t \neq t_k, \\ \\ \Delta x(t_k) = I_k(x(t_k)), \ k = \pm 1, \pm 2, ..., \end{cases} \tag{3.57}$$

where $t \in \mathbb{R}$, $t_0 \in \mathbb{R}$, $\{t_k\} \in \mathcal{B}$, $f : \mathbb{R} \times \Omega \times \mathbb{R}^n \to \mathbb{R}^n$, $K : \mathbb{R} \times \mathbb{R} \times \Omega \to \mathbb{R}^n$, $I_k : \Omega \to \mathbb{R}^n$, $k = \pm 1, \pm 2, ...$

Introduce the following conditions:

H3.16. The function $f \in C[\mathbb{R} \times \Omega \times \mathbb{R}^n, \mathbb{R}^n)]$, and $f(t, 0, 0) = 0$ for $t \in \mathbb{R}$.
H3.17. $I_k \in C[\Omega, \mathbb{R}^n]$, and $I_k(0) = 0$ for $k = \pm 1, \pm 2....$
H3.18. $E + I_k : \Omega \to \Omega$, $k = \pm 1, \pm 2...$, where E is the identity in $\mathbb{R}^{n \times n}$.
H3.19. The function $K \in C[\mathbb{R} \times \mathbb{R} \times \Omega, \mathbb{R}^n]$, and $K(t_0, 0, 0) = 0$.

We shall denote by $x(t) = x(t; t_0, x_0)$ the solution of (3.57) with the initial condition $x(t_0^+; t_0, x_0) = x_0$.

Definition 3.6. The function $K \in C[\mathbb{R} \times \mathbb{R} \times \Omega, \mathbb{R}^n]$ is said to be *integro-almost periodic uniformly for* $x \in PC[\mathbb{R}, \Omega]$, if for every sequence of real numbers $\{s_m'\}$ there exists a subsequence $\{s_i\}$, $s_i = s_{m_i}'$ such that the sequence

$$\left\{ \int_{t_0}^{t+s_i} K(t + s_i, \xi, x(\xi))d\xi \right\}, \quad t_0 \in \mathbb{R}$$

converges uniformly with respect to $i \to \infty$.

Introduce the following conditions:

H3.20. The function $f(t, x, y)$ is almost periodic in t uniformly with respect to $x \in \Omega$, $y \in \mathbb{R}^n$.

H3.21. The sequence $\{I_k(x)\}$, $k = \pm 1, \pm 2, ...$ is almost periodic uniformly with respect to $x \in \Omega$.

H3.22. The function $K(t, s, x)$ is integro-almost periodic in t uniformly with respect to $x \in PC[\mathbb{R}, \Omega]$.

H3.23. The set of sequences $\{t_k^j\}$, $t_k^j = t_{k+j} - t_k$, $k = \pm 1, \pm 2, ...$, $j = \pm 1, \pm 2, ...$ is uniformly almost periodic, and $\inf_k t_k^1 = \theta > 0$.

Let conditions H3.16–H3.23 hold, and let $\{s_m'\}$ be an arbitrary sequence of real numbers. Then, there exists a subsequence $\{s_i\}$, $s_i = s_{m_i}'$, so that analogous to the process in Chap. 1, the system (3.57) moves to the system

$$\begin{cases} \dot{x}(t) = f^s(t, x(t), \int_{t_0}^t K^s(t, \xi, x(\xi))d\xi), \ t \neq t_k^s, \\ \Delta x(t_k^s) = I_k(x(t_k^s)), \ k = \pm 1, \pm 2, ..., \end{cases} \tag{3.58}$$

and in this case, the set of systems in the form (3.58) we shall denote by $H(f, K, I_k, t_k)$.

We shall use the following definition for the stability of the zero solution of (3.57).

Definition 3.7. [93] The zero solution $x(t) \equiv 0$ of (3.57) is said to be:

(a) *Uniformly stable*, if

$$(\forall \varepsilon > 0)(\exists \delta > 0)(\forall t_0 \in \mathbb{R})(\forall x_0 \in B_\delta)(\forall t \geq t_0) : \ ||x(t; t_0, x_0)|| < \varepsilon;$$

(b) *Uniformly attractive*, if

$$(\exists \lambda > 0)(\forall \varepsilon > 0)(\exists T > 0)(\forall t_0 \in \mathbb{R})(\forall x_0 \in B_\lambda)$$

$$(\forall t \geq t_0 + T) : \ ||x(t; t_0, x_0)|| < \varepsilon;$$

(c) *Uniformly asymptotically stable*, if it is uniformly stable and uniformly attractive.

In this section, we shall use, also, the classes V_1, V_2 and W_0, defined in Chap. 1. Let $V \in V_1$, $t \neq t_k$, $x \in PC[\mathbb{R}, \Omega]$, $y \in PC[\mathbb{R}, \Omega]$.

Introduce the function

$$D^+V(t, x(t), y(t))$$

$$= \lim_{\delta \to 0^+} \sup \frac{1}{\delta} \Big\{ V\Big(t + \delta, x(t) + \delta f(t, x(t), \int_{t_0}^t K(t, \xi, x(\xi))d\xi\Big)$$

$$+ \delta f\Big(t, y(t), \int_{t_0}^t K(t, \xi, y(\xi))d\xi\Big)\Big) - V(t, x(t), y(t)) \Big\}. \qquad (3.59)$$

By analogy for $W \in W_0$, $t \neq t_k^s$, $x \in PC[\mathbb{R}, \Omega]$, we shall introduce the function

$$D^+W(t, x(t)) = \lim_{\delta \to 0^+} \sup \frac{1}{\delta} \Big\{ W\Big(t$$

$$+ \delta, x(t) + \delta f^s\Big(t, x(t), \int_{t_0}^t K^s(t, \xi, x(\xi))d\xi\Big)\Big) - W(t, x(t)) \Big\}.$$

Now, we shall proof the main result for the system (3.57).

Theorem 3.7. *Let the following conditions hold:*

1. *Conditions H3.16–H3.23 are met.*
2. *There exist functions $V \in V_2$ and $a, b \in K$ such that*

$$V\big(t^+, x + I_k(x), y + I_k(y)\big) \leq V(t, x, y), \ x, y \in \Omega, \ t = t_k, \ k = \pm 1, \pm 2, ...,$$

$$a(||x - y||) \leq V(t, x, y) \leq b(||x - y||), \ t \in \mathbb{R}, \ x, y \in \Omega, \qquad (3.60)$$

and the inequality

$$D^+V(t, x(t), y(t)) \leq -cV(t, x(t), y(t)), \ t \neq t_k, \ k = \pm 1, \pm 2, ..., \quad (3.61)$$

is valid for $t > t_0$, $x, y \in PC[\mathbb{R}, \Omega]$, for which $V(s, x(s), y(s)) \leq V(t, x(t), y(t))$, $s \in [t_0, t]$, $c > 0$.
3. *There exists a solution $x(t; t_0, x_0)$ of (3.57) such that $||x(t; t_0, x_0)|| < \alpha_1$, where $t \geq t_0$, $\alpha_1 < \alpha$.*

Then for the system (3.57) there exists a unique almost periodic solution $\omega(t)$ such that:

(a) $||\omega(t)|| \leq \alpha_1$.
(b) $\omega(t)$ is uniformly asymptotically stable.
(c) $H(\omega, t_k) \subset H(f, K, I_k, t_k)$.

Proof. Let $\{s_i\}$ be an arbitrary sequence of real numbers, so that $s_i \to \infty$ as $i \to \infty$ and $\{s_i\}$ moves the system (3.57) to a system from $H(f, K, I_k, t_k)$.

For any real number β, let $i_0 = i_0(\beta)$ be the smallest value of i, such that $s_{i_0} + \beta \geq t_0$. Since $\|x(t; t_0, x_0)\| \leq \alpha_1$, where $\alpha_1 < \alpha$, for all $t \geq t_0$, then $x(t + s_i; t_0, x_0) \in B_{\alpha_1}$ for $t \geq \beta$, $i \geq i_0$.

Let U, $U \subset (\beta, \infty)$ be a compact and for any $\varepsilon > 0$, choose an integer $n_0(\varepsilon, \beta) \geq i_0(\beta)$, so large that for $l \geq i \geq n_0(\varepsilon, \beta)$ and $t \in (\beta, \infty)$, it follows

$$b(2\alpha_1)e^{-c(\beta + s_i - t_0)} < \frac{a(\varepsilon)}{2}, \tag{3.62}$$

$$\left\| f\left(t + s_l, x(t), \int_{t_0}^{t+s_l} K(t + s_l, \xi, x(\xi))d\xi\right) \right.$$

$$\left. - f(t + s_i, x(t), \int_{t_0}^{t+s_i} K(t + s_i, \xi, x(\xi))d\xi) \right\| < \frac{a(\varepsilon)c}{2H_1}, \tag{3.63}$$

where $H_1 > 0$ is the Lipschitz constant of the function $V(t, x, y)$.

Now, we shall consider the function $V(\sigma, x(\sigma), x(\sigma + s_l - s_i))$.

For $\sigma > t_0$, $x(\sigma) \in PC[(t_0, \infty), \mathbb{R}^n]$, $x(\sigma + s_l - s_i) \in PC[(t_0, \infty), \mathbb{R}^n]$ and

$$V(s, x(s), x(s + s_l - s_i)) \geq V(\sigma, x(\sigma), x(\sigma + s_l - s_i)),$$

$s \in [t_0, \sigma]$, from condition 2 of the theorem, we obtain

$$D^+ V(\sigma, x(\sigma), x(\sigma + s_l - s_i)) \leq -cV(\sigma, x(\sigma), x(\sigma + s_l - s_i))$$

$$+ H_1 \left\| f\left(\sigma + s_l - s_i, x(\sigma + s_l - s_i), \int_{t_0}^{\sigma + s_l - s_i} K(\sigma + s_l - s_i, \xi, x(\xi))d\xi\right) \right.$$

$$\left. - f\left(\sigma, x(\sigma + s_l - s_i), \int_{t_0}^{\sigma} K(\sigma, \xi, x(\xi))d\xi\right) \right\|$$

$$\leq -cV\left(\sigma, x(\sigma), x(\sigma + s_l - s_i)\right) + \frac{a(\varepsilon)c}{2}. \tag{3.64}$$

From (3.64) it follows that the conditions of Lemma 3.2 are fulfilled. Consequently from (3.62), for $l \geq n_0(\varepsilon, \beta) \geq i_0(\beta)$ and for $t \in U$, it follows

$$V(t + s_i, x(t + s_i), x(t + s_l))$$

$$\leq e^{-c(t + s_i - t_0)} V(t_0, x(t_0), x(t_0 + s_l - s_i)) + \frac{a(\varepsilon)}{2} < a(\varepsilon).$$

Then using (3.60), we have

$$\|x(t + s_i) - x(t + s_l)\| < \varepsilon.$$

Then there exists a function $\omega(t)$, $t \in (\beta, \infty)$ such that $x(t+s_i) - \omega(t) \to 0$ for $i \to \infty$.

Since β is arbitrary chosen, it follows that $\omega(t)$ is defined uniformly on $t \in U$.

Next, we shall show that $\omega(t)$ is a solution of (3.58).

Since $x(t; t_0, x_0)$ is a solution of (3.57), we have

$$\|\dot{x}(t+s_i) - \dot{x}(t+s_l)\| \leq \left\| f\left(t+s_i, x(t+s_i), \int_{t_0}^{t+s_i} K(t+s_i, \xi, x(\xi))d\xi\right) \right.$$

$$\left. - f\left(t+s_l, x(t+s_i), \int_{t_0}^{t+s_i} K(t+s_l, \xi, x(\xi))d\xi\right) \right\|$$

$$+ \left\| f\left(t+s_l, x(t+s_i), \int_{t_0}^{t+s_i} K(t+s_l, \xi, x(\xi))d\xi\right) \right.$$

$$\left. - f\left(t+s_l, x(t+s_l), \int_{t_0}^{t+s_l} K(t+s_l, \xi, x(\xi))d\xi\right) \right\|,$$

for $t + s_j \neq t_k$, $j = i, l$, $k = \pm 1, \pm 2, \ldots$.

As $x(t+s_i) \in B_{\alpha_1}$ for large i for each compact subset of $\mathbb{R}$ there exists an $n_1(\varepsilon) > 0$ such that if $l \geq i \geq n_1(\varepsilon)$, then

$$\left\| f\left(t+s_i, x(t+s_i), \int_{t_0}^{t+s_i} K(t+s_i, \xi, x(\xi))d\xi\right) \right.$$

$$\left. - f\left(t+s_l, x(t+s_i), \int_{t_0}^{t+s_i} K(t+s_l, \xi, x(\xi))d\xi\right) \right\| < \frac{\varepsilon}{2}.$$

Since $x(t+s_j) \in B_{\beta(\alpha)}$, $j = i, l$ and from Lemma 1.6 it follows that there exists $n_2(\varepsilon) > 0$ such that if $l \geq i \geq n_2(\varepsilon)$, then

$$\left\| f\left(t+s_l, x(t+s_i), \int_{t_0}^{t+s_l} K(t+s_i, \xi, x(\xi))d\xi\right) \right.$$

$$\left. - f\left(t+s_l, x(t+s_l), \int_{t_0}^{t+s_l} K(t+s_l, \xi, x(\xi))d\xi\right) \right\| < \frac{\varepsilon}{2}.$$

Then for $l \geq i \geq n(\varepsilon)$, $n(\varepsilon) = max\{n_1(\varepsilon), n_2(\varepsilon)\}$, we obtain

$$\|\dot{x}(t+s_i) - \dot{x}(t+s_l)\| \leq \varepsilon,$$

$t + s_j \neq t_k^s$, $j = i, l$ and $k = \pm 1, \pm 2, \ldots$ which shows that $\lim\limits_{i \to \infty} \dot{x}(t+s_i)$ exists uniformly on all compact subsets of $\mathbb{R}$.

Therefore, $\lim_{i\to\infty} \dot{x}(t + s_i) = \dot{\omega}(t)$, and

$$
\begin{aligned}
\dot{\omega}(t) = \lim_{i\to\infty} &\left(f\Big(t + s_i, x(t + s_i), \int_{t_0}^{t+s_i} K(t + s_i, \xi, x(\xi))d\xi\Big) \right. \\
&- f\Big(t + s_i, \omega(t), \int_{t_0}^{t+s_i} K(t + s_i, \xi, \omega(\xi))d\xi\Big) \\
&\left. + f\Big(t + s_i, \omega(t), \int_{t_0}^{t+s_i} K(t + s_i, \xi, \omega(\xi))d\xi\Big) \right) \\
= &\, f^s\Big(t, \omega(t), \int_{t_0}^{t} K^s(t, \xi, \omega(\xi))d\xi\Big),
\end{aligned}
\tag{3.65}
$$

where $t \neq t_k^s$, $t_k^s = \lim_{i\to\infty} t_{k+i(s)}$.

On the other hand, for $t + s_i = t_k^s$, it follows

$$
\begin{aligned}
\omega(t_k^{s+}) - \omega(t_k^{s-}) &= \lim_{i\to\infty}\left(x(t_k^s + s_i + 0) - (x(t_k^s + s_i - 0)\right) \\
&= \lim_{i\to\infty} I_k^s(x(t_k^s + s_i)) = I_k^s(\omega(t_k^s)).
\end{aligned}
\tag{3.66}
$$

From (3.65) and (3.66) it follows that $\omega(t)$ is a solution of (3.58).

We shall show that $\omega(t)$ is an almost periodic function.

Let the sequence $\{s_i\}$ moves the system (3.57) to $H(f, K, I_k, t_k)$. For any $\varepsilon > 0$ there exists $m_0(\varepsilon) > 0$ such that if $l \geq i \geq m_0(\varepsilon)$, then

$$
e^{-cs_i} b(2\alpha_1) < \frac{a(\varepsilon)}{4},
$$

and

$$
\begin{aligned}
\Big\| f\Big(\sigma + s_i, x(\sigma), &\int_{t_0}^{t+s_i} K(t + s_i, \xi, x(\xi))d\xi\Big) \\
&- f\Big(\sigma + s_l, x(\sigma + s_l), \int_{t_0}^{t+s_l} K(t + s_l, \xi, x(\xi))d\xi\Big) \Big\| < \frac{a(\varepsilon)}{4H_1},
\end{aligned}
$$

where $x \in PC[(t_0, \infty), \mathbb{R}^n]$, $c = const > 0$.

For each fixed $t \in \mathbb{R}$ let $\tau_\varepsilon = \frac{a(\varepsilon)}{4H_1}$ be a translation number of the function f such that $t + \tau_\varepsilon \geq 0$.

We shall consider the function

$$
V\big(\tau_\varepsilon + \sigma, \omega(\sigma), \omega(\sigma + s_l - s_i)\big),
$$

where $t \leq \sigma \leq t + s_i$.

Then

$$D^+V(\tau_\varepsilon + \sigma, \omega(\sigma), \omega(\sigma + s_l - s_i))$$

$$\leq -cV(\tau_\varepsilon + \sigma, \omega(\sigma), \omega(\sigma + s_l - s_i)) + H_1\Big\|f^s\big(\sigma, \omega(\sigma),$$

$$\times \int_{t_0}^\sigma K^s(\sigma, \xi, \omega(\xi))d\xi\big)$$

$$- f^s\big(\tau_\varepsilon + \sigma, \omega(\sigma), \int_{t_0}^{\tau_\varepsilon + \sigma} K^s(\tau_\varepsilon + \sigma, \xi, \omega(\xi))d\xi\big)\Big\|$$

$$+ H_1\Big\|f^s\big(\sigma + s_l - s_i, \omega(\sigma + s_l - s_i),$$

$$\times \int_{t_0}^{\sigma + s_l - s_i} K^s(\sigma + s_l - s_i, \xi, \omega(\xi))d\xi\big)$$

$$- f^s\big(\tau_\varepsilon + \sigma, \omega(\sigma + s_l - s_i), \int_{t_0}^{\tau_\varepsilon + \sigma} K^s(\tau_\varepsilon + \sigma, \xi, \omega(\xi))d\xi\big)\Big\|$$

$$\leq -cV(\tau_\varepsilon + \sigma, \omega(\sigma), \omega(\sigma + s_l - s_i)) + \frac{3a(\varepsilon)}{4}. \tag{3.67}$$

On the other hand,

$$V(\tau_\varepsilon + t_k^s, \omega(t_k^s) + I_k^s(\omega(t_k^s)), \omega(t_k^s + s_l - s_i) + I_k^s(\omega(t_k^s + s_l - s_i)))$$

$$\leq V(\tau_\varepsilon + t_k^s, \omega(t_k^s), \omega(t_k^s + s_l - s_i)). \tag{3.68}$$

From (3.67), (3.68) and Lemma 3.2, it follows

$$V(\tau_\varepsilon + t + s_i, \omega(t + s_i), \omega(t + s_l))$$

$$\leq e^{-cs_i}V(\tau_\varepsilon + t, \omega(t), \omega(t + s_i - s_l)) + \frac{3a(\varepsilon)}{4} < a(\varepsilon). \tag{3.69}$$

Then from (3.69) for $l \geq i \geq m_0(\varepsilon)$, we have

$$\|\omega(t + s_i) - \omega(t + s_l)\| < \varepsilon. \tag{3.70}$$

From definitions of the sequence $\{s_i\}$ and system (3.58) for $l \geq i \geq m_0(\varepsilon)$ it follows that $\rho(t_k + s_i, t_k + s_l) < \varepsilon$, where $\rho(.,.)$ is an arbitrary distance on $\mathcal{B}$.

Then from (3.70) and the last inequality we obtain that the sequence $\omega(t + s_i)$ is convergent uniformly to the function $\omega(t)$.

The assertions (a) and (c) of the theorem follow immediately. We shall proof the assertion (b).

Let $\bar{\omega}(t)$ be an arbitrary solution of (3.58).

Set

$$u(t) = \overline{w}(t) - w(t), g^s\left(t, u(t), \int_{t_0}^{t} K^s(t, \xi, u(\xi))d\xi\right)$$

$$= f^s\left(t, u(t) + w(t), \int_{t_0}^{t} K^s(t, \xi, u(\xi) + w(\xi))d\xi\right)$$

$$- f^s\left(t, w(t), \int_{t_0}^{t} K^s(t, \xi, w(\xi))d\xi\right),$$

$$B_k^s(u) = I_k^s(u + w) - I_k^s(u).$$

We shall consider the system

$$\begin{cases} \dot{u} = g^s\left(t, u(t), \int_{t_0}^{t} K^s(t, \xi, u(\xi))d\xi\right), \ t \neq t_k^s, \\ \\ \Delta u(t_k^s) = B_k^s(u(t_k^s)), \ k = \pm 1, \pm 2, ..., \end{cases} \tag{3.71}$$

and let $W(t, u(t)) = V(t, w(t), w(t) + u(t))$.

Then, from Lemma 3.3 it follows that the zero solution $u(t) = 0$ of (3.71) is uniformly asymptotically stable and hence, $w(t)$ is uniformly asymptotically stable.   □

3.3   Impulsive Differential Equations with Time-varying Delays

In this part we apply the comparison principle to the problem of existence of almost periodic solutions of impulsive differential equations with time-varying delays. The impulses are in the fixed moments of time and since the solutions of such systems are piecewise continuous functions the investigations are carried out by using minimal subset [64, 132, 145, 177] of a suitable space of piecewise continuous functions, by the elements of which the derivatives of Lyapunov functions are estimated.

Let $\varphi_0 \in PC[\mathbb{R}, \Omega]$ and $|\varphi_0|_\infty = \sup_{t \in \mathbb{R}} ||\varphi_0(t)||$.

Consider the following system of impulsive differential equations with time-varying delay

$$\begin{cases} \dot{x}(t) = f(t, x(t), x(t - \eta(t))), \ t \neq t_k, \\ \Delta x(t_k) = I_k(x(t_k)), \ t_k \geq t_0, \ k = \pm 1, \pm 2, ..., \end{cases} \tag{3.72}$$

where $t \in \mathbb{R}$, $f : \mathbb{R} \times \Omega \times \Omega \to \mathbb{R}^n$, $\eta : \mathbb{R} \to \mathbb{R}^+$, $\{t_k\} \in \mathcal{B}$, $I_k : \Omega \to \mathbb{R}^n$, $k = \pm 1, \pm 2, \dots$.

Introduce the following conditions:

H3.24. The function $f \in C[\mathbb{R} \times \Omega \times \Omega \to \mathbb{R}^n]$, and $f(t, 0, 0) = 0$, $t \in \mathbb{R}$.

H3.25. The function $\eta \in C[\mathbb{R}, \mathbb{R}^+]$, and $t - \eta(t) \geq 0$, $t \in \mathbb{R}$.

H3.26. The functions $I_k \in C[\Omega, \mathbb{R}^n]$, and $I_k(0) = 0$, $k = \pm 1, \pm 2, \dots$.

H3.27. $E + I_k : \Omega \to \Omega$, $k = \pm 1, \pm 2, \dots$ where E is the identity in $\mathbb{R}^{n \times n}$.

Let $t_0 \in \mathbb{R}$. Denote by $x(t) = x(t; t_0, \varphi_0)$, $\varphi_0 \in PC[\mathbb{R}, \Omega]$, the solution of the system (3.72) with initial conditions:

$$\begin{cases} x(t) = \varphi_0(t), \ t \in (-\infty, t_0], \\ x(t_0^+; t_0, \varphi_0) = \varphi_0(t_0). \end{cases}$$

We assume that the solution $x(t) = x(t; t_0, \varphi_0)$ of (3.72) with an initial function $\varphi_0 \in PC[\mathbb{R}, \Omega]$ exists, and from [177] it follows that $x(t)$ is a piecewise continuous function with points of discontinuity at the moments t_k, $k = \pm 1, \pm 2, \dots$ at which it is continuous from the left.

Introduce the following conditions:

H3.28. The function $f(t, x, y)$ is almost periodic in t uniformly with respect to $x \in \Omega$, $y \in \Omega$.

H3.29. The function $\eta(t)$ is almost periodic in the sense of Bohr.

H3.30. The sequence of functions $\{I_k(x)\}$, $k = \pm 1, \pm 2, \dots$ is almost periodic uniformly with respect to $x \in \Omega$.

H3.31. The function $\varphi_0 \in PC[\mathbb{R}, \Omega]$ is almost periodic.

H3.32. The set of sequences $\{t_k^j\}$, $t_k^j - t_{k+j} - t_k$, $k = \pm 1, \pm 2, \dots$, $j = \pm 1, \pm 2, \dots$ is uniformly almost periodic, and $inf_k t_k^1 = \theta > 0$.

Let conditions H3.24–H3.32 hold, and let $\{s_m'\}$ be an arbitrary sequence of real numbers. Then there exists a subsequence $\{s_n\}$, $s_n = s_{m_n}{}'$, such that the system (3.72) moves to the system

$$\begin{cases} \dot{x}(t) = f^s(t, x(t), x(t - \eta^s(t))), \ t \neq t_k^s, \\ \Delta x(t_k^s) = I_k^s(x(t_k^s)), \ k = \pm 1, \pm 2, \dots, \end{cases} \tag{3.73}$$

and by analogy, we shall denote the set of these systems by $H(f, \eta, I_k, t_k)$.

Definition 3.8. [177] The zero solution $x(t) \equiv 0$ of system (3.72) is said to be:

(a) *Uniformly stable*, if

$$(\forall \varepsilon > 0)(\exists \delta > 0)(\forall \varphi_0 \in PC[\mathbb{R}, \Omega] : |\varphi_0|_\infty < \delta)(\forall t_0 \in \mathbb{R})$$

$$(\forall t \geq t_0) : \ ||x(t; t_0, \varphi_0)|| < \varepsilon;$$

(b) *Uniformly attractive*, if

$$(\exists \lambda > 0)(\forall \varepsilon > 0)(\exists T > 0)(\forall t_0 \in \mathbb{R})(\forall \varphi_0 \in PC[\mathbb{R}, \Omega] : |\varphi_0|_\infty < \lambda)$$
$$(\forall t \geq t_0 + T) : \ ||x(t; t_0, \varphi_0)|| < \varepsilon;$$

(c) *Uniformly asymptotically stable*, if it is uniformly stable and uniformly attractive.

We shall use the classes V_1, V_2 and W_0 from Chap. 1. Let $V \in V_1$, $t \neq t_k$, $k = \pm 1, \pm 2, \ldots$, $x \in PC[\mathbb{R}, \Omega]$, $y \in PC[\mathbb{R}, \Omega]$.

Introduce

$$D^+ V(t, x(t), y(t)) = \lim_{\delta \to 0^+} sup \frac{1}{\delta} \Big\{ V\big(t + \delta, x(t) + \delta f(t, x(t), x(t - \eta(t))), y(t) $$
$$+ \delta f(t, y(t), y(t - \eta(t))) \big) - V(t, x(t), y(t)) \Big\}. \qquad (3.74)$$

By analogy, introduce:

$$D^+ W(t, x(t)) = \lim_{\delta \to 0^+} sup \frac{1}{\delta} \Big\{ W\big(t + \delta, x(t) + \delta f^s(t, x(t), x(t $$
$$- \eta^s(t))) \big) - W(t, x(t)) \Big\}.$$

For $V \in V_1$ and for some $t \geq t_0$, define the following set:

$$\Omega_1 = \{x, y \in PC[\mathbb{R}, \Omega] : \ V(s, x(s), y(s)) \leq V(t, x(t), y(t)), \ s \in (-\infty, t]\}.$$

In the proof of the main results we shall use the following lemmas. The proofs of these lemmas are similar to the proofs of similar lemmas in [177].

Lemma 3.5. *Let the following conditions hold:*

1. *The function $g : \mathbb{R} \times \mathbb{R}^+ \to \mathbb{R}^+$ is continuous in each of the sets $(t_{k-1}, t_k] \times \mathbb{R}^+$, $k = \pm 1, \pm 2, \ldots$ and $g(t, 0) = 0$ for $t \in \mathbb{R}$.*
2. *$B_k \in C[\mathbb{R}^+, \mathbb{R}^+]$, $B_k(0) = 0$ and $\psi_k(u) = u + B_k(u)$, $k = \pm 1, \pm 2, \ldots$ are nondecreasing with respect to u.*
3. *The maximal solution $u^+ : J^+(t_0, u_0) \to \mathbb{R}^+$ of (1.16) is defined in $[t_0, \infty)$.*
4. *The solutions $x(t) = x(t; t_0, \varphi_0)$, $y(t) = y(t; t_0, \phi_0)$, $\phi_0 \in PC[\mathbb{R}, \Omega]$, of (3.72) are such that x, $y \in PC[\mathbb{R}, \Omega] \cap PC^1[\mathbb{R}, \Omega]$.*
5. *The function $V \in V_1$ is such that $V(t_0^+, \varphi_0, \phi_0) \leq u_0$,*

$$V(t^+, x + I_k(x), y + I_k(y)) \leq \psi_k(V(t, x, y)), \ x, y \in \Omega, t = t_k, k = \pm 1, \pm 2, \ldots,$$

and the inequality

$$D^+V(t, x(t), y(t)) \leq g(t, V(t, x(t), y(t))), \ t \neq t_k, \ k = \pm 1, \pm 2, \dots$$

is valid for $t \in [t_0, \infty)$, $x \in \Omega_1$.

Then

$$V(t, x(t; t_0, \varphi_0), y(t; t_0, \phi_0)) \leq u^+(t; t_0, u_0),$$

as $t \geq t_0$.

Lemma 3.6. *Let the following conditions hold:*

1. *Conditions H3.24–H3.32 are met.*
2. *For any equation from the set* $H(f, \eta, I_k, t_k)$ *there exist functions* $W \in W_0$, $a, b \in K$ *such that*

$$a(||x||) \leq W(t, x) \leq b(||x||), \ (t, x) \in \mathbb{R} \times \Omega,$$

$$W(t^+, x + I_k^s(x)) \leq W(t, x), \ x \in \mathbb{R}^n, \ t = t_k^s, \ k = \pm 1, \pm 2, \dots,$$

and the inequality

$$D^+W(t, x(t)) \leq -cW(t, x(t)), \ t \neq t_k^s, \ k = \pm 1, \pm 2, \dots$$

is valid for $t \geq t_0$, $x \in PC[\mathbb{R}, \mathbb{R}^n]$ *for which* $W(s, x(s)) \leq W(t, x(t))$, $s \in [t_0, t]$.

Then the zero solution of (3.73) is uniformly asymptotically stable.
Now we shall prove the main theorem in this paragraph.

Theorem 3.8. *Let the following conditions hold:*

1. *Conditions H3.24–H3.32 are met.*
2. *There exist functions* $V \in V_2$ *and* $a, b \in K$ *such that*

$$a(||x - y||) \leq V(t, x, y) \leq b(||x - y||), \ (t, x, y) \in \mathbb{R} \times \Omega \times \Omega, \quad (3.75)$$

$$V(t^+, x + I_k(x), y + I_k(y)) \leq V(t, x, y), \ x, y \in \Omega, \ t = t_k, \ k = \pm 1, \pm 2, \dots,$$

and the inequality

$$D^+V(t, x(t), y(t)) \leq -cV(t, x(t), y(t)), \ t \neq t_k, \ k = \pm 1, \pm 2, \dots, \quad (3.76)$$

is valid for $t \geq t_0$, $(x, y) \in \Omega_1$, *where* $c > 0$.
3. *There exists a solution* $x(t; t_0, \varphi_0)$ *of (3.72) such that*

$$||x(t; t_0, \varphi_0)|| < \alpha_1,$$

where $t \geq t_0$, $\alpha_1 < \alpha$.

Then, in B_α, $\alpha > 0$ for the system (3.72) there exists a unique almost periodic solution $\omega(t)$ such that:

1. $||\omega(t)|| \le \alpha_1$.
2. $\omega(t)$ *is uniformly asymptotically stable.*
3. $H(\omega, t_k) \subset H(f, \eta, I_k, t_k)$.

Proof. Let $\{s_i\}$ be any sequence of real numbers such that $s_i \to \infty$ as $i \to \infty$, and $\{s_i\}$ moves the system (3.72) in a system from the set $H(f, \eta, I_k, t_k)$.

For any real number β, let $i_0 = i_0(\beta)$ be the smallest value of i, such that $s_{i_0} + \beta \ge t_0$. Since $||x(t; t_0, \varphi_0)|| < \alpha_1$, where $\alpha_1 < \alpha$, for all $t \ge t_0$, then $x(t + s_i; t_0, \varphi_0) \in B_{\alpha_1}$ for $t \ge \beta$ and $i \ge i_0$.

Let U, $U \subset (\beta, \infty)$ be a compact. Then, for any $\varepsilon > 0$, choose an integer $n_0(\varepsilon, \beta) \ge i_0(\beta)$, so large that for $l \ge i \ge n_0(\varepsilon, \beta)$ and $t \in (\beta, \infty)$, it follows

$$b(2\alpha_1)e^{-c(\beta + s_i - t_0)} < \frac{a(\varepsilon)}{2}, \tag{3.77}$$

$$||f(t + s_l, x(t), x(t - \eta(t))) - f(t + s_i, x(t), x(t - \eta(t)))|| < \frac{a(\varepsilon)c}{2H_1}, \tag{3.78}$$

where $a, b \in K$, $x \in PC[\mathbb{R}, \Omega]$, $c > 0$ and $H_1 > 0$ is the Lipschitz constant of the function $V(t, x, y)$.

We shall consider the function $V(\sigma, x(\sigma), x(\sigma + s_l - s_i))$, and for $\sigma > t_0$, $\big(x(\sigma), x(\sigma + s_l - s_i)\big) \in \Omega_1$ from (3.76), we obtain

$$D^+ V(\sigma, x(\sigma), x(\sigma + s_l - s_i)) \le -cV(\sigma, x(\sigma), x(\sigma + s_l - s_i))$$
$$+ H_1 ||f(\sigma + s_l - s_i, x(\sigma + s_l - s_i), x(\sigma + s_l - s_i - \eta(\sigma + s_l - s_i)))$$
$$- f(\sigma, x(\sigma + s_l - s_i), x(\sigma + s_l - s_i - \eta(\sigma + s_l - s_i)))||$$
$$\le -cV(\sigma, x(\sigma), x(\sigma + s_l - s_i)) + \frac{a(\varepsilon)c}{2}. \tag{3.79}$$

On the other hand, from (3.77), (3.79) and Lemma 3.5, it follows

$$V(t + s_i, x(t + s_i), x(t + s_l))$$

$$\le e^{-c(t + s_i - t_0)} V(t_0, x(t_0), x(t_0 + s_l - s_i)) + \frac{a(\varepsilon)c}{2} < a(\varepsilon).$$

Then from (3.75), we have

$$||x(t + s_i) - x(t + s_l)|| < \varepsilon,$$

for $l \ge i \ge n_0(\varepsilon, \beta)$, $t \in U$.

Consequently, there exists a function $\omega(t)$, such that $x(t + s_i) - \omega(t) \to \infty$ for $i \to \infty$, and since β is arbitrary, it follows that $\omega(t)$ is defined uniformly on $t \in U$.

Next, we shall show that $\omega(t)$ is a solution of (3.73).

Since $x(t; t_0, \varphi_0)$ is a solution of (3.72), we have

$$
\begin{aligned}
||\dot{x}(t + s_i) - \dot{x}(t + s_l)|| &\leq ||f(t + s_i, x(t + s_i), x(t + s_i - \eta(t + s_i))) \\
&\quad - f(t + s_l, x(t + s_i), x(t + s_i - \eta(t + s_i)))|| \\
&\quad + ||f(t + s_l, x(t + s_i), x(t + s_i - \eta(t + s_i))) \\
&\quad - f(t + s_l, x(t + s_l), x(t + s_l - \eta(t + s_l)))||,
\end{aligned}
$$

for $t + s_j \neq t_k$, $j = i, l$ and $k = \pm 1, \pm 2, \dots$.

As $x(t + s_i) \in B_{\alpha_1}$ for large s_i for each compact subset of $\mathbb{R}$ there exists $n_1(\varepsilon) > 0$ such that if $l \geq i \geq n_1(\varepsilon)$, then

$$
\begin{aligned}
||f(t + s_i, x(t + s_i), x(t + s_i - \eta(t + s_i)) \\
- f(t + s_l, x(t + s_i), x(t + s_i - \eta(t + s_i))|| < \frac{\varepsilon}{2}.
\end{aligned}
$$

Since $x(t + s_j) \in B_{\beta(\alpha)}$, $j = i, l$ and from Lemma 1.6 it follows that there exists $n_2(\varepsilon) > 0$ such that if $l \geq i \geq n_2(\varepsilon)$, then

$$
\begin{aligned}
||f(t + s_l, x(t + s_i), x(t + s_i - \eta(t + s_i))) \\
- f(t + s_l, x(t + s_l), x(t + s_l - \eta(t + s_l)))|| < \frac{\varepsilon}{2}.
\end{aligned}
$$

Then for $l \geq i \geq n(\varepsilon)$, $n(\varepsilon) = max\{n_1(\varepsilon), n_2(\varepsilon)\}$, we obtain

$$
||\dot{x}(t + s_i) - \dot{x}(t + s_l)|| < \varepsilon,
$$

where $t + s_j \neq t_k^s$, $j = i, l$ and $k = \pm 1, \pm 2, \dots$, which shows that $\lim_{i \to \infty} \dot{x}(t + s_i)$ exists uniformly on all compact subsets of $\mathbb{R}$.

Let now $\lim_{i \to \infty} \dot{x}(t + s_i) = \dot{\omega}(t)$, and

$$
\begin{aligned}
\dot{\omega}(t) &= \lim_{i \to \infty} \Big[f(t + s_i, x(t + s_i), x(t + s_i - \eta(t + s_i))) \\
&\quad - f(t + s_i, \omega(t), \omega(t + s_i - \eta(t + s_i))) \\
&\quad + f(t + s_i, \omega(t), \omega(t + s_i - \eta(t + s_i))) \Big] \\
&= f^s(t, \omega(t), \omega(t - \eta^s(t))), \quad\quad\quad\quad\quad\quad (3.80)
\end{aligned}
$$

where $t \neq t_k^s$ and $t_k^s = \lim_{i \to \infty} t_{k+i(s)}$.

On the other hand, for $t + s_i = t_k^s$ it follows

$$\omega(t_k^{s+}) - \omega(t_k^{s-}) = \lim_{i \to \infty} \left(x(t_k^s + s_i + 0) - x(t_k^s + s_i - 0) \right)$$

$$= \lim_{i \to \infty} I_k^s(x(t_k^s + s_i)) = I_k^s(\omega(t_k^s)). \tag{3.81}$$

From (3.80) and (3.81) we get that $\omega(t)$ is a solution of (3.73).

We have to show that $\omega(t)$ is an almost periodic function.

Let the sequence $\{s_i\}$ moves the system (3.72) to $H(f, \eta, I_k, t_k)$. For any $\varepsilon > 0$ there exists $m_0(\varepsilon) > 0$ such that if $l \geq i \geq m_0(\varepsilon)$, then

$$e^{-cs_i} b(2\alpha_1) < \frac{a(\varepsilon)}{4},$$

and

$$\|f(\sigma + s_i, x(\sigma + s_i), x(\sigma + s_i - \eta(\sigma + s_i)))$$

$$- f(\sigma + s_l, x(\sigma + s_l), x(\sigma + s_l - \eta(\sigma + s_l)))\| < \frac{a(\varepsilon)}{4H_1},$$

where $a, b \in K$.

For each fixed $t \in \mathbb{R}$, let $\tau_\varepsilon = \dfrac{a(\varepsilon)}{4H_1}$ be a translation number of the function f such that $t + \tau_\varepsilon \geq 0$.

Now, we consider the function

$$V(\tau_\varepsilon + \sigma, \omega(\sigma), \omega(\sigma + s_l - s_i)),$$

where $t \leq \sigma \leq t + s_i$.

Then

$$D^+ V(\tau_\varepsilon + \sigma, \omega(\sigma), \omega(\sigma + s_l - s_i))$$

$$\leq -cV(\tau_\varepsilon + \sigma, \omega(\sigma), \omega(\sigma + s_l - s_i)) + H_1 \|f^s(\sigma, \omega(\sigma), \omega(\sigma - \eta^s(\sigma)))$$

$$- f^s(\tau_\varepsilon + \sigma, \omega(\sigma), \omega(\tau_\varepsilon + \sigma - \eta^s(\tau_\varepsilon + \sigma)))\|$$

$$+ H_1 \|f^s(\sigma + s_l - s_i, \omega(\sigma + s_l - s_i), \omega(\sigma + s_l - s_i - \eta^s(\sigma + s_l - s_i)))$$

$$- f^s(\tau_\varepsilon + \sigma, \omega(\sigma + s_l - s_i), \omega(\sigma + s_l - s_i - \eta^s(\sigma + s_l - s_i)))\|$$

$$\leq -cV(\tau_\varepsilon + \sigma, \omega(\sigma), \omega(\sigma + s_l - s_i)) + \frac{3a(\varepsilon)c}{4}. \tag{3.82}$$

On the other hand,

$$V(\tau_\varepsilon + t_k^s, \omega(t_k^s) + I_k^s(\omega(t_k^s)), \omega(t_k^s + s_l - s_i) + I_k^s(\omega(t_k^s + s_l - s_i)))$$
$$\leq V(\tau_\varepsilon + t_k^s, \omega(t_k^s), \omega(t_k^s + s_l - s_i)). \tag{3.83}$$

Then from (3.82), (3.83) and Lemma 3.5 it follows

$$V(\tau_\varepsilon + t + s_i, \omega(t + s_i), \omega(t + s_l))$$
$$\leq e^{-cs_i} V(\tau_\varepsilon + t, \omega(t), \omega(t + s_i - s_l)) + \frac{3a(\varepsilon)}{4} < a(\varepsilon). \tag{3.84}$$

From (3.84) for $l \geq i \geq m_0(\varepsilon)$, we have

$$||\omega(t + s_i) - \omega(t + s_l)|| < \varepsilon. \tag{3.85}$$

Now, from the definition of the sequence $\{s_i\}$ for $l \geq i \geq m_0(\varepsilon)$, it follows that
$$\rho(t_k + s_i, t_k + s_l) < \varepsilon,$$
where $\rho(.,.)$ is an arbitrary distance in $\mathcal{B}$.

From (3.85) and the last inequality, we obtain that the sequence $\omega(t + s_i)$ is convergent uniformly to the function $\omega(t)$.

The assertions (a) and (c) of the theorem follow immediately. We shall prove the assertion (b).

Let $\overline{\omega}(t)$ be an arbitrary solution of (3.73).

Set

$$u(t) = \overline{\omega}(t) - \omega(t),$$
$$g^s(t, u(t)) = f^s(t, u(t) + \omega(t), u(t) + \omega(t - \eta^s(t)))$$
$$- f^s(t, \omega(t), \omega(t - \eta^s(t))),$$
$$B_k^s(u) = I_k^s(u + \omega) - I_k^s(u).$$

Now, we consider the system

$$\begin{cases} \dot{u} = g^s(t, u(t)), \ t \neq t_k^s, \\ \Delta u(t_k^s) = B_k^s(u(t_k^s)), \ k = \pm 1, \pm 2, ..., \end{cases} \tag{3.86}$$

and let $W(t, u(t)) = V(t, \omega(t), \omega(t) + u(t))$.

Then from Lemma 3.6 it follows that the zero solution $u(t) = 0$ of system (3.86) is uniformly asymptotically stable for $t_0 \geq 0$, and hence $\omega(t)$ is uniformly asymptotically stable. □

Example 3.2. To apply the results of Theorem 3.8, we shall consider the following system

$$\begin{cases} \dot{u} = r(t)u(t)\big(1 - u(t - \eta(t)K^{-1})\big), \ t \neq t_k, \\ \Delta u(t_k) = K\gamma_k\big(u(t_k)K^{-1} - 1\big), \ k = \pm 1, \pm 2, ..., \end{cases} \qquad (3.87)$$

where $r, \eta \in C[\mathbb{R}, \mathbb{R}^+]$ and $r(t) > 0$, $0 \leq \eta(t) \leq \eta_0$, $\gamma_k \in C[\mathbb{R}, \mathbb{R}], k = \pm 1, \pm 2, ...$, $K = const > 0$, $\{t_k\} \in \mathcal{B}$.

This system presents the simulation in mathematical ecology of the dynamics of the population $u(t)$, where $K > 0$ is the capacity of the environment, γ_k are functions which characterize the magnitude of the impulsive effect at the moments t_k.

The systems (3.87), where $\gamma_k(u) = 0$, $u \geq 0$, $k = \pm 1, \pm 2, ...$ has been studied in [198].

Let $N(t) = u(t)K^{-1} - 1$ and from (3.87), we obtain the system

$$\begin{cases} \dot{N}(t) = -r(t)\big(1 + N(t)\big)N(t - \eta(t)), \ t \neq t_k, \\ \Delta N(t_k) = \gamma_k(N(t_k)), \ k = \pm 1, \pm 2, ... \end{cases} \qquad (3.88)$$

Under the standard type of initial conditions, $1 + N(0) > 0$, $1 + N(s) \geq 0$ for $s \in (-\infty, 0]$ it is easy to see that the solution of (3.88) satisfies $1 + N(t) > 0$ for $t \geq 0$.

Theorem 3.9. *Let the following conditions hold:*

1. *Conditions H3.29 and H3.32 are met.*
2. *The function $r(t)$ is almost periodic in sense of Bohr and*

$$\int_{t-\eta(t)}^{t} r(s)ds = \infty.$$

3. *The sequence of functions $\{\gamma_k(u)\}$, $\gamma_k \in C[\mathbb{R}, \mathbb{R}]$, is almost periodic uniformly on $u \in \mathbb{R}$, and $\gamma_k(0) = 0$, $-1 < \gamma_k(u) \leq 0$, $k = \pm 1, \pm 2, ...$*
4. *There exists a bounded solution of (3.88).*
5. *The inequalities*

$$N(t - \eta(t))\big(1 + N(t)\big) \geq N(t), \ t \neq t_k, \ k = \pm 1, \pm 2, ...,$$

$$M(t - \eta(t))\big(1 + M(t)\big) \geq M(t), \ t \neq t_k, \ k = \pm 1, \pm 2, ...,$$

$$2N(t)\big(N(t) + \gamma_k^2(N(t))\big) \leq \gamma_k(t), \ t = t_k, \ k = \pm 1, \pm 2, ...,$$

are valid for $t \geq 0$ and $(N, M) \in \Omega_2$ where

$$\Omega_2 = \Big\{(N, M) . \ N, M \subset PC[\mathbb{R}, \mathbb{R}^+], \ N^2(s)$$

$$+ M^2(s) \leq N^2(t) + M^2(t), s \in (-\infty, t], \ t \geq 0\Big\}.$$

Then for the system (3.88) there exists a unique almost periodic solution.

Proof. Let $V(t, N, M) = N^2 + M^2$. Then, from the conditions of the theorem, it follows that

$$D^+ V(t, N(t), M(t))$$

$$= 2N(t)\big(-r(t)\big)\Big(1 + N(t)\Big)N(t - \eta(t)) + 2M(t)[-r(t)]\Big(1 + M(t)\Big)$$

$$\times M(t - \eta(t))$$

$$\leq -2r(t)V(t, N(t), M(t)) \leq 0, \ t \neq t_k, \ k = \pm 1, \pm 2, ...,$$

and

$$V\Big(t_k^+, N(t_k) + \gamma_k(N(t_k)), M(t_k) + \gamma_k(M(t_k))\Big)$$

$$= \Big[N(t_k) + \gamma_k(N(t_k))\Big]^2 + \Big[M(t_k) + \gamma_k(M(t_k))\Big]^2$$

$$\leq V(t_k, N(t_k), M(t_k)), \ k = \pm 1, \pm 2, ...$$

Thus, all conditions of Theorem 3.8 are satisfied, and the conclusion of Theorem 3.9 follows. $\square$

3.4 Impulsive Functional Differential Equations

In the present part, the existence and stability of almost periodic solutions of nonlinear systems of impulsive functional differential equations are considered. Some known results are improved and generalized [19, 28, 142, 144, 197].

Let $\varphi_0 \in PC[\mathbb{R}, \Omega]$ and $|\varphi_0|_\infty = \sup\limits_{t \in \mathbb{R}} ||\varphi_0(t)||$.

We shall consider the system of impulsive functional differential equations

$$\begin{cases} \dot{x}(t) = f(t, x_t), \ t \geq t_0, \ t \neq t_k, \\ \Delta x(t_k) = I_k(x(t_k)), \ k = \pm 1, \pm 2, ..., \end{cases} \tag{3.89}$$

where $t_0 \in \mathbb{R}$, $t \in \mathbb{R}$, $f : \mathbb{R} \times PC[\mathbb{R}, \Omega] \to \mathbb{R}^n$, $\{t_k\} \in \mathcal{B}$, $I_k : \Omega \to \mathbb{R}^n$, $k = \pm 1, \pm 2, ...$ and for $t > t_0$, $x_t \in PC[\mathbb{R}, \Omega]$ is defined by the relation $x_t = x(t + s)$, $-\infty \leq s \leq 0$.

Denote by $x(t) = x(t; t_0, \varphi_0)$, $\varphi_0 \in PC[\mathbb{R}, \Omega]$ the solution of the system (3.89) with initial conditions:

$$\begin{cases} x(t; t_0, \varphi_0) = \varphi_0(t - t_0), \ t \leq t_0, \\ x(t_0^+; t_0, \varphi_0) = \varphi_0(0), \end{cases} \tag{3.90}$$

and by $J^+(t_0, \varphi_0)$—the maximal interval of type $[t_0, \beta)$ in which the solution $x(t; t_0, \varphi_0)$ is defined.

The solution $x(t) = x(t; t_0, \varphi_0)$ of the initial value problem (3.89), (3.90) is characterized by the following [177]:

(a) For $t \leq t_0$, $t_0 \in [t_{k_0}, t_{k_1})$, $t_{k_0} < t_{k_1}, t_{k_i} \in \{t_k\}$, $i = 0, 1$, the solution $x(t)$ satisfies the initial conditions (3.90).
(b) For $t_0 < t \leq t_{k_1}$, $x(t)$ coincides with the solution of the problem

$$\begin{cases} \dot{x}(t) = f(t, x_t), \ t > t_0, \\ x_{t_0} = \varphi_0(s), \ -\infty < s \leq 0. \end{cases}$$

At the moment $t = t_{k_1}$ the mapping point $(t, x(t; t_0, \varphi_0))$ of the extended phase space jumps momentarily from the position $(t_{k_1}, x(t_{k_1}; t_0, \varphi_0))$ to the position $(t_{k_1}, x(t_{k_1}; t_0, \varphi_0) + I_{k_1}(x(t_{k_1}; t_0, \varphi_0)))$.
(c) For $t_{k_1} < t \leq t_{k_2}$, $t_{k_2} \in \{t_k\}$ the solution $x(t)$ coincides with the solution of

$$\begin{cases} \dot{y}(t) = f(t, y_t), \ t > t_{k_1}, \\ y_{t_{k_1}} = \varphi_1, \ \varphi_1 \in PC[\mathbb{R}, \Omega], \end{cases}$$

where

$$\varphi_1(t - t_{k_1}) = \begin{cases} \varphi_0(t - t_{k_1}), \ t \in (-\infty, t_{k_1}], \\ x(t; t_0, \varphi_0) + I_{k_1}(x(t; t_0, \varphi_0)), \ t = t_{k_1}. \end{cases}$$

At the moment $t = t_{k_2}$ the mapping point $(t, x(t))$ jumps momentarily, etc.

Thus in interval $J^+(t_0, \varphi_0)$ the solution $x(t; t_0, \varphi_0)$ of the problem (3.89), (3.90) is a piecewise continuous function with points of discontinuity of the first kind at the moments $t = t_k$, $k = \pm 1, \pm 2, ...$, at which it is continuous from the left.

Introduce the following conditions:

H3.33. The function $f \in C[\mathbb{R} \times PC[\mathbb{R}, \Omega], \mathbb{R}^n]$, $f(t, 0) = 0$, $t \in \mathbb{R}$.
H3.34. The function $f(t, \varphi)$ is Lipschitz continuous with respect to $\varphi \in PC[\mathbb{R}, \Omega]$ uniformly on $t \in \mathbb{R}$.
H3.35. The functions $I_k \in C[\Omega, \mathbb{R}^n]$, $I_k(0) = 0$.
H3.36. $(E + I_k) : \Omega \to \Omega$, $k = \pm 1, \pm 2, ...$, where E is the identity in $\mathbb{R}^{n \times n}$.

We shall use the next lemma in the prove of the main results.

Lemma 3.7 ([178]). *Let the conditions H3.33–H3.36 are met.*
Then, $J^+(t_0, \varphi_0) = [t_0, \infty)$.

Introduce the following conditions:

H3.37. The function $f(t, \varphi)$ is almost periodic in $t \in \mathbb{R}$ uniformly with respect to $\varphi \in PC[\mathbb{R}, \Omega]$.

H3.38. The sequence $\{I_k(x)\}$, $k = \pm 1, \pm 2, \ldots$ is almost periodic uniformly with respect to $x \in \Omega$.

H3.39. The function $\varphi_0 \in PC[\mathbb{R}, \Omega]$ is almost periodic.

H3.40. The set of sequences $\{t_k^j\}$, $t_k^j = t_{k+j} - t_k$, $k = \pm 1, \pm 2, \ldots$, $j = \pm 1, \pm 2, \ldots$ is uniformly almost periodic, and $\inf_k t_k^1 = \theta > 0$.

Let the conditions H3.33–H3.40 hold, and let $\{s_m'\}$ be an arbitrary sequence of real numbers. Then, there exist a subsequence $\{s_n\}$, $s_n = s_{m_n}'$ such that the system (3.89) moves to the system

$$\begin{cases} \dot{x}(t) = f^s(t, x_t), \ t \geq t_0, \ t \neq t_k^s, \\ \Delta x(t_k^s) = I_k^s(x(t_k^s)), \ k = \pm 1, \pm 2, \ldots, \end{cases} \qquad (3.91)$$

and the set of systems in the form (3.91) we shall denote by $H(f, I_k, t_k)$.

We shall use the classes V_1, V_2 and W_0 which are defined in Chap. 1. Let $V \in V_1$, $t \neq t_k$, $x \in PC[\mathbb{R}, \Omega]$, $y \in PC[\mathbb{R}, \Omega]$.

Introduce

$$D^+ V(t, x(t), y(t)) = \lim_{\delta \to 0^+} \sup \frac{1}{\delta} \Big\{ V(t + \delta, x(t) + \delta f(t, x_t), y(t) + \delta f(t, y_t))$$

$$- V(t, x(t), y(t)) \Big\}. \qquad (3.92)$$

By analogy, let $W \in W_0$, $t \neq t_k^s$, $x \in PC[\mathbb{R}, \Omega]$.
Introduce:

$$D^+ W(t, x(t)) = \lim_{\delta \to 0^+} \sup \frac{1}{\delta} \Big\{ W(t + \delta, x(t) + \delta f^s(t, x_t)) - W(t, x(t)) \Big\}.$$

For $V \in V_1$ and for some $t \geq t_0$, define the following set:

$$\Omega_1 = \big\{ (x, y) : \ x, y \in PC[\mathbb{R}, \Omega] : \ V(s, x(s), y(s))$$

$$\leq V(t, x(t), y(t)), \ s \in (-\infty, t] \big\}.$$

In the next theorem are considered the conditions for the existence of almost periodic solutions of the system (3.89).

Theorem 3.10. *Let the following conditions hold:*

1. *Conditions H3.33–H3.40 are met.*
2. *There exist functions $V \in V_2$ and $a, b \in K$ such that*

$$a(||x - y||) \leq V(t, x, y) \leq b(||x - y||), \ (t, x, y) \in \mathbb{R} \times \Omega \times \Omega, \qquad (3.93)$$

$$V\left(t^+, x + I_k(x), y + I_k(y)\right) \leq V(t, x, y), \ x, y \in \Omega, \ t = t_k, \ k = \pm 1, \pm 2, ...,$$

and the inequality

$$D^+ V(t, x(t), y(t)) \leq -cV(t, x(t), y(t)), \ t \neq t_k, \ k = \pm 1, \pm 2, ..., \qquad (3.94)$$

is valid for $t > t_0, \ (x, y) \in \Omega_1,$ *where* $c > 0$.

3. *There exists a solution* $x(t; t_0, \varphi_0)$ *of* (3.89) *such that*

$$||x(t; t_0, \varphi_0)|| \leq \alpha_1,$$

where $t \geq t_0, \ \alpha_1 < \alpha$.

Then, in $B_\alpha, \ \alpha > 0$ *for the system* (3.89) *there exists a unique almost periodic solution* $\omega(t)$ *such that:*

1. $||\omega(t)|| \leq \alpha_1$.
2. $\omega(t)$ *is uniformly asymptotically stable.*
3. $H(\omega, t_k) \subset H(f, I_k, t_k)$.

Proof. Let $\{s_i\}$ be any sequence of real numbers such that $s_i \to \infty$ as $i \to \infty$ and $\{s_i\}$ moves the system (3.89) to a system in $H(f, I_k, t_k)$.

For any real number β, let $i_0 = i_0(\beta)$ be the smallest value of i, such that $s_{i_0} + \beta \geq t_0$. Since $||x(t; t_0, \varphi_0)|| < \alpha_1, \ \alpha_1 < \alpha$ for all $t \geq t_0$, then $x(t + s_i; t_0, \varphi_0) \in B_{\alpha_1}$ for $t \geq \beta, \ i \geq i_0$.

Let $U, \ U \subset (\beta, \infty)$ be a compact. Then, for any $\varepsilon > 0$, choose an integer $n_0(\varepsilon, \beta) \geq i_0(\beta)$, so large that for $l \geq i \geq n_0(\varepsilon, \beta)$ and $t \in \mathbb{R}$, it follows

$$b(2\alpha_1)e^{-c(\beta + s_i - t_0)} < \frac{a(\varepsilon)}{2}, \qquad (3.95)$$

$$||f(t + s_l, x_t) - f(t + s_i, x_t)|| < \frac{a(\varepsilon)c}{2H_1}, \qquad (3.96)$$

where $x \in PC[\mathbb{R}, \Omega], \ c > 0$ and H_1 is Lipschitz constant of the function $V(t, x, y)$.

Consider the function $V(\sigma, x(\sigma), x(\sigma + s_l - s_i))$, and for $\sigma > t_0, \ (x(\sigma), x(\sigma + s_l - s_i)) \in \Omega_1$ from (3.94), (3.95), we obtain

$$D^+ V(\sigma, x(\sigma), x(\sigma + s_l - s_i)) \leq -cV(\sigma, x(\sigma), x(\sigma + s_l - s_i))$$

$$+ H_1 ||f(\sigma + s_l - s_i, x_{\sigma + s_l - s_i}) - f(\sigma, x_{\sigma + s_l - s_i})||$$

$$\leq -cV(\sigma, x(\sigma), x(\sigma + s_l - s_i)) + \frac{a(\varepsilon)c}{2}. \qquad (3.97)$$

On the other hand, from (3.97) and Lemma 3.5, it follows that for $l \geq n_0(\varepsilon, \beta) \geq i_0(\beta)$ and $t \in U$

$$V(t + s_i, x(t + s_i), x(t + s_l))$$

$$\leq e^{-c(t + s_i - t_0)} V(t_0, x(t_0), x(t_0 + s_l - s_i)) + \frac{a(\varepsilon)}{2} < a(\varepsilon).$$

Then from the condition (3.93) and $t \in U$, we have

$$||x(t + s_i) - x(t + s_l)|| < \varepsilon.$$

Consequently, there exists a function $\omega(t)$, such that $x(t + s_i) - \omega(t) \to \infty$ for $i \to \infty$. Since β is arbitrary, it follows that $\omega(t)$ is defined uniformly on $t \in \mathbb{R}$.

Next, we shall show that $\omega(t)$ is a solution of (3.91).

Since $x(t; t_0, \varphi_0)$ is solution of (3.89), we have

$$||\dot{x}(t + s_i) - \dot{x}(t + s_l)|| \leq ||f(t + s_i, x_{t+s_i}) - f(t + s_l, x_{t+s_i})||$$

$$+ ||f(t + s_l, x_{t+s_i}) - f(t + s_l, x_{t+s_l})||,$$

for $t + s_j \neq t_k$, $j = i, k$; $k = \pm 1, \pm 2, \ldots$.

As $x(t + s_i) \in B_{\alpha_1}$ for large s_i for each compact subset of $\mathbb{R}$ there exists an $n_1(\varepsilon) > 0$ such that if $l \geq i \geq n_1(\varepsilon)$, then

$$||f(t + s_i, x_{t+s_i}) - f(t + s_l, x_{t+s_i})|| < \frac{\varepsilon}{2}.$$

Since $x(t + s_j) \in B_{\beta(\alpha)}$, $j = i, l$ and from Lemma 1.6 it follows that there exists $n_2(\varepsilon) > 0$ such that if $l \geq i \geq n_2(\varepsilon)$, then

$$||f(t + s_l, x_{t+s_i}) - f(t + s_l, x_{t+s_l})|| < \frac{\varepsilon}{2}.$$

For $l \geq i \geq n(\varepsilon)$, $n(\varepsilon) = max\{n_1(\varepsilon), n_2(\varepsilon)\}$, we obtain

$$||\dot{x}(t + s_i) - \dot{x}(t + s_l)|| \leq \varepsilon,$$

where $t + s_j \neq t_k^s$ which shows that $\lim_{i \to \infty} \dot{x}(t + s_i)$ exists uniformly on all compact subsets of $\mathbb{R}$.

Let now $\lim_{i \to \infty} \dot{x}(t + s_i) = \dot{\omega}(t)$, and

$$\dot{\omega}(t) = \lim_{i \to \infty} \left[f(t + s_i, x_{t+s_i}) - f(t + s_i, \omega(t)) + f(t + s_i, \omega(t)) \right]$$

$$= f^s(t, \omega(t)), \tag{3.98}$$

where $t \neq t_k^s$, $t_k^s = \lim_{i \to \infty} t_{k+i(s)}$.

On the other hand, for $t + s_i = t_k^s$ it follows

$$w(t_k^{s+}) - w(t_k^{s-}) = \lim_{i \to \infty} (x(t_k^s + s_i + 0) - x(t_k^s + s_i - 0))$$

$$= \lim_{i \to \infty} I_k^s(x(t_k^s + s_i)) = I_k^s(w(t_k^s)). \qquad (3.99)$$

From (3.98) and (3.99), we get that $w(t)$ is a solution of (3.91).

We shall show that $w(t)$ is an almost periodic function.

Let the sequence $\{s_i\}$ moves the system (3.89) to $H(f, I_k, t_k)$. For any $\varepsilon > 0$ there exists $m_0(\varepsilon) > 0$ such that if $l \geq i \geq m_0(\varepsilon)$, then

$$e^{-cs_i} b(2\alpha_1) < \frac{a(\varepsilon)}{4},$$

and

$$||f(\sigma + s_i, x_{\sigma+s_i}) - f(\sigma + s_l, x_{\sigma+s_l})|| < \frac{a(\varepsilon)}{4H_1},$$

where $x \in PC[\mathbb{R}, \Omega]$, $c > 0$.

For each fixed $t \in \mathbb{R}$ let $\tau_\varepsilon = \frac{a(\varepsilon)}{4H_1}$ be a translation number of f such that $t + \tau_\varepsilon \geq 0$.

Consider the function

$$V(\tau_\varepsilon + \sigma, w(\sigma), w(\sigma + s_l - s_i)),$$

where $t \leq \sigma \leq t + s_i$.

Then

$$D^+ V(\tau_\varepsilon + \sigma, w(\sigma), w(\sigma + s_l - s_i))$$

$$\leq -cV(\tau_\varepsilon + \sigma, w(\sigma), w(\sigma + s_l - s_i)) + H_1 ||f^s(\sigma, w(\sigma)) - f^s(\tau_\varepsilon + \sigma, w(\sigma))||$$

$$+ H_1 ||f^s(\sigma + s_l - s_i, w(\sigma + s_l - s_i)) - f^s(\tau_\varepsilon + \sigma, w(\sigma + s_l - s_i))||$$

$$\leq -cV(\tau_\varepsilon + \sigma, w(\sigma), w(\sigma + s_l - s_i)) + \frac{3a(\varepsilon)}{4}. \qquad (3.100)$$

On the other hand,

$$V(\tau_\varepsilon + t_k^s, w(t_k^s) + I_k^s(w(t_k^s)), w(t_k^s + s_l - s_i) + I_k^s(w(t_k^s + s_l - s_i)))$$

$$\leq V(\tau_\varepsilon + t_k^s, w(t_k^s), w(t_k^s + s_l - s_i)). \qquad (3.101)$$

From (3.100), (3.101) and Lemma 3.5 it follows

$$V(\tau_\varepsilon + t + s_i, w(t + s_i), w(t + s_l)))$$

$$\leq e^{-cs_i} V(\tau_\varepsilon + t, w(t), w(t + s_i - s_l))) + \frac{3a(\varepsilon)}{4} < a(\varepsilon). \qquad (3.102)$$

Then from (3.102) for $l \geq i \geq m_0(\varepsilon)$, we have

$$\|w(t + s_i) - w(t + s_l)\| < \varepsilon. \tag{3.103}$$

Now, from definitions of the sequence $\{s_i\}$ and system (3.99) for $l \geq i \geq m_0(\varepsilon)$ it follows that $\rho(t_k+s_i, t_k+s_l) < \varepsilon$, where $\rho(.,.)$ is an arbitrary distance in $\mathcal{B}$.

Then from (3.103) and the last inequality we obtain that the sequence $w(t + s_i)$ is convergent uniformly to the function $w(t)$.

The assertions (a) and (c) of the theorem follow immediately. We shall prove the assertion (b).

Let $\overline{w}(t)$ be an arbitrary solution of (3.91).

Set

$$u(t) = \overline{w}(t) - w(t),$$

$$g^s(t, u(t)) = f^s(t, u(t) + w(t)) - f^s(t, w(t)),$$

$$B_k^s(u) = I_k^s(u + w) - I_k^s(u).$$

Now we consider the system

$$\begin{cases} \dot{u} = g^s(t, u(t)), \ t \neq t_k^s, \\ \Delta u(t_k^s) = B_k^s(u(t_k^s)), \ k = \pm 1, \pm 2, ..., \end{cases} \tag{3.104}$$

and let $W(t, u(t)) = V(t, w(t), w(t) + u(t))$. Then, from Lemma 3.5 it follows that the zero solution $u(t) = 0$ of (3.104) is uniformly asymptotically stable, and consequently $w(t)$ is uniformly asymptotically stable. □

Example 3.3. We shall consider the next scalar impulsive functional differential equation

$$\begin{cases} \dot{x}(t) = -a(t)x(t) + \displaystyle\int_{-\infty}^{t} c(t - s)x(s)ds + f(t), \ t > t_0, \ t \neq t_k, \\ \Delta x(t_k) = b_k x(t_k), \ k = \pm 1, \pm 2, ..., \end{cases} \tag{3.105}$$

where $a, c, f \in C[\mathbb{R}, \mathbb{R}]$ are almost periodic in the sense of Bohr. The function $f(t)$ is Lipschitz continuous in $\mathbb{R}$, $\{b_k\}$, $b_k \geq 0$, $k = \pm 1, \pm 2, ...$ is an almost periodic sequence of real number and the condition H3.40 for the sequence $\{t_k\} \in \mathcal{B}$ holds.

Let

$$-a(t) + M \int_0^\infty |c(u)|du \leq -\lambda,$$

where $\lambda > 0$ and $M = \displaystyle\prod_{k=\pm 1, \pm 2, ...} (1 + b_k)$.

For the function $V(t, x, y) = |x| + |y|$ and (3.105) the conditions 2 and 3 of Theorem 3.10 hold, and from [110] it follows that for the (3.105) there exists a uniformly bounded solution.

Then, all the conditions of Theorem 3.10 hold, and consequently, for the (3.105) there exists a uniformly asymptotically stable almost periodic solution.

3.5   Uncertain Impulsive Dynamical Equations

In the present part sufficient conditions for the existence of almost periodic solutions of uncertain impulsive dynamical equations are obtained.

The investigations are carried out by means the concepts of uniformly positive definite matrix functions and Hamilton–Jacobi–Riccati inequalities.

We shall consider the following system of uncertain impulsive dynamical equations

$$\begin{cases} \dot{x} = f(t, x) + g(t, x), \ t \neq t_k, \\ \Delta x(t_k) = I_k(x(t_k)) + J_k(x(t_k)), \ k = \pm 1, \pm 2, ..., \end{cases} \tag{3.106}$$

where $t \in \mathbb{R}$, $\{t_k\} \in \mathcal{B}$, $f, g : \mathbb{R} \times \Omega \to \mathbb{R}^n$, $I_k, J_k : \Omega \to \mathbb{R}^n$, $k = \pm 1, \pm 2, ...$

The functions $g(t, x)$, $J_k(x)$ represent a structural uncertainty or a uncertain perturbation in the system (3.106) and are characterized by

$$g \in U_g = \big\{ g : \ g(t, x) = e_g(t, x) . \delta_g(t, x), \ ||\delta_g(t, x)|| \leq ||m_g(t, x)|| \big\},$$

and

$$J_k \in U_J = \big\{ J_k : \ J_k(x) = e_k(x) . \delta_k(x), \ ||\delta_k(x)|| \leq ||m_k(x)|| \big\}, \ k = \pm 1, \pm 2, ...,$$

where $e_g : \mathbb{R} \times \Omega \to \mathbb{R}^{n \times m}$, and $e_k : \Omega \to \mathbb{R}^{n \times m}$ are known matrix functions, whose entries are smooth functions of the state, and δ_g, δ_k are unknown vector-valued functions, whose norms are bounded, respectively, by the norms of vector-valued functions $m_g(t, x), m_k(x)$, respectively. Here $m_g : \mathbb{R} \times \Omega \to \mathbb{R}^m$, $m_k : \Omega \to \mathbb{R}^m$, $k = \pm 1, \pm 2, ...$ are given functions.

We denote by $x(t) = x(t; t_0, x_0)$, the solution of (3.106) with initial condition $t_0 \in \mathbb{R}$, $x(t_0^+) = x_0$.

Introduce the following conditions:

H3.41. The functions $f(t, x)$ and $e_g(t, x)$ are almost periodic in t uniformly with respect to $x \in \Omega$.

H3.42. The sequences $\{I_k(x)\}$ and $\{e_k(x)\}$, $k = \pm 1, \pm 2, ...$ are almost periodic uniformly with respect to $x \in \Omega$.

H3.43. The set of sequences $\{t_k^j\}$, $t_k^j = t_{k+j} - t_k$, $k = \pm 1, \pm 2, ..., j = \pm 1, \pm 2, ...$ is uniformly almost periodic, and $\inf_k t_k^1 = \theta > 0$.

Let the conditions H3.41–H3.43 hold, and let $\{s_m'\}$ be an arbitrary sequence of real numbers. Then there exists a subsequence $\{s_n\}$, $s_n = s_{m_n}'$ such that the system (3.106) moves to the system

$$\begin{cases} \dot{x} = f^s(t,x) + g^s(t,x), \ t \neq t_k^s, \\ \Delta x(t_k^s) = I_k^s(x(t_k^s)) + J_k^s(x(t_k^s)), \ k = \pm 1, \pm 2, ..., \end{cases} \tag{3.107}$$

The set of all systems at the form (3.107) we shall denote by $H(f, g, I_k, J_k, t_k)$.

Introduce the following condition.

H3.44. $f(t,0) = 0$, $\delta_g(t,0) = 0$, $I_k(0) = 0$ and $\delta_k(0) = 0$, for all $t \in \mathbb{R}$ and $k = \pm 1, \pm 2, ...$

We shall note that from the last condition it follows that $x(t) = 0$ is a solution of the system (3.106).

Definition 3.9 ([108]). The uncertain impulsive dynamical system (3.106) is said to be *uniformly robustly stable, uniformly robustly attractive, uniformly robustly asymptotically stable*, if for any $g \in U_g$, $J_k \in U_J$, $k = \pm 1, \pm 2, ...$ the trivial solution $x(t) = 0$ is uniformly stable, uniformly attractive, uniformly asymptotically stable, respectively.

Next we shall use the classes V_1, V_2 and W_0, which are defined in Chap. 1. Let $V \in V_1$, $t \neq t_k$, $x \in PC[\mathbb{R}, \Omega]$, $y \in PC[\mathbb{R}, \Omega]$.

Introduce

$$D^+V(t, x(t), y(t)) - \lim_{\delta \to 0^+} \sup \frac{1}{\delta} \Big\{ V(t + \delta, x(t) + \delta f(t, x(t)), y(t) + \delta f(t, y(t)))$$

$$- V(t, x(t), y(t)) \Big\}. \tag{3.108}$$

For the proof of the main results we shall use the following nominal system of the system (3.106)

$$\begin{cases} \dot{x} = f(t,x), \ t \neq t_k, \\ \Delta x(t_k) = I_k(x(t_k)), \ k = \pm 1, \pm 2, ... \end{cases} \tag{3.109}$$

Definition 3.10. The matrix function $X : \mathbb{R} \to \mathbb{R}^{n \times n}$ is said to be:

(a) *A positive define matrix function*, if for any $t \in \mathbb{R}$, $X(t)$ is a positive define matrix.

(b) *A positive define matrix function bounded above*, if it is a positive definite matrix function, and there exists a positive real number $M > 0$ such that

$$\lambda_{max}(X(t)) \leq M, \ t \in R,$$

where $\lambda_{max}(X(t))$ is the maximum eigenvalue.

(c) *An uniformly positive define matrix function*, if it is a positive definite matrix function, and there exists a positive real number $m > 0$ such that

$$\lambda_{min}(X(t)) \geq m, \ t \in \mathbb{R},$$

where $\lambda_{min}(X(t))$ is the minimum eigenvalue.

The proof of the following lemma is obvious.

Lemma 3.8. *Let $X(t)$ be a positive define matrix function, and $Y(t)$ be a symmetric matrix.*

Then for any $x \in \mathbb{R}^n$, $t \in \mathbb{R}$ the following inequality holds

$$x^T Y(t) x \leq \lambda_{max}(X^{-1}(t)Y(t)) x^T X(t) x. \tag{3.110}$$

We shall use the next lemma.

Lemma 3.9 ([108]). *Let $\Sigma(t)$ be a diagonal matrix function.*

Then for any positive scalar function $\lambda(t)$ and for any $\xi, \eta \in \mathbb{R}^n$, the following inequality holds

$$2\xi^T \Sigma(t) \eta \leq \lambda^{-1}(t) \xi^T \xi + \lambda(t) \eta^T \eta. \tag{3.111}$$

Now we shall prove the main theorem.

Theorem 3.11. *Let the following conditions hold:*

1. *Conditions H3.41–H3.44 are met.*
2. *There exist functions $V \in V_2$ and $a, b \in K$ such that*

$$a(||x - y||) \leq V(t, x, y) \leq b(||x - y||), \ (t, x, y) \in \mathbb{R} \times \Omega \times \Omega.$$

3. *There exist positive define matrix functions $G_{1k} : \mathbb{R} \times \mathbb{R}^n \times \mathbb{R}^n \to \mathbb{R}^{1 \times m}$, $G_{2k} : \mathbb{R} \times \mathbb{R}^n \times \mathbb{R}^n \to \mathbb{R}^{m \times m}$ and for $t \in \mathbb{R}$, $k = \pm 1, \pm 2, ...$, $x, y \in PC^1[\mathbb{R}, \Omega]$, $z \in \mathbb{R}^m$ it follows*

$$V\Big(t, x(t) + I_k(x(t)) + e_k(x(t))z, y(t) + I_k(y(t)) + e_k(y(t))z\Big)$$

$$\leq V\Big(t, x(t) + I_k(x(t)), y(t) + I_k(y(t))\Big) + G_{1k}(t, x(t), y(t))z$$

$$+ z^T G_{2k}(t, x(t), y(t))z. \tag{3.112}$$

4. *There exist positive constants* χ_k, $k = \pm 1, \pm 2, \ldots$ *such that*

$$V\left(t_k^+, x(t_k) + I_k(x(t_k)), y(t_k) + I_k(y(t_k))\right)$$

$$+ \chi_k^{-1} G_{1k} G_{1k}^T + \left(\chi_k + \lambda_{max}(G_{2k})\right) m_k^T m_k$$

$$\leq V(t_k, x(t_k), y(t_k)), \tag{3.113}$$

where $G_{1k} = G_{1k}(t_k, x(t_k), y(t_k))$, $G_{2k} = G_{2k}(t_k, x(t_k), y(t_k))$, $m_k(x(t_k)) = m_k$.

5. *There exists a constant* $c > 0$ *and scalar functions* $\lambda_k \in C[\mathbb{R}^n, \mathbb{R}^+]$ *such that for* $t \in \mathbb{R}$, $t \neq t_k$, $k = \pm 1, \pm 2, \ldots$, $x, y \in \Omega$ *it follows*

$$\frac{\partial V}{\partial t} + \left(\frac{\partial V}{\partial x} + \frac{\partial V}{\partial y}\right) f + \frac{\lambda_k^2}{2}\left(\frac{\partial V}{\partial x} + \frac{\partial V}{\partial y}\right) e_g e_g^T \left(\frac{\partial V}{\partial x} + \frac{\partial V}{\partial y}\right)^T + \frac{1}{2\lambda_k^2} m_g^T m_g$$

$$\leq -cV(t, x, y). \tag{3.114}$$

6. *There exists a solution* $x(t; t_0, x_0)$ *of* (3.106) *such that*

$$\|x(t; t_0, x_0)\| < \alpha_1, \ where \ \alpha_1 < \alpha, \ \alpha > 0.$$

Then, in B_α *for the system* (3.106) *there exists a unique almost periodic solution* $\omega(t)$ *such that*:

1. $\|\omega(t)\| \leq \alpha_1$.
2. $\omega(t)$ *is uniformly robustly asymptotically stable.*
3. $H(\omega, t_k) \subset H(f, g, I_k, J_k, t_k)$.

Proof. From (3.110), (3.111), (3.112), for $t = t_k$, $k = \pm 1, \pm 2, \ldots$, we have

$$V\left(t_k^+, x(t_k) + I_k(x(t_k)) + J_k(x(t_k)), y(t_k) + I_k(y(t_k)) + J_k(y(t_k))\right)$$

$$\leq V\left(t_k^+, x(t_k) + I_k(x(t_k)), y(t_k) + I_k(y(t_k))\right) + G_{1k}\delta(x(t_k))$$

$$+ \delta(x(t_k))^T G_{2k}\delta(x(t_k))$$

$$\leq V\left(t_k^+, x(t_k) + I_k(x(t_k)), y(t_k) + I_k(y(t_k))\right)$$

$$+ \chi_k^{-1} G_{1k} G_{1k}^T + \left(\chi_k + \lambda_{max}(G_{2k})\right) m_k^T m_k$$

$$\leq V(t_k, x(t_k), y(t_k)). \tag{3.115}$$

On the other hand, for $t \neq t_k$, $k = \pm 1, \pm 2, \ldots$, from (3.108) and (3.114), we get

$$D^+V(t, x(t), y(t)) = \frac{\partial V}{\partial t} + \left(\frac{\partial V}{\partial x} + \frac{\partial V}{\partial y}\right)(f + g)$$

$$= \frac{\partial V}{\partial t} + \left(\frac{\partial V}{\partial x} + \frac{\partial V}{\partial y}\right)f + \left(\frac{\partial V}{\partial x} + \frac{\partial V}{\partial y}\right)e_g \delta_g$$

$$= \frac{\partial V}{\partial t} + \left(\frac{\partial V}{\partial x} + \frac{\partial V}{\partial y}\right)f + \frac{\lambda_k^2}{2}\left(\frac{\partial V}{\partial x} + \frac{\partial V}{\partial y}\right)e_g e_g^T\left(\frac{\partial V}{\partial x} + \frac{\partial V}{\partial y}\right)^T$$

$$+ \frac{1}{2\lambda_k^2}m_g m_g^T - \frac{1}{2}\left\{\lambda_k\left(\frac{\partial V}{\partial x} + \frac{\partial V}{\partial y}\right)e_g - \frac{1}{\lambda_k}\delta_g^T\right\}\left\{\lambda_k e_g^T\left(\frac{\partial V}{\partial x} + \frac{\partial V}{\partial y}\right)^T e_g$$

$$- \frac{1}{\lambda_k}\delta_g\right\} - \frac{1}{2\lambda_k^2}\left\{m_g^T m_g - \delta_g^T \delta_g\right\}$$

$$\leq \frac{\partial V}{\partial t} + \left(\frac{\partial V}{\partial x} + \frac{\partial V}{\partial y}\right)f + \frac{\lambda_k^2}{2}\left(\frac{\partial V}{\partial x} + \frac{\partial V}{\partial y}\right)e_g e_g^T\left(\frac{\partial V}{\partial x} + \frac{\partial V}{\partial y}\right) + \frac{1}{2\lambda_k^2}m_g^T m_g$$

$$\leq -cV(t, x(t), y(t)). \tag{3.116}$$

Then from (3.115), (3.116) and conditions of the theorem it follows that for the system (3.106) the conditions of Theorem 3.2 are satisfied, and hence, the proof of the theorem is complete.   □

Now, we shall consider the linear system of uncertain impulsive dynamical equations

$$\begin{cases} \dot{x} = A(t)x + B(t)x, \ t \neq t_k, \\ \Delta x(t_k) = A_k(t_k)x(t_k) + B_k(t_k)x(t_k), \ k = \pm 1, \pm 2, ..., \end{cases} \tag{3.117}$$

where $t \in \mathbb{R}$, $\{t_k\} \in \mathcal{B}$, $A, A_k : \mathbb{R} \to \mathbb{R}^{n \times n}$, $k = \pm 1, \pm 2, ...$ are known matrix functions, and $B, B_k : \mathbb{R} \in \mathbb{R}^{n \times n}$, $k = \pm 1, \pm 2, ...$ are interval matrix functions, i.e. $B(t) \in IN[P(t), Q(t)] = \Big\{B(t) \in \mathbb{R}^{n \times n} : B(t) = (b_{ij}(t)), \ p_{ij}(t) \leq b_{ij}(t) \leq q_{ij}(t), \ i, j = 1, 2, ..., n\Big\}$. $B_k(t_k) \in IN[P_k(t_k), Q_k(t)]$, $k = \pm 1, \pm 2, ...,$ where $P(t) = (p_{ij}(t))$, $Q(t) = (q_{ij}(t))$, and $P_k(t), Q_k(t)$, $k = \pm 1, \pm 2, ...$ are known matrices.

Introduce the following conditions:

H3.45. The matrix functions $A(t)$, $P(t)$, $Q(t)$ are almost periodic.
H3.46. The sequences $A_l(t_k)$, $P_l(t_k)$, $Q_l(t_k)$, $l = \pm 1, \pm 2, ...$, $k = \pm 1, \pm 2, ...$ are almost periodic for any $k = \pm 1, \pm 2, ...$

Lemma 3.10. *[108] Let $B(t) \in IN[P(t), Q(t)]$, where $P(t)$, $Q(t)$ be known matrices.*

Then $B(t)$ can be written

$$B(t) = B_0(t) + E(t)\Sigma(t)F(t),$$

where:

$$B_0(t) = \frac{1}{2}(P(t) + Q(t)),$$

$$\Sigma(t) = diag\{\varepsilon_{11}(t), ..., \varepsilon_{1n}(t), ..., \varepsilon_{n1}(t), ..., \varepsilon_{nn}(t)\} \in \mathbb{R}^{n^2 \times n^2},$$

$$||\varepsilon_{ij}(t)|| \leq 1, \ i, j = 1, 2, ..., n,$$

$$H(t) = (h_{ij}(t)) = \frac{1}{2}(Q(t) - P(t)), \ h_{ij}(t) \geq 0, \ t \in \mathbb{R}, \ i, j = 1, 2, ..., n,$$

$$E(t) = \left(\sqrt{h_{11}(t)}e_1, ..., \sqrt{h_{1n}(t)}e_1, ..., \sqrt{h_{n1}(t)}e_n, ..., \sqrt{h_{nn}(t)}e_n\right) \in \mathbb{R}^{n \times n^2},$$

$$F(t) = \left(\sqrt{h_{11}(t)}e_1, ..., \sqrt{h_{1n}(t)}e_n, ..., \sqrt{h_{n1}(t)}e_1, ..., \sqrt{h_{nn}(t)}e_n\right)^T \in \mathbb{R}^{n^2 \times n},$$

$$e_i(0, ..., 0, 1, 0, ..., 0)^T \in \mathbb{R}^n, \ i = 1, 2, ..., n.$$

By Lemma 3.10, we rewrite the system (3.117) in the form

$$\begin{cases} \dot{x} = A_0(t)x + E(t)\Sigma(t)F(t)x, \ t \neq t_k, \\ \Delta x(t_k) = \tilde{A}_k(t_k)x(t_k) + \tilde{E}_k(t_k)\tilde{\Sigma}_k(t_k)\tilde{F}_k(t_k)x(t_k), \end{cases} \tag{3.118}$$

where

$$k = \pm 1, \pm 2, ...,$$

$$A_0(t) = A(t) + B_0(t), \ \tilde{A}_k(t) = A_k(t) + \tilde{B}_{k0}(t),$$

$$B_k(t) = \tilde{B}_{k0}(t) + \tilde{E}_k(t)\tilde{\Sigma}_k(t)\tilde{F}_k(t), \ k = \pm 1, \pm 2, ...,$$

and B_0, $\tilde{E}_k(t)$, $\tilde{\Sigma}_k(t)$, $\tilde{F}_k(t)$ are defined in Lemma 3.10.

Now, we shall prove the next theorem.

Theorem 3.12. *Let the following conditions hold:*

1. *Conditions H3.41, H3.43 and H3.45, H3.46 are met.*
2. *There exist scalar functions $\lambda(t) > 0$, $\alpha(t) > 0$ and an uniformly positive matrix function $X(t)$ bounded above such that:*

   (a) *$X(t)$ is differentiable at $t \neq t_k$ and the Riccati inequality holds:*

   $$\dot{X} + XA_0 + A_0^T X + \lambda^{-1}XEE^T X + \lambda F^T F \leq -\alpha X, \tag{3.119}$$

   for $t \neq t_k$, $k = \pm 1, \pm 2, ...,$
   (b) *There exist some $r_k \in \mathbb{R}$ and positive constants χ_k, $k = \pm 1, \pm 2, ...$ such that*

$$\int_{t_k}^{t_{k+1}} \alpha(s)ds + \ln \beta_k \le -r_k, \quad k = \pm 1, \pm 2, ..., \qquad (3.120)$$

where

$$\beta_k = \lambda_{max}\Big\{ X^{-1}(t_k)\Big[(E + A_k^T(t_k))$$
$$+ \chi_k^{-1} X(t_k)\tilde{E}_k(t_k)\tilde{E}_k^T(t_k)X(t_k)\big(E + A_k(t_k)\big)$$
$$+ \Big(\chi_k + \lambda_{max}\tilde{E}_k^T(t_k)X(t_k)\tilde{E}_k(t_k)\tilde{F}_k^T(t_k)\Big)\tilde{F}_k(t_k)\Big]\Big\} \le 1,$$

where E is an identity in $\mathbb{R}^{n \times n}$.
3. *There exists a solution $x(t; t_0, x_0)$ of (3.117) such that*

$$||x(t; t_0, x_0)|| < \nu_1, \text{ where } \nu_1 < \nu, \ \nu > 0.$$

Then, if $\displaystyle\sum_{k=1}^{\infty} r_k = \infty$ for the system (3.117), there exists an almost periodic solution $\omega(t)$ such that:

1. $||\omega(t)|| \le \nu_1$.
2. $H(\omega, t_k) \subset H(A, B, A_k, B_k, t_k)$.
3. $\omega(t)$ *is uniformly robustly asymptotically stable.*

Proof. Let $V(t, x, y) = (x + y)^T X(t)(x + y)$. Then $V \in V_2$, and

$$\lambda_{min}(X(t))\Big(||x(t)||^2 + ||y(t)||^2\Big) \le V \le \lambda_{max}(X(t))\Big(||x(t)||^2 + ||y(t)||^2\Big),$$

where $(t, x(t), y(t)) \in \mathbb{R} \times B_\nu \times B_\nu$.

The matrix $X(t)$ is an uniformly positive define matrix function and is bounded above. Then, we have positive numbers $M \ge m > 0$ such that

$$m \le \lambda_{min}(X(t)) \le \lambda_{max}(X(t)) \le M,$$

and for $a(t) = mt^2$, $b(t) = Mt^2$, $t \in \mathbb{R}$, $a, b \in K$, it follows that

$$a(||x(t) - y(t)||) \le V(t, x(t), y(t)) \le b(||x(t) - y(t)||). \qquad (3.121)$$

Similar to the proofs of (3.115) and (3.116), from (3.119) and (3.120) we get

$$V\Big(t_k^+, x(t_k) + A_k(x(t_k)) + B_k(x(t_k)), y(t_k) + A_k(y(t_k)) + B_k(y(t_k))\Big)$$
$$\le \beta_k V(t_k, x(t_k), y(t_k)) \le V(t_k, x(t_k), y(t_k)), \qquad (3.122)$$

and

$$D^+V(t, x(t), y(t)) \leq -cV(t, x(t), y(t)), \qquad (3.123)$$

where $t \neq t_k$, $k = \pm 1, \pm 2, ...,$ $0 < c \leq \alpha(t)$.

Then, from (3.121), (3.112) and (3.123) it follows that for the system (3.117), the conditions of Theorem 3.11 hold, and the proof of Theorem 3.12 is complete. $\qquad \square$

Chapter 4
Applications

In this chapter, we shall consider the some applications to the real world problems to illustrate the theory developed in the previous chapters.

Section 4.1 will offer some impulsive biological models. We shall consider conditions for the existence of almost periodic solutions for an impulsive Lasota–Wazewska model, an impulsive model of hematopoiesis, and an impulsive delay logarithmic population model.

Section 4.2 will deal with conditions for the existence of almost periodic solutions of different kinds of n-species Lotka–Volterra type impulsive models.

Section 4.3 we shall present impulsive neural networks. By means of Lyapunov functions sufficient conditions for the existence of almost periodic solutions will be established.

4.1   Biological Models

4.1.1   An Impulsive Lasota–Wazewska Model

The main problem of this paragraph is to study the following generalized system of impulsive differential equations with delay of Lasota–Wazewska type:

$$\begin{cases} \dot{x}(t) = -\alpha(t)x(t) + \displaystyle\sum_{i=1}^{n} \beta_i(t)e^{-\gamma_i(t)x(t-h)}, \ t \neq t_k, \\ \Delta x(t_k) = \alpha_k x(t_k) + \nu_k, \ k = \pm 1, \pm 2, \ldots, \end{cases} \tag{4.1}$$

where $t \in \mathbb{R}$, $\{t_k\} \in \mathcal{B}$, $\alpha(t)$, $\beta_i(t)$, $\gamma_i(t) \in C[\mathbb{R}, \mathbb{R}^+]$, $i = 1, 2, \ldots, n$, $h = const > 0$, and the constants $\alpha_k \in \mathbb{R}$, $\nu_k \in \mathbb{R}$, $k = \pm 1, \pm 2, \ldots$.

We shall note that in the special cases when α, β, γ are positive constants, the differential equations with delay and without impulses in the form

$$\dot{u} = -\alpha u(t) + \beta e^{-\gamma u(t-h)},$$

G.T. Stamov, *Almost Periodic Solutions of Impulsive Differential Equations*, Lecture Notes in Mathematics 2047, DOI 10.1007/978-3-642-27546-3_4, © Springer-Verlag Berlin Heidelberg 2012

are considered by Wazewska-Czyzewska and Lasota [186]. The aim is an
investigation of the development and survival of the red corpuscles in the
organisms.

The late investigations in this area are in the work of Kulenovic and Ladas
[89] for the oscillations of the last equations and studying of the equation

$$\dot{u} = -\mu u(t) + \sum_{i=1}^{n} p_i e^{-r_i u(t-h)},$$

where μ, p_i and r_i are positive constants.

Let $t_0 \in \mathbb{R}$. Introduce the following notation:

$PC(t_0)$ is the space of all functions $\phi : [t_0 - h, t_0] \to \Omega$ having points
of discontinuity at $\theta_1, \theta_2, \ldots, \theta_s \in (t_0 - h, t_0)$ of the first kind and are left
continuous at these points.

Let ϕ_0 be an element of $PC(t_0)$. Denote by $x(t) = x(t; t_0, \phi_0)$, $x \in \Omega$, the
solution of system (4.1), satisfying the initial conditions:

$$\begin{cases} x(t; t_0, \phi_0) = \phi_0(t), \ t_0 - h \ \leq \ t \leq \ t_0, \\ x(t_0^+; t_0, \phi_0) = \phi_0(t_0). \end{cases} \tag{4.2}$$

Together with (4.1), we consider the linear system

$$\begin{cases} \dot{x}(t) = -\alpha(t)x(t), \ t \neq t_k, \\ \Delta x(t_k) = \alpha_k x(t_k), \ k = \pm 1, \pm 2, \ldots. \end{cases} \tag{4.3}$$

Introduce the following conditions:

H4.1.   The function $\alpha(t)$ is almost periodic in the sense of Bohr, and there
exists a positive constant α such that $\alpha \leq \alpha(t)$.

H4.2.   The sequence $\{\alpha_k\}$ is almost periodic and $-1 < \alpha_k \leq 0$, $k = \pm 1, \pm 2, \ldots$.

H4.3.   The set of sequences $\{t_k^j\}$, $t_k^j = t_{k+j} - t_k$, $k = \pm 1, \pm 2, \ldots$, $j = \pm 1, \pm 2, \ldots$ is uniformly almost periodic, and there exists $\theta > 0$ such that
$\inf_k t_k^1 = \theta > 0$.

Now, we shall consider the equations

$$\dot{x}(t) = -\alpha(t)x(t), \ \ t_{k-1} < t \leq t_k$$

and their solutions

$$x(t) = x(s)exp\{-\int_{0}^{t} \alpha(\sigma)d\sigma\}$$

for $t_{k-1} < s \leq t \leq t_k$, $k = \pm 1, \pm 2, \ldots$.

Then by the definition of the Cauchy matrix for the linear equation (1.13) at Chap. 1, we obtain for (4.3) the matrix

$$W(t,s) = \begin{cases} exp\{ -\int_s^t \alpha(\sigma)d\sigma \}, \ t_{k-1} < s \le t \le t_k, \\ \prod_{i=m}^{k+1}(1+\alpha_i)exp\{ -\int_s^t \alpha(\sigma)d\sigma \}, \\ t_{m-1} < s \le t_m < t_k < t \le t_{k+1}. \end{cases}$$

Then the solutions of (4.3) are in the form

$$x(t; t_0, x_0) = W(t, t_0)x_0, \ t_0, x_0 \in \mathbb{R}.$$

Introduce the following conditions:

H4.4. The functions $\beta_i(t)$ are almost periodic in the sense of Bohr, and

$$0 < \sup_{t \in \mathbb{R}} |\beta_i(t)| < B_i, \ B_i > 0, \ \beta_i(0) = 0, \ i = 1, 2, \ldots, n.$$

H4.5. The functions $\gamma_i(t)$, $i = 1, 2, \ldots, m$ are almost periodic in the sense of Bohr, and

$$0 < \sup_{t \in \mathbb{R}} |\gamma_i(t)| < G_i, \ G_i > 0, \ \gamma_i(0) = 0, \ i = 1, 2, \ldots, n.$$

H4.6. The sequence $\{\nu_k\}$, $k = \pm 1, \pm 2, \ldots$ is almost periodic.

In the proof of the main theorem we shall use the following lemma the proof of which is similar to the proof of Lemma 1.7.

Lemma 4.1. *Let conditions H4.1–H4.6 hold. Then for each $\varepsilon > 0$ there exist ε_1, $0 < \varepsilon_1 < \varepsilon$, a relatively dens sets $\overline{T}$ of real numbers, and a set P of integer numbers such that the following relations are fulfilled:*

(a) $|\alpha(t+\tau) - \alpha(t)| < \varepsilon, \ t \in \mathbb{R}, \ \tau \in \overline{T}.$
(b) $|\beta_i(t+\tau) - \beta_i(t)| < \varepsilon, \ t \in \mathbb{R}, \ \tau \in \overline{T}.$
(c) $|\gamma_i(t+\tau) - \gamma_i(t)| < \varepsilon, \ t \in \mathbb{R}, \ \tau \in \overline{T}.$
(d) $|\alpha_{k+q} - \alpha_k| < \varepsilon, \ q \in P, \ k = \pm 1, \pm 2, \ldots.$
(e) $|\nu_{k+q} - \nu_k| < \varepsilon, \ q \in P, \ k = \pm 1, \pm 2, \ldots.$
(f) $|t_k^q - r| < \varepsilon_1, \ q \in P, \ r \in \overline{T}, \ k = \pm 1, \pm 2, \ldots.$

Lemma 4.2. *Let conditions H4.1–H4.3 hold.*
   Then:

1. *For the Cauchy matrix $W(t,s)$ of system (4.3) it follows*

$$|W(t,s)| \leq e^{-\alpha(t-s)}, \ t \geq s, \ t,s \in \mathbb{R}.$$

2. *For any $\varepsilon > 0$, $t \in \mathbb{R}$, $s \in \mathbb{R}$, $t \geq s$, $|t - t_k| > \varepsilon$, $|s - t_k| > \varepsilon$, $k = \pm 1, \pm 2, \ldots$ there exists a relatively dense set $\overline{T}$ of ε-almost periods of the function $\alpha(t)$ and a positive constant Γ such that for $\tau \in \overline{T}$ it follows*

$$|W(t+\tau, s+\tau) - W(t,s)| \leq \varepsilon \Gamma e^{-\frac{\alpha}{2}(t-s)}.$$

Proof. Since the sequence $\{\alpha_k\}$ is almost periodic, then it is bounded and from H4.2 it follows that $(1 + \alpha_k) \leq 1$.

From the presentation of $W(t,s)$ and last inequality it follows that

$$|W(t,s)| \leq e^{-\alpha(t-s)}, \ t \geq s, \ t,s \in \mathbb{R}.$$

Consider the sets $\overline{T}$ and P determined by Lemma 4.1, and let $\tau \in \overline{T}$. Then for the matrix $W(t+\tau, s+\tau)$, we have

$$\frac{\partial W}{\partial t} = -\alpha(t)W(t+\tau, s+\tau) + \big(\alpha(t) - \alpha(t+\tau)\big)W(t+\tau, s+\tau), \ t \neq t'_k,$$

$$\Delta W(t'_k, s) = \alpha_k W(t_k + \tau, s+\tau) + \big(\alpha_{k+q} - \alpha_k\big)W(t'_k + \tau, s+\tau),$$

where $t'_k = t_k - q$, $q \in P$, $k = \pm 1, \pm 2, \ldots$.
   Then

$$W(t+\tau, s+\tau) - W(t,s)$$

$$= \int_s^t W(t,\sigma)\big(\alpha(\sigma) - \alpha(\sigma+\tau)\big)W(\sigma+\tau, s+\tau)d\sigma$$

$$+ \sum_{s<t'_k<t} W(t, t'^+_k)\big(\alpha_{k+q} - \alpha_k\big)W(t'_k + \tau, s+\tau), \qquad (4.4)$$

and again from Lemma 4.1 it follows that if $|t - t'_k| > \varepsilon$, then

$$t_{h+q} < t + \tau < t'_{h+q+1}.$$

From(4.4), we obtain

$$|W(t + \tau, s + \tau) - W(t, s)| < \varepsilon(t - s)e^{-\alpha(t-s)} + \varepsilon i(s, t)e^{-\alpha(t-s)} \qquad (4.5)$$

for $|t - t_k'| > \varepsilon$, $|s - t_k'| > \varepsilon$, where $i(s, t)$ is the number of points t_k' in the interval (s, t).

Now from Lemma 4.2, (4.5) and the obvious inequality

$$\frac{t - s}{2} \leq e^{-\frac{\alpha}{2}(t-s)},$$

we obtain

$$|W(t + \tau, s + \tau) - W(t, s)| < \varepsilon \Gamma e^{-\frac{\alpha}{2}(t-s)},$$

where $\Gamma = \dfrac{2}{\alpha}\left(1 + N + \dfrac{\alpha}{2}N\right)$. $\square$

We shall proof the main theorem of this part.

Theorem 4.1. *Let the following conditions hold:*

1. *Conditions H4.1–H4.6 are met.*
2. *The following inequality holds*

$$\sum_{i=1}^{n} B_i < \alpha.$$

Then:

(1)   There exists a unique almost periodic solution $x(t)$ of (4.1).
(2)   The solution $x(t)$ is exponentially stable.

Proof. We denote by AP the set of all almost periodic functions $\varphi(t)$, $\varphi \in PC[\mathbb{R}, \mathbb{R}^+]$, satisfying the inequality $|\varphi|_\infty < K$, where

$$K = \frac{1}{\alpha}\sum_{i=1}^{n} B_i + \sup_{k=\pm 1, \pm 2, \ldots} |\nu_k| \frac{1}{1 - e^{-\alpha}}.$$

Here we denote

$$|\varphi|_\infty = \sup_{t \in \mathbb{R}} |\varphi(t)|.$$

We define in AP an operator S such that if $\varphi \in AP$,

$$S\varphi = \int_{-\infty}^{t} W(t, s) \sum_{i=1}^{n} \beta_i(s)e^{-\gamma_i(s)\varphi(s-h)} ds + \sum_{t_k < t} W(t, t_k)\nu_k. \qquad (4.6)$$

For an arbitrary $\varphi \in AP$ it follows

$$|S\varphi|_\infty = \int_{-\infty}^t |W(t,s)| \sum_{i=1}^n |\beta_i(s)| e^{-\gamma_i(s)\varphi(s-h)} ds + \sum_{t_k<t} |W(t,t_k)||\nu_k|$$

$$< \frac{1}{\alpha} \sum_{i=1}^n B_i + \sup_{k=\pm 1, \pm 2, \ldots} |\nu_k| \frac{1}{1-e^{-\alpha}} = K. \tag{4.7}$$

On the other hand, let $\tau \in \overline{T}$, $q \in P$ where the sets $\overline{T}$ and P are determined in Lemma 4.1. Then

$$|S\varphi(t+\tau) - S\varphi(t)|_\infty$$

$$\leq \int_{-\infty}^t |W(t+\tau, s+\tau) - W(t,s)| \sum_{i=1}^n |\beta_i(s+\tau)| e^{-\gamma_i(s+\tau)\varphi(s+\tau-h)} ds$$

$$+ \int_{-\infty}^t |W(t,s)| \left| \sum_{i=1}^n |\beta_i(s+\tau)| e^{-\gamma_i(s+\tau)\varphi(s+\tau-h)} \right.$$

$$\left. - \sum_{i=1}^n |\beta_i(s)| e^{-\gamma_i(s)\varphi(s-h)} \right| ds + \sum_{t_k<t} |W(t+\tau, t_{k+q}) - W(t,t_k)||\nu_{k+q}|$$

$$+ \sum_{t_k<t} |W(t,t_k)||\nu_{k+q} - \nu_k| \leq C_1 \varepsilon, \tag{4.8}$$

where

$$C_1 = \frac{1}{\alpha} \sum_{i=1}^n B_i \left\{ 2\Gamma + \sum_{i=1}^n (B_i + \alpha G_i) \right\} + \varepsilon \left(\Gamma \sup_{k=\pm 1, \pm 2, \ldots} |\nu_k| + \frac{1}{1-e^{-\alpha}} \right).$$

From (4.7) and (4.8), we obtain that $S\varphi \in AP$.
Let $\varphi \in AP$, $\psi \in AP$.
We get

$$|S\varphi - S\psi| \leq \int_{-\infty}^t |W(t,s)| \sum_{i=1}^n |\beta_i(s)| \left| e^{-\gamma_i(s)\varphi(s-h)} - e^{-\gamma_i(s)\psi(s-h)} \right| ds$$

$$\leq \frac{1}{\alpha} \sum_{i=1}^n B_i |\varphi - \psi|_\infty. \tag{4.9}$$

Then from (4.9) and the conditions of Theorem 4.1 it follows that S is a contracting operator in AP. So, there exists a unique almost periodic solution of (4.1).

Now, let $y(t)$ be another solution of (4.1) with the initial conditions

$$\begin{cases} y(t; t_0, \varpi_0) = \varpi_0, \ t_0 - h \leq t \leq t_0, \\ y(t_0^+; t_0, \varpi_0) = \varpi_0, \end{cases}$$

where $\varpi_0 \in PC(t_0)$.
  Then

$$y(t) - x(t) = W(t, t_0)(\varpi_0 - \phi_0)$$

$$+ \int_{t_0}^{t} W(t, s) \sum_{i=1}^{n} \beta_i(s) \left(e^{-\gamma_i(s)x(s-h)} - e^{-\gamma_i(s)y(s-h)} \right) ds.$$

Consequently,

$$|y(t) - x(t)| \leq e^{-\alpha(t-t_0)}|\varpi_0 - \phi_0| + \int_{t_0}^{t} e^{-\alpha(t-s)} \sum_{i=1}^{n} B_i |y(s) - x(s)| ds.$$

Set $u(t) = |y(t) - x(t)|e^{\alpha t}$ and from Gronwall–Bellman's inequality and Theorem 1.9, we have

$$|y(t) - x(t)| \leq |\varpi_0 - \phi_0| \exp \left\{ -(\alpha - \sum_{i=1}^{m} B_i)(t - t_0) \right\}.$$

and the proof of Theorem 4.1 is complete. $\qquad\qquad\qquad\qquad\square$

Example 4.1. We consider the linear impulsive delay differential equation in the form

$$\begin{cases} \dot{x}(t) = -\alpha(t)x(t) + \beta(t)e^{-\gamma(t)x(t-h)}, \ t \neq t_k, \\ \Delta x(t_k) = \alpha_k x(t_k), \end{cases} \qquad (4.10)$$

where $t \in \mathbb{R}$, α, β, $\gamma \in C[\mathbb{R}, \mathbb{R}^+]$, $h > 0$, $\{t_k\} \in \mathcal{B}$, and the constants $\alpha_k \in \mathbb{R}$, $k = \pm 1, \pm 2, \ldots$.

Corollary 4.1. *Let the following conditions hold:*

1. The functions $\alpha(t)$, $\beta(t)$, $\gamma(t)$ are almost periodic.
2. Conditions H4.2 and H4.3 are met.

Then if $\sup\limits_{t \in \mathbb{R}} \alpha(t) > \sup\limits_{t \in \mathbb{R}} \beta(t)$, there exists a unique almost periodic exponentially stable solution $x(t)$ of (4.10).

Proof. The proof follows from Theorem 4.1. $\qquad\qquad\qquad\qquad\square$

Remark 4.1. The results in this part show that by means of appropriate impulsive perturbations we can control the almost periodic dynamics of these equations.

4.1.2   An Impulsive Model of Hematopoiesis

In this part of the paragraph the existence and asymptotic stability of positive almost periodic solution for nonlinear impulsive delay model of hematopoiesis is investigated.

The nonlinear delay differential equation

$$\dot{h}(t) = -\alpha h(t) + \frac{\beta}{1 + h^n(t - \omega)}, \quad t \ge t_0, \qquad (4.11)$$

where $\alpha > 0$, $\beta > 0$, $\omega > 0$, $n \in \mathbb{N}$ has been proposed by Mackey and Glass [112] as an appropriate model of hematopoiesis that describes the process of production of all types of blood cells generated by a remarkable self-regulated system that is responsive to the demands put upon it. In medical terms, $h(t)$ denotes the density of mature cells in blood circulation at time t and τ is the time delay between the production of immature cells in the bone marrow and their maturation for release in circulating bloodstream. It is assumed [190] that the cells are lost from the circulation at a rate α and the flux of the cells into the circulation from the stem cell compartment depends on the density of mature cells at the previous time $t - \tau$.

In the real world phenomena, the parameters can be nonlinear functions. The variation of the environment, however, plays an important role in many biological and ecological dynamical systems. In particular, the effects of a periodically varying environment are important for evolutionary theory as the selective forces on systems in a fluctuating environment differ from those in a stable environment. Thus, the assumption of periodicity of the parameters are a way of incorporating the periodicity of the environment. It has been suggested by Nicholson [123] that any periodical change of climate tends to impose its period upon oscillations of internal origin or to cause such oscillations to have a harmonic relation to periodic climatic changes.

On the other hand, some dynamical systems which describe real phenomena are characterized by the fact that at certain moments in their evolution they undergo rapid changes. Most notably this takes place due to certain seasonal effects such as weather, resource availability, food supplies, mating habits, etc. These phenomena are best described by the system of impulsive differential equations of the form

$$\begin{cases} \dot{h}(t) = -\alpha(t)h(t) + \dfrac{\beta(t)}{1 + h^n(t - \omega)}, & t \neq t_k, \\ \Delta h(t_k) = \gamma_k h(t_k) + \delta_k, & k \in \mathbb{N}, \end{cases} \quad (4.12)$$

where t_k represent the instants $h(t)$ at which the density suffers an increment of δ_k units. The density of mature cells in blood circulation decreases at prescribed instants t_k by some medication and it is proportional to the density at that time t_k.

Let $\{t_k\} \in \mathcal{B}$, $\sigma > 0$ and $\sigma \leq t_1 < t_2 < \ldots$. For a given initial function $\xi \in PC[[\sigma - \omega, \sigma], \mathbb{R}^+]$, it is well known [177] that the system (4.12) has a unique solution $h(t) = h(t; \sigma, \xi)$ defined on $[\sigma - \omega, \infty)$, and satisfying the initial conditions

$$\begin{cases} h(t; \sigma, \xi) = \xi(t), & \sigma - \omega \leq t \leq \sigma, \\ h(\sigma^+; \sigma, \xi) = \xi(\sigma). \end{cases} \quad (4.13)$$

As we are interested in solutions of biomedical significance, we restrict our attention to positive ones. To say that impulsive delay differential equations have positive almost periodic solutions, one need to adopt the definitions of almost periodicity only for $t \in \mathbb{R}^+$ and $k = 1, 2, \ldots$.

Related to the system (4.12), we consider the linear system

$$\begin{cases} \dot{h}(t) = -\alpha(t)h(t), & t \neq t_k, \\ \Delta h(t_k) = \gamma_k h(t_k), & k = 1, 2, \ldots. \end{cases} \quad (4.14)$$

The system (4.14) with an initial condition $h(t_0) = h_0$ has a unique solution represented by the form

$$h(t; t_0, h_0) - H(t, t_0)h_0,$$

where $H(t, s)$ is the Cauchy matrix of (4.14) defined as follows:

$$H(t, s) = \begin{cases} \exp\left\{ -\displaystyle\int_s^t \alpha(\sigma)d\sigma \right\}, & t_{k-1} < s \leq t \leq t_k, \\ \displaystyle\prod_{i=m}^{k+1}(1 + \gamma_i)exp\left\{ -\displaystyle\int_s^t \alpha(\sigma)d\sigma \right\}, \\ t_{m-1} < s \leq t_m < t_k < t \leq t_{k+1}. \end{cases} \quad (4.15)$$

Introduce the following conditions:

H4.7. The function $\alpha \in C[\mathbb{R}^+, \mathbb{R}^+]$ is almost periodic in the sense of Bohr, and there exists a constant μ such that $\alpha(t) \geq \mu > 0$.

H4.8. The sequence $\{\gamma_k\}$ is almost periodic, and $-1 < \gamma_k \leq 0$, $k = 1, 2, \ldots$.

H4.9.   The function $\beta \in C[\mathbb{R}^+, \mathbb{R}^+]$ is almost periodic in the sense of
Bohr, and

$$0 < \sup_{t \in \mathbb{R}} |\beta(t)| < \nu, \ \nu > 0, \ \beta(0) = 0.$$

H4.10.   The sequence $\{\delta_k\}, k = 1, 2, \ldots$ is almost periodic and $\sup_{k=1,2,\ldots} |\delta_k|$
$\leq \kappa.$

In the proof of the main theorem we shall use the following lemmas.

Lemma 4.3. *Let conditions H4.3 and H4.7–H4.10 hold.*
Then for each $\varepsilon > 0$ there exists ε_1, $0 < \varepsilon_1 < \varepsilon$, a relatively dense sets $\overline{T}$ of
positive real numbers, and a set P of natural numbers such that the following
relations are fulfilled:
(a) $|\alpha(t + \tau) - \alpha(t)| < \varepsilon, \ t \in \mathbb{R}^+, \ \tau \in \overline{T}.$
(b) $|\beta(t + \tau) - \beta(t)| < \varepsilon, \ t \in \mathbb{R}^+, \ \tau \in \overline{T}.$
(c) $|\gamma_{k+q} - \gamma_k| < \varepsilon, \ q \in P, \ k = 1, 2, \ldots.$
(d) $|\delta_{k+q} - \delta_k| < \varepsilon, \ q \in P, \ k = 1, 2, \ldots.$
(e) $|t_k^q - r| < \varepsilon_1, \ r \in P, \ \tau \in \overline{T}, \ k = 1, 2, \ldots.$

Lemma 4.4. *Let conditions H4.3 and H4.7–H4.10 hold.*
Then:

1. *For the Cauchy matrix $H(t, s)$ of system (4.14) there exists a positive*
constant μ such that

$$H(t, s) \leq e^{-\mu(t-s)}, \ t \geq s, \ t, s \in \mathbb{R}^+.$$

2. *For each $\varepsilon > 0$, $t \in \mathbb{R}^+$, $s \in \mathbb{R}^+$, $t \geq s$, $|t - t_k| > \varepsilon$, $|s - t_k| > \varepsilon$, $k =$*
$1, 2, \ldots$ there exists a relatively dense set $\overline{T}$ of $\varepsilon-$ almost periods of the
function $\alpha(t)$ such that for $\tau \in \overline{T}$ it follows

$$|H(t + \tau, s + \tau) - H(t, s)| \leq \varepsilon M e^{-\frac{\mu}{2}(t-s)},$$

where $M = \dfrac{2}{\mu}\left(1 + N + \dfrac{\mu}{2}N\right).$

We shall prove the next theorem.

Theorem 4.2. *Let the following conditions hold:*

1. *Conditions H4.3 and H4.7–H4.10 are met.*
2. *The following inequality is fulfilled*

$$\nu < \mu. \tag{4.16}$$

Then:

1. *There exists a unique positive almost periodic solution $h(t)$ of (4.12).*
2. *The solution $h(t)$ is exponentially stable.*

Proof. Let $AP \subset PC[\mathbb{R}^+, \mathbb{R}^+]$ denote the set of all positive almost periodic functions $\varphi(t)$ with $|\varphi|_\infty \leq K$, where

$$K = \frac{1}{\mu}\nu + \frac{2}{1 - e^{-\mu}}\kappa N,$$

and $|\varphi|_\infty = \sup_{t \in \mathbb{R}^+} |\varphi(t)|$.

In AP we define an operator S such that if $\varphi \in AP$, we have

$$S\varphi = \int_{-\infty}^t H(t,s)\beta(s)\frac{1}{1 + \varphi^n(s - \omega)}ds + \sum_{t_k < t} H(t,t_k)\delta_k. \qquad (4.17)$$

One can easily check that if $\varphi \in AP$, then

$$|S\varphi| = \int_{-\infty}^t H(t,s)|\beta(s)|\frac{1}{1 + \varphi^n(s - \omega)}ds + \sum_{t_k < t} H(t,t_k)|\delta_k|$$

$$< \frac{1}{\mu}\nu + \frac{2}{1 - e^{-\mu}}\kappa N = K. \qquad (4.18)$$

Now let $\omega \in T$, $q \in P$, where the sets $\overline{T}$ and P are defined as in Lemma 4.3, it follows that

$$|S\varphi(t + \tau) - S\varphi(t)|$$

$$\leq \int_{-\infty}^t \left| H(t + \tau, s + \tau) - H(t,s) \right| |\beta(s + \tau)| \frac{1}{1 + \varphi^n(s + \tau - \omega)}ds$$

$$+ \int_{-\infty}^t H(t,s) \left| |\beta(s + \tau)| \frac{1}{1 + \varphi^n(s + \tau - \omega)} - |\beta(s)| \frac{1}{1 + \varphi^n(s - \omega)} \right| ds$$

$$+ \sum_{t_k < t} \left| H(t + \tau, t_k + q) - H(t, t_k) \right| |\delta_{k+q}|$$

$$+ \sum_{t_k < t} H(t, t_k) \left| \delta_{k+q} - \delta_k \right|,$$

or

$$|S\varphi(t+\tau) - S\varphi(t)|$$

$$\le \int_{-\infty}^{t} \left| H(t+\tau, s+\tau) - H(t,s) \right| |\beta(s+\tau)| \frac{1}{1+\varphi^n(s+\tau-\omega)} ds$$

$$+ \int_{-\infty}^{t} H(t,s) \left\{ |\beta(s+\tau) - \beta(s)| \frac{1}{1+\varphi^n(s+\tau-\omega)} \right.$$

$$+ |\beta(s)| \left| \frac{1}{1+\varphi^n(s+\tau-\omega)} - \frac{1}{1+\varphi^n(s-\omega)} \right| ds$$

$$+ \sum_{t_k < t} \left| H(t+\tau, t_{k+q}) - H(t, t_k) \right| |\delta_{k+q}|$$

$$+ \sum_{t_k < t} H(t, t_k) |\delta_{k+q} - \delta_k| \right\} \le \varepsilon C_1, \tag{4.19}$$

where

$$C_1 = \frac{2}{\mu}\nu M + \frac{1}{\mu}(1+\nu) + \kappa M \frac{2N}{1-e^{-\frac{\mu}{2}}} + \frac{2N}{1-e^{-\mu}}.$$

In virtue of (4.18) and (4.19), we deduce that $S\varphi \in AP$.
Let $\varphi, \psi \in AP$ and then

$$|S\varphi - S\psi| \le \int_{-\infty}^{t} H(t,s)|\beta(s)| \left| \frac{1}{1+\varphi^n(s-\omega)} - \frac{1}{1+\psi^n(s-\omega)} \right| ds$$

$$\le \frac{1}{\mu}\nu|\varphi - \psi|_\infty. \tag{4.20}$$

From (4.20) and the condition (4.16) it follows that S is a contraction mapping on AP. Then there exists a unique fixed point $h \in D$ such that $Sh = h$. This implies that (4.12) has a unique positive almost periodic solution $h(t)$.

Let now $g(t)$ be an arbitrary solution of (4.12) supplemented with initial conditions

$$\begin{cases} g(t) = \zeta(t), \ \zeta \in PC[[\sigma - \omega, \sigma], \mathbb{R}^+], \\ g(\sigma^+; \sigma, \zeta) = \zeta(\sigma), \end{cases}$$

and $h(t)$ be the unique positive almost periodic solution of (4.12) with initial conditions (4.13).

It follows that

$$h(t) - g(t) = H(t, \sigma)(\xi - \zeta)$$

$$+ \int_{\sigma}^{t} H(t,s)\beta(s) \left(\frac{1}{1+h^n(s-\omega)} - \frac{1}{1+g^n(s-\omega)} \right) ds.$$

Consequently,

$$|h(t) - g(t)| \le e^{-\mu(t-\sigma)}|\xi - \zeta| + \int_{\sigma}^{t} e^{-\mu(t-s)}\nu|h(s) - g(s)|ds.$$

Setting $u(t) = |h(t) - g(t)|e^{\mu t}$, applying Gronwall–Bellman's inequality and Theorem 1.9 we end up with the expression

$$\left|h(t) - g(t)\right| \le |\xi - \zeta|e^{-(\mu-\nu)(t-\sigma)}.$$

The assumption (4.6) implies that the unique positive almost periodic solution of equation (4.12) is exponentially stable.   □

4.1.3   An Impulsive Delay Logarithmic Population Model

By employing the contraction mapping principle and applying Gronwall–Bellman's inequality, sufficient conditions are established to prove the existence and exponential stability of positive almost periodic solution for an impulsive delay logarithmic population model.

The following single species logarithmic population model

$$\dot{x}(t) = x(t)\Big[\lambda - \alpha \ln x(t) - \beta \ln x(t - \omega)\Big],$$

where α, β, λ, ω are positive constants has been proposed by Gopalsamy [59] and Kirlinger [82] and is then generalized by Liu [107] to the non autonomous case

$$\dot{x}(t) = x(t)\Big[\lambda(t) - \alpha(t) \ln x(t) - \beta(t) \ln x(t - \omega(t))\Big]. \tag{4.21}$$

In this part we shall study the existence and exponential stability of positive almost periodic solutions of (4.21) accompanied with impulses. Namely, system of the form

$$\begin{cases} \dot{x}(t) = x(t)\Big[\lambda(t) - \alpha(t) \ln x(t) - \beta(t) \ln x(t - \omega(t))\Big], \ t \ne t_k, \\ x(t_k^+) = [x(t_k)]^{1+\gamma_k}e^{\delta_k}, \ k = \pm 1, \pm 2, \dots, \end{cases} \tag{4.22}$$

where the moments t_k, $\{t_k\} \in \mathcal{B}$ represent the instants at which the density suffers an increment of e^{δ_k} units.

Our first observation is that under the invariant transformation $x(t) = e^{y(t)}$, system (4.22) reduces to

$$\begin{cases} \dot{y}(t) = -\alpha(t)y(t) - \beta(t)y(t - \omega(t)) + \lambda(t), \ t \ne t_k, \\ \Delta y(t_k) = \gamma_k y(t_k) + \delta_k, \ k = \pm 1, \pm 2, \dots. \end{cases} \tag{4.23}$$

Clearly, the transformation $x(t) = e^{y(t)}$ preserves the asymptotic properties of equation (4.22).

Let $\omega = \max\limits_{t \in \mathbb{R}} \omega(t)$ and $\sigma \in \mathbb{R}$, and let $\xi \in PC[[\sigma - \omega, \sigma], \mathbb{R}^+]$, then from [183] it follows that system (4.23) has a unique solution $y(t) = y(t; \sigma, \xi)$ defined on $[\sigma - \omega, \infty)$, and satisfies initial conditions

$$\begin{cases} y(t; \sigma, \xi) = \xi(t), \ \ \sigma - \omega \le t \le \sigma, \\ y(\sigma^+; \sigma, \xi) = \xi(\sigma). \end{cases} \tag{4.24}$$

As we are interested in solutions of biomedical significance, we again restrict our attention to positive ones.

Related to system (4.23), we consider the linear system

$$\begin{cases} \dot{y}(t) = -\alpha(t)y(t), \ t \ne t_k, \\ \Delta y(t_k) = \gamma_k y(t_k), \ k = \pm 1, \pm 2, \dots \end{cases} \tag{4.25}$$

That system (4.25) with an initial condition $y(t_0) = y_0$ has a unique solution in the form

$$y(t; t_0, y_0) = Y(t, t_0)y_0,$$

where Y is the Cauchy matrix of (4.25) defined as follows:

$$Y(t, s) = \begin{cases} exp\Big\{ -\displaystyle\int_s^t \alpha(\sigma)d\sigma \Big\}, \ t_{k-1} < s \le t \le t_k, \\ \displaystyle\prod_{i=m}^{k+1}(1 + \gamma_i)exp\Big\{ -\displaystyle\int_s^t \alpha(\sigma)d\sigma \Big\}, \\ t_{m-1} < s \le t_m < t_k < t \le t_{k+1}. \end{cases} \tag{4.26}$$

Introduce the following conditions:

H4.11. The function $\alpha \in C[\mathbb{R}, \mathbb{R}^+]$ is almost periodic in the sense of Bohr, and there exists a constant $\mu > 0$ such that $\mu \le \alpha(t)$.

H4.12. The sequence $\{\gamma_k\}$ is almost periodic, and $-1 < \gamma_k \le 0$, $k = \pm 1, \pm 2, \dots$.

H4.13. The function $\beta \in C[\mathbb{R}, \mathbb{R}^+]$ is almost periodic in the sense of Bohr, and

$$0 < \sup_{t \in \mathbb{R}} |\beta(t)| < \nu, \ \nu > 0, \ \beta(0) = 0.$$

H4.14. The function $\lambda \in C[\mathbb{R}, \mathbb{R}^+]$ is almost periodic in the sense of Bohr, and

$$0 < \sup_{t \in \mathbb{R}} |\lambda(t)| < \eta, \ \eta > 0, \ \lambda(0) = 0.$$

H4.15. The sequence $\{\delta_k\}, k = 1, 2, \dots$ is almost periodic and $\sup\limits_{k=1,2,\dots} |\delta_k| \le \kappa$.

The proofs of the next lemmas are similar to the proofs of similar lemmas in Chap. 1.

Lemma 4.5. *Let conditions H4.3 and H4.11–H4.15 hold.*
Then for each $\varepsilon > 0$ there exists ε_1, $0 < \varepsilon_1 < \varepsilon$, a relatively dense sets $\overline{T}$ of positive real numbers, and a set P of natural numbers such that the following relations are fulfilled:

(a) $|\alpha(t + \tau) - \alpha(t)| < \varepsilon,\ t \in \mathbb{R},\ \tau \in \overline{T}.$
(b) $|\lambda(t + \tau) - \lambda(t)| < \varepsilon,\ t \in \mathbb{R},\ \tau \in \overline{T}.$
(c) $|\beta(t + \tau) - \beta(t)| < \varepsilon,\ t \in \mathbb{R}^+,\ \tau \in \overline{T}.$
(d) $|\gamma_{k+q} - \gamma_k| < \varepsilon,\ q \in P,\ k = 1, 2, \ldots .$
(e) $|\delta_{k+q} - \delta_k| < \varepsilon,\ q \in P,\ k = 1, 2, \ldots .$
(f) $|t_k^q - r| < \varepsilon_1,\ r \in P,\ \tau \in \overline{T},\ k = 1, 2, \ldots .$

Lemma 4.6. *Let conditions H4.3 and H4.11–H4.15 hold.*
Then:

1. *For the Cauchy matrix $Y(t, s)$ of system (4.25) there exists a positive constant μ such that*

$$Y(t, s) \leq e^{-\mu(t-s)},\ t \geq s,\ t, s \in \mathbb{R}.$$

2. *For each $\varepsilon > 0$, $t \in \mathbb{R}$, $s \in \mathbb{R}$, $t \geq s$, $|t - t_k| > \varepsilon$, $|s - t_k| > \varepsilon$, $k = \pm 1, \pm 2, \ldots$ there exists a relatively dense set $\overline{T}$ of ε–almost periods of the function $\alpha(t)$ and a positive constant M such that for $\tau \in \overline{T}$ it follows*

$$|Y(t + \tau) - Y(t)| \leq \varepsilon M e^{-\frac{\mu}{2}(t-s)},$$

where $M = \dfrac{2}{\mu}\left(1 + N + \dfrac{\mu}{2}\right).$

The proof of the next theorem is as the same way like the Theorems 4.1 and 4.2 using Lemmas 4.5 and 4.6.

Theorem 4.3. *Let the following conditions hold:*

1. *Conditions H4.3 and H4.12–H4.15 are met.*
2. *The following inequality is fulfilled*

$$\nu < \mu.$$

Then:

1. *There exists a unique positive almost periodic solution $x(t)$ of (4.22).*
2. *The solution $x(t)$ is exponentially stable.*

4.2   Population Dynamics

4.2.1   *Impulsive* n-*dimensional Lotka–Volterra Models*

During the past few decades, a lot of works has been done of Lotka–Volterra models, see [1–4, 7, 29, 37, 54, 59, 76, 78, 82, 88, 99–103, 117, 125, 130, 131, 135, 177, 181, 182, 185, 187, 189, 196, 198, 200] and the references cited therein.

The classical n-species Lotka–Volterra model can be expressed as follows:

$$\dot{u}_i(t) = u_i(t)\Big[r_i(t) - \sum_{j=1}^{n} a_{ij}u_j(t)\Big], \ i = 1, 2, \ldots, n,$$

where $u_i(t)$ represents the density of species i at the moment $t \in \mathbb{R}$, $u_i \in \mathbb{R}$, $r_i(t)$ is the reproduction rate function, and $a_{ij}(t)$ are functions which describe the effect of the j-th population upon the i-th population, which is positive if it enhances, and negative if it inhibits the growth.

These kinds of systems are of great interest not only for population dynamics or in chemical kinetics, but they are important in ecological modeling and all fields of science, from plasma physics to neural nets.

If at certain moments of time the evolution of the process is subject to sudden changes, then the population numbers vary by jumps. Therefore, it is important to study the behavior of the solutions of Lotka–Volterra systems with impulsive perturbations.

In this part of Sect. 4.2, we shall investigate the existence of almost periodic solutions of following n-species Lotka–Volterra type impulsive system,

$$\begin{cases} \dot{u}_i(t) = u_i(t)\Big[r_i(t) - a_i(t)u_i(t) - \displaystyle\sum_{j=1, j\neq i}^{n} a_{ij}(t)u_j(t)\Big], \ t \neq t_k, \\ \Delta u_i(t_k) = d_{ik}u_i(t_k), \ k = \pm 1, \pm 2, \ldots, \end{cases} \quad (4.27)$$

where $i = 1, 2, \ldots, n$, $t \in \mathbb{R}$, $\{t_k\} \in \mathcal{B}$, $n \geq 2$, r_i, $a_i \in C[\mathbb{R}, \mathbb{R}]$ and $a_{ij} \in C[\mathbb{R}, \mathbb{R}]$, $j = 1, 2, \ldots, n$, $i \neq j$, the constants $d_{ik} \in \mathbb{R}$, $k = \pm 1, \pm 2, \ldots$.

In mathematical ecology, the system (4.27) denotes a model of the dynamics of an n-species system in which each individual competes with all others of the system for a common resource. The numbers $u_i(t_k)$ and $u_i(t_k^+)$ are respectively, the population densities of species i before and after impulsive perturbation at the moment t_k. The constants d_{ik} characterize the magnitude of the impulsive effect on the species i at the moments t_k.

Let $t_0 \in \mathbb{R}$, $u_0 = col(u_{10}, u_{20}, \ldots, u_{n0})$, $u_{i0} \in \mathbb{R}$ for $1 \leq i \leq n$.

Denote by $u(t) = u(t; t_0, u_0)$, $u(t) = col(u_1(t), u_2(t), \ldots, u_n(t))$ the solution of (4.27) with the initial condition

$$u(t_0^+; t_0, u_0) = u_0. \tag{4.28}$$

The solution $u(t) = u(t; t_0, u_0)$ of problem (4.27), (4.28) is a piece-wise continuous function with points of discontinuity at the moments t_k, $k = \pm 1, \pm 2, \ldots$ at which it is continuous from the left, i.e. the following relations are valid:

$$u_i(t_k^-) = u_i(t_k),$$

$$u_i(t_k^+) = u_i(t_k) + d_{ik}u_i(t_k), \ \ k = \pm 1, \pm 2, \ldots, \ 1 \le i \le n.$$

In this paragraph for a given a nonnegative function $g(t)$ which is defined on $\mathbb{R}$, we set

$$g_M = \sup_{t \in \mathbb{R}} g(t), \ \ g_L = \inf_{t \in \mathbb{R}} g(t).$$

Introduce the following conditions:

H4.16.   The functions $r_i(t)$, $a_i(t)$, $1 \le i \le n$ and $a_{ij}(t)$, $1 \le i, j \le n$, $i \ne j$ are almost periodic, nonnegative, and $r_{iL} > 0$, $r_{iM} < \infty$, $a_{iL} > 0$, $a_{iM} < \infty$, $a_{ijL} \ge 0$, $a_{ijM} < \infty$ for $1 \le i, j \le n$, $i \ne j$.

H4.17.   The sequences $\{d_{ik}\}$, $1 \le i \le n$, $k = \pm 1, \pm 2, \ldots$ are almost periodic, and $-1 < d_{ik} \le 0$.

H4.18.   The set of sequences $\{t_k^j\}$, $t_k^j = t_{k+j} - t_k$, $k, j = \pm 1, \pm 2, \ldots$ is uniformly almost periodic, and there exists $\theta > 0$ such that $\inf_k t_k^1 = \theta > 0$.

Let conditions H4.16–H4.18 hold, and let $\{s_m'\}$ be an arbitrary sequence of real numbers. Then there exist a subsequence $\{s_l\}$, $s_l = s_{m_l}'$ such that the system (4.27) by the process described in Chap. 1 gives rise to the limiting system

$$
\begin{cases}
\dot{u}_i(t) = u_i(t)\left[r_i^s(t) - a_i^s(t)u_i(t) - \displaystyle\sum_{j=1, j \ne i}^{n} a_{ij}^s(t)u_j(t)\right], \ t \ne t_k^s, \\
\Delta u_i(t_k^s) = d_{ik}^s u_i(t_k^s)), \ \ k = \pm 1, \pm 2, \ldots.
\end{cases}
\tag{4.29}
$$

In the proof of the main results we shall use the following definitions and lemmas.

Let $u_0 = col(u_{10}, u_{20}, \ldots, u_{n0})$ and $v_0 = col(v_{10}, v_{20}, \ldots, v_{n0})$, u_{i0}, $v_{i0} \in \mathbb{R}$, $1 \le i \le n$, and let

$$u(t) = col(u_1(t), u_2(t), \ldots, u_n(t)), \ v(t) = col(v_1(t), v_2(t), \ldots, v_n(t))$$

be two solutions of (4.27) with initial conditions

$$u(t_0^+, t_0, u_0) = u_0, \ \ v(t_0^+, t_0, v_0) = v_0.$$

Definition 4.1 ([177]). The solution $u(t)$ of (4.27) is said to be:

(a)   *stable*, if

$$(\forall \varepsilon > 0)(\forall t_0 \in \mathbb{R})(\exists \delta > 0)(v_0 \in \mathbb{R}^n : ||u_0 - v_0|| \le \delta)(\forall t \ge t_0) :$$

$$||u(t; t_0, u_0) - v(t; t_0, v_0)|| < \varepsilon;$$

(b)   *globally asymptotically stable*, if it is stable and

$$\lim_{t \to \infty} ||u(t; t_0, u_0) - v(t; t_0, v_0)|| = 0;$$

(c)   *globally exponentially stable*, if

$$(\exists c > 0)(\forall \alpha > 0)(\exists \gamma > 0)(\forall t_0 \in \mathbb{R})(\forall v_0 \in \mathbb{R}^n : ||u_0 - v_0|| \le \alpha)(\forall t \ge t_0) :$$

$$||u(t; t_0, u_0) - v(t; t_0, v_0)|| < \gamma ||u_0 - v_0|| \exp\{-c(t - t_0)\}.$$

Definition 4.2 ([177]). The solution $u(t) = (u_1(t), u_2(t), \ldots, u_n(t))$ of system (4.27) is said to be a *strictly positive solution*, if for $i = 1, 2, \ldots, n$,

$$0 < \inf_{t \in \mathbb{R}} u_i(t) \le \sup_{t \in \mathbb{R}} u_i(t) < \infty.$$

We shall prove the next lemmas.

Lemma 4.7. *Let the following conditions hold:*

1. *Conditions H4.16–H4.18 are met.*
2. *The function $U(t) = col(U_1(t), U_2(t), \ldots, U_n(t))$ is the maximal solution of the system*

$$\begin{cases} \dot{U}_i(t) = U_i(t)\Big[r_{iM} - a_{iL}U_i(t)\Big], \ t \ne t_k, \\ \Delta U_i(t_k) = d_{Mk}U_i(t_k), \ k = \pm 1, \pm 2, \ldots, \end{cases} \qquad (4.30)$$

   where $d_{Mk} = max\{d_{ik}\}$ for $1 \le i \le n$ and $k = \pm 1, \pm 2, \ldots$.
3. *The function $V(t) = col(V_1(t), V_2(t), \ldots, V_n(t))$ is the minimal solution of the system*

$$\begin{cases} \dot{V}_i(t) = V_i(t)\Big[r_{iL} - a_{iM}V_i(t) - \sum_{j=1, j \ne i}^{n} a_{ijM} \sup_{t \in \mathbb{R}} U_j(t)\Big], \ t \ne t_k, \\ \Delta V_i(t_k) = d_{Lk}V_i(t_k), \ k = \pm 1, \pm 2, \ldots, \end{cases} \qquad (4.31)$$

   where $d_{Lk} = min\{d_{ik}\}$ for $1 \le i \le n$ and $k = \pm 1, \pm 2, \ldots$.
4. *For each $1 \le i \le n$,*

$$V_i(t_0^+) \le u_i(t_0^+) \le U_i(t_0^+).$$

Then

$$V_i(t) \leq u_i(t) \leq U_i(t) \qquad (4.32)$$

for $t \in \mathbb{R}$ *and* $1 \leq i \leq n$.

Proof. We follow [4] and for (4.27) have that

$$\begin{cases} \dot{u}_i(t) \leq u_i(t)\big[r_{iM} - a_{iL}u_i(t)\big], \ t \neq t_k, \\ \Delta u_i(t_k) \leq d_{Mk}u_i(t_k), \ k = \pm 1, \pm 2, \ldots, \end{cases}$$

and

$$\begin{cases} \dot{u}_i(t) \geq u_i(t)\Big[r_{iL} - a_{iM}u_i(t) - \displaystyle\sum_{j=1, j \neq i}^{n} a_{ijM} \sup_{t \in \mathbb{R}} u_j(t)\Big], \ t \neq t_k, \\ \Delta u_i(t_k) \geq d_{Lk}u_i(t_k), \ k = \pm 1, \pm 2, \ldots. \end{cases}$$

Then from the differential inequalities for piecewise continuous functions $V_i(t)$, $U_i(t)$ and $u_i(t)$, we obtain that inequality (4.32) is valid for $t \in \mathbb{R}$ and $1 \leq i \leq n$. $\qquad\square$

Lemma 4.8. *Let the following conditions hold:*

1. *Conditions H4.16–H4.18 are met.*
2. *The solution* $u(t) = col(u_1(t), u_2(t), \ldots, u_n(t))$ *of (4.27) is such that* $u_i(t_0^+) > 0$, $1 \leq i \leq n$.

Then:

1. $u_i(t) > 0$, $1 \leq i \leq n$, $t \in \mathbb{R}$.
2. *For* $t \in \mathbb{R}$ *and* $1 \leq i \leq n$

$$min\Big\{u_i(t_0^+), \Big[r_{iL} - \sum_{j=1, j \neq i}^{n} \frac{a_{ijM}r_{jM}}{a_{jL}}\Big]\Big/a_{iM}\Big\}$$

$$\leq u_i(t) \leq max\Big\{u_i(t_0^+), \frac{r_{iM}}{a_{iL}}\Big\}.$$

Proof. Let we again follow [4] and then the solution $u(t; t_0, u_0)$, $t_0 \in (t_{-1}, t_1]$ of (4.27) is defined by the equality

$$u(t; t_0, u_0) = \begin{cases} \ldots\ldots\ldots\ldots\ldots\ldots\ldots\ldots, \\ u^{-1}(t; t_{-1}, u^{-1+}), \ t_{-1} < t \leq t_0, \\ u^0(t; t_0, u^{0+}), \ t_0 < t \leq t_1, \\ \ldots\ldots\ldots\ldots\ldots\ldots\ldots\ldots \\ u^k(t; t_k, u^{k+}), \ t_k < t \leq t_{k+1}, \\ \ldots\ldots\ldots\ldots\ldots\ldots\ldots\ldots, \end{cases}$$

where $u^k(t; t_k, u^{k+})$ is a solution of the equation without impulses $\dot{u}(t) = p(t)u(t)$, $p(t) = col(p_1(t), p_2(t), \ldots, p_n(t))$,

$$p_i(t) = r_i(t) - a_i(t)u_i(t) - \sum_{j=1, j\neq i}^{n} a_{ij}(t)u_j(t),$$

in the interval $(t_k, t_{k+1}]$, $k = \pm 1, \pm 2, \ldots$ for which

$$u_i^{k+} = (1 + d_{ik})u_i^k(t_k; t_{k-1}, u_i^{k-1} + 0), \ k = \pm 1, \pm 2, \ldots, \ 1 \leq i \leq n,$$

and $u^{0+} = u_0$.

Thus,

$$u_i(t) = u_i(t_0^+)exp\left\{ \int_{t_0}^{t} p_i(s)\, ds \right\} \prod_{t_0 < t_k < t} (1 + d_{ik})$$

for $1 \leq i \leq n$, so $u_i(t) > 0$ for $t \in \mathbb{R}$.

Now from Lemma 4.7 we get that the inequalities (4.32) hold for all $t \in \mathbb{R}$ and $1 \leq i \leq n$.

We shall prove that

$$\left[r_{iL} - \sum_{j=1, j\neq i}^{n} \frac{a_{ijM}r_{jM}}{a_{jL}} \right] \Big/ a_{iM} \leq V_i(t) \leq U_i(t) \leq \frac{r_{iM}}{a_{iL}},$$

for all $t \in \mathbb{R}$ and $1 \leq i \leq n$.

First, we shall prove

$$U_i(t) \leq \frac{r_{iM}}{a_{iL}}, \tag{4.33}$$

for all $t \in \mathbb{R}$ and $1 \leq i \leq n$.

If $t \in \mathbb{R}$ and for some i with $1 \leq i \leq n$, $U_i(t) > \dfrac{r_{iM}}{a_{iL}}$, then for $t \in [t_{k-1}, t_k)$, $k = \pm 1, \pm 2, \ldots$ we will have

$$\dot{U}_i(t) < U_i(t)\left[r_{iM} - a_{iL}U_i(t) \right] < 0.$$

This proves that (4.33) holds for all i, $1 \leq i \leq n$ as long as $U_i(t)$ is defined.

The inequality $\left[r_{iL} - \displaystyle\sum_{j=1, j\neq i}^{n} \dfrac{a_{ijM}r_{jM}}{a_{jL}} \right] / a_{iM} \leq V_i(t)$ is proves by analogous way.

Hence, for $t \in \mathbb{R}$, we have

$$min\left\{ u_i(t_0^+), \left[r_{iL} - \sum_{j=1, j\neq i}^{n} \frac{a_{ijM}r_{jM}}{a_{jL}} \right] / a_{iM} \right\}$$

$$\leq u_i(t) \leq max\left\{ u_i(t_0^+), \frac{r_{iM}}{a_{iL}} \right\}. \qquad \square$$

For the proof of the main result in this part we shall consider system (4.29) and then discuss the almost periodic solutions of the system (4.27). For simplification, we rewrite (4.27) in the form

$$\begin{cases} \dot{u} = f(t, u), \ t \neq t_k, \\ \Delta u(t_k) = D_k u(t_k), \ k = \pm 1, \pm 2, \dots . \end{cases} \tag{4.34}$$

$\square$

Lemma 4.9. *Let the following conditions hold:*

1. *Conditions H4.16–H4.18 are met.*
2. *$\{s'_m\}$ is an arbitrary sequence of real numbers.*
3. *For the system (4.29) there exist strictly positive almost periodic solutions. Then the system (4.27) has a unique strictly positive almost periodic solution.*

Proof. In (4.34) from H4.16–H4.18 it follows that $f(t, x)$ is an almost periodic function with respect to $t \in \mathbb{R}$ and $x \in B_\alpha$, $\alpha > 0$ and D_k is almost periodic sequence with respect to $k = \pm 1, \pm 2, \dots$.

Let $\phi(t)$ be a strictly positive solution of (4.34), and let the sequences of real numbers α' and β' be such that for their common subsequences $\alpha \subset \alpha'$, $\beta \subset \beta'$. Then we have $\theta_{\alpha+\beta} f(t, u) = \theta_\alpha \theta_\beta f(t, u)$ and $\theta_{\alpha+\beta}\phi(t), \theta_\alpha\theta_\beta\phi(t)$ exist uniformly on the compact set $\mathbb{R} \times \mathcal{B}$, and are solutions of the system

$$\begin{cases} \dot{u} = f^{\alpha+\beta}(t, u), \ t \neq t_k^{\alpha+\beta}, \\ \Delta u(t_k^{\alpha+\beta}) = D_k^{\alpha+\beta} u(t_k^{\alpha+\beta}), \ k = \pm 1, \pm 2, \dots . \end{cases}$$

Therefore, $\theta_{\alpha+\beta}\phi(t) = \theta_\alpha\theta_\beta\phi(t)$ and then from Lemma 2.23 it follows that $\phi(t)$ is an almost periodic solution of system (4.27). $\square$

In the proof of the next theorem we shall use the sets G_k, $k = \pm 1, \pm 2, \dots$, $G = \cup_{k=-\infty}^{\infty} G_k$, and the class of function V_0 defined in Chap. 1. $\square$

Theorem 4.4. *Let the following conditions hold:*

1. *Conditions H4.16–H4.18 are met.*
2. *There exist nonnegative functions $\delta_\nu(t)$, $1 \leq \nu \leq n$ such that*

$$a_\nu(t) - \sum_{i=1, i\neq\nu}^{n} a_{i\nu}(t) \geq \delta_\nu(t), \ t \neq t_k, \ k = \pm 1, \pm 2, \dots, \tag{4.35}$$

for $t \in \mathbb{R}$.

Then:

1. *For the system (4.27) there exists a unique strictly positive almost periodic solution.*

2. *If there exists a constant $c \geq 0$ such that*

$$\int_{t_0}^{t} \delta(t) ds = c(t - t_0),$$

where $\delta(t) = min(\delta_1(t), \delta_2(t), \ldots, \delta_n(t))$ then the almost periodic solution is globally exponentially stable.

Proof. From the construction of (4.29) it follows that for an arbitrary sequence of real numbers $\{s'_m\}$ there exists a subsequence $\{s_l\}$, $s_l < s_{l+1}$ and $s_l \to \infty$ for $l \to \infty$ such that

$$r_i(t + s_l) \to r_i^s(t),\ a_i(t + s_l) \to a_i^s(t),\ a_{ij}(t + s_l) \to a_{ij}^s(t),\ l \to \infty,$$

uniformly on $t \in \mathbb{R}$, $t \neq t_k$, and there exists a subsequence $\{k_l\}$ of $\{l\}$, $k_l \to \infty$, $l \to \infty$ such that $t_{k_l} \to t_k^s$, $d_{ik_l} \to d_{ik}^s$.

Then for the system

$$\begin{cases} \dot{u}_i(t) = u_i(t) \Big[r_i^s(t) - a_i^s(t) u_i(t) - \displaystyle\sum_{j=1, j \neq i}^{n} a_{ij}^s(t) u_j(t) \Big], \ t \neq t_k^s, \\ \Delta u_i(t_k^s) = d_{ik_n}^s u_i(t_k^s), \ k = \pm 1, \pm 2, \ldots, \end{cases} \tag{4.36}$$

the conditions of Lemma 4.8 hold.

Then, if $u^s(t)$ is a solution of system (4.36) it follows that

$$0 < \inf_{t \in \mathbb{R}} u_i(t) \leq \sup_{t \in \mathbb{R}} u_i(t) < \infty, \ i = 1, 2, \ldots, n. \tag{4.37}$$

From (4.37) we have that for every system in the form (4.29) there exists at least one strictly positive solution.

Now suppose that (4.29) has two arbitrary strictly positive solutions $u^s(t) = (u_1^s(t), u_2^s(t), \ldots, u_n^s(t))$, $v^s(t) = (v_1^s(t), v_2^s(t), \ldots, v_n^s(t))$.

Consider a Lyapunov function

$$V^s(t, u^s(t), v^s(t)) = \sum_{i=1}^{n} \left| \ln \frac{u_i^s}{v_i^s} \right|, \ t \in \mathbb{R}.$$

Then for $t \in \mathbb{R}$, $t \neq t_k$

$$D^+ V^s(t, u^s(t), v^s(t)) = \sum_{i=1}^{n} \left[\frac{\dot{u}_i^s(t)}{u_i^s(t)} - \frac{\dot{v}_i^s(t)}{v_i^s(t)} \right] sgn(u_i^s(t) - v_i^s(t))$$

$$\leq \sum_{l=1}^{n} \Big(-a_l^s(t) |u_l^s(t) - v_l^s(t)| + \sum_{i=1, i \neq l}^{n} a_{il}^s(t) |u_i^s(t) - v_i^s(t)| \Big).$$

Thus, in view of hypothesis (4.35), we obtain

$$D^+ V^s(t, u^s(t), v^s(t)) \leq -\delta^s(t) m^s(t), \ t \in \mathbb{R}, \ t \neq t_k^s, \qquad (4.38)$$

where $\delta_\nu(t + s_l) \to \delta^s(t)$, $l \to \infty$, $\nu = 1, 2, \ldots, n$,

$$\delta^s(t) = min(\delta_1^s(t), \delta_2^s(t), \ldots, \delta_n^s(t)), \quad m^s(t) = \sum_{i=1}^{n} |u_i^s - v_i^s|.$$

On the other hand for $t = t_k^s$ we have

$$V^s(t_k^{s+}, u^s(t_k^{s+}), v^s(t_k^{s+})) = \sum_{i=1}^{n} \left| \ln \frac{u_i^s(t_k^{s+})}{v_i^s(t_k^{s+})} \right| = \sum_{i=1}^{n} \left| \ln \frac{(1 + d_{ik}^s) u_i^s(t_k^s)}{(1 + d_{ik}^s) v_i^s(t_k^s)} \right|$$

$$= V^s(t_k^s, u^s(t_k^s), v^s(t_k^s)). \qquad (4.39)$$

From (4.38) and (4.39), we get

$$D^+ V^s(t, u^s(t), v^s(t)) \leq 0, \ t \in \mathbb{R}, \ t \neq t_k,$$
$$\Delta V^s(t_k^s, u^s(t_k^s), v^s(t_k^s)) = 0,$$

and hence

$$V^s(t, u^s(t), v^s(t)) \leq V^s(t_0, u^s(t_0), v^s(t_0))$$

for all $t \geq t_0$, $t_0 \in \mathbb{R}$.

From the last inequality, (4.38) and (4.39), we have

$$\int_{t_0}^{t} \delta^s(t) m^s(t) ds \leq V^s(t_0) - V^s(t), \ t \geq t_0.$$

Consequently,

$$\int_{t_0}^{\infty} |u_i^s(t) - v_i^s(t)| < \infty, \ i = 1, 2, \ldots, n,$$

and $u_i^s(t) - v_i^s(t) \to 0$ for $t \to \infty$.

Let $\mu^s = \inf_{t \in \mathbb{R}} \{u_i^s, v_i^s, \ i = 1, 2, \ldots, n\}$.

From the definition of $V^s(t)$ we have

$$V^s(t, u^s(t), v^s(t)) = \sum_{i=1}^{n} \left| \ln u_i^s - \ln v_i^s \right|$$

$$\leq \frac{1}{\mu^s} \sum_{i=1}^{n} |u_i^s - v_i^s|.$$

Hence, $V^s(t) \to 0$, $t \to -\infty$, $V^s(t)$ is a nonincreasing nonnegative function on $\mathbb{R}$, and from (4.38), we obtain

$$V^s(t) = 0, \quad t \in \mathbb{R}. \tag{4.40}$$

Now, from (4.40) and the boundedness of the right-hand side of (4.27) it follows that $u^s \equiv v^s$ for all $t \in \mathbb{R}$ and $i = 1, 2, \ldots, n$. Then for an arbitrary sequence of real numbers $\{s_n\}$ the system (4.29) has a unique strictly positive almost periodic solution $u(t)$.

From Lemma 4.8 analogously it follows that system (4.27) has a unique strictly positive almost periodic solution.

Now, consider again the Lyapunov function

$$V(t) = V(t, u(t), v(t)) = \sum_{i=1}^{n} \left| \ln \frac{u_i(t)}{v_i(t)} \right|,$$

where $v(t) = (v_1(t), v_2(t), \ldots, v_n(t))$ is an arbitrary solution of (4.27) with the initial condition $v(t_0^+) = v_0$.

By Mean Value Theorem it follows that for any closed interval contained in $(t_{k-1}, t_k]$, $k = \pm 1, \pm 2, \ldots$ there exist positive numbers r and R such that for $1 \le i \le n$, $r \le u_i(t)$, $v_i(t) \le R$, and

$$\frac{1}{R} \left| u_i(t) - v_i(t) \right| \le \left| \ln u_i(t) - \ln v_i(t) \right| \le \frac{1}{r} \left| u_i(t) - v_i(t) \right|. \tag{4.41}$$

Hence, we obtain

$$V(t_0^+, u_0, v_0) = \sum_{i=1}^{n} \left| \ln u_i(t_0^+) - \ln v_i(t_0^+) \right| \le \frac{1}{r} ||u_0 - v_0||. \tag{4.42}$$

On the other hand,

$$D^+ V(t, u(t), v(t)) \le -\delta(t) m(t)$$
$$\le -\delta(t) r V(t, u(t), v(t)), \quad t \in \mathbb{R}, \ t \ne t_k, \tag{4.43}$$

and for $t \in \mathbb{R}$, $t = t_k$,

$$V(t_k^+, u(t_k^+), v(t_k^+)) = \sum_{i=1}^{n} \left| \ln \frac{u_i(t_k^+)}{v_i(t_k^+)} \right|$$
$$= \sum_{i=1}^{n} \left| \ln \frac{(1+d_{ik})u_i(t_k)}{(1+d_{ik})v_i(t_k)} \right| = V(t_k, u(t_k), v(t_k)). \tag{4.44}$$

From (4.41), (4.43) and (4.44) it follows

$$V(t, u(t), v(t)) \le V(t_0^+, u_0, v_0) \exp\left\{ -r \int_{t_0}^t \delta(s)ds \right\}. \qquad (4.45)$$

Therefore, from (4.41), (4.44) and (4.45), we deduce the inequality

$$\sum_{i=1}^n |u_i(t) - v_i(t)| \le \frac{R}{r} ||u_0 - v_0|| e^{-rc(t-t_0)},$$

where $t \ge t_0$.

The last inequality shows that the unique almost periodic solution $u(t)$ of system (4.27) is globally exponentially stable. $\square$

Example 4.2. Consider a three-dimensional impulsive Lotka–Volterra system in the form

$$\begin{cases} \dot{u}_1(t) = u_1(t)\big[6 - 9u_1(t) - (3 - \sin\sqrt{2}t)u_2(t) - (3 - \sin\sqrt{2}t)u_3(t)\big], \ t \ne t_k, \\ \dot{u}_2(t) = u_2(t)\big[7 - (2 - \sin\sqrt{2}t)u_1(t) - (10 + \cos\sqrt{3}t)u_2(t) \\ \qquad\qquad -(2 - \cos\sqrt{3}t)u_3(t)\big], \ t \ne t_k, \\ \dot{u}_3(t) = u_3(t)\big[6 + \sin t - 4u_1(t) - (3 - \sin\sqrt{3}t)u_2(t) - 9u_3(t)\big], \ t \ne t_k, \\ \Delta u_i(t_k) = d_{ik}u_i(t_k), \ k = \pm 1, \pm 2, \dots, \ i = 1, 2, 3, \end{cases}$$

where conditions H4.17 and H4.18 hold.

Then, we have that

$$a_1 - a_{21} - a_{31} = 9 - (2 - \sin\sqrt{2}t) - 4 \ge 2 = \delta_1,$$

$$a_2 - a_{12} - a_{32} = 10 + \cos\sqrt{3}t - (3 - \sin\sqrt{2}t) - (3 - \sin\sqrt{3}t) \ge 1 = \delta_2,$$

$$a_3 - a_{13} - a_{23} = 9 - (3 - \sin\sqrt{2}t) - (2 - \cos\sqrt{3}t) \ge 2 = \delta_3.$$

For $\delta(t) = min(\delta_1, \delta_2, \delta_3) = 1$ and $c = 1$ all conditions of Theorem 4.4 are satisfied and the three-dimensional system considered has a unique strictly positive almost periodic solution which is globally exponentially stable.

4.2.2 *Impulsive Lotka–Volterra Models with Dispersions*

In the present part we shall investigate the existence of almost periodic processes of ecological systems which are presented with nonautonomous n-dimensional impulsive Lotka–Volterra competitive systems with dispersions and fixed moments of impulsive perturbations.

Consider the system

$$
\begin{cases}
\dot{u}_i(t) = u_i(t)\left[r_i(t) - a_i(t)u_i(t) - \sum_{j=1, j\neq i}^{n} a_{ij}(t)u_j(t)\right] + \\
\qquad + \sum_{j=1}^{n} b_{ij}(t)\big(u_j(t) - u_i(t)\big),\ t \neq t_k, \\
\Delta u_i(t_k) = d_k u_i(t_k),\ k = \pm 1, \pm 2, \dots,
\end{cases}
\tag{4.46}
$$

where $t \in \mathbb{R}$, $\{t_k\} \in \mathcal{B}$, $i = 1, 2, \dots, n$, $n \geq 2$, the functions r_i, $a_i \in C[\mathbb{R}, \mathbb{R}]$, $1 \leq i \leq n$, $a_{ij} \in C[\mathbb{R}, \mathbb{R}]$, $i \neq j$, $b_{ij} \in C[\mathbb{R}, \mathbb{R}]$, $1 \leq i, j \leq n$, the constants $d_k \in \mathbb{R}$.

The solution $u(t) = u(t; t_0, u_0)$ of problem (4.46), (4.28) is a piecewise continuous function with points of discontinuity at the moments t_k, $k = \pm 1, \pm 2, \dots$ at which it is continuous from the left, i.e. the following relations are valid:

$$
u_i(t_k^-) = u_i(t_k),
$$

$$
u_i(t_k^+) = u_i(t_k) + d_k u_i(t_k),\ k = \pm 1, \pm 2, \dots,\ 1 \leq i \leq n.
$$

Introduce the following conditions:

H4.19.　The functions $b_{ij}(t)$ are almost periodic, nonnegative, continuous and $b_{ijL} \geq 0$, $b_{ijM} < \infty$ for $1 \leq i, j \leq n$.

H4.20.　The sequence $\{d_k\}$, $k = \pm 1, \pm 2, \dots$, is almost periodic and $-1 < d_k \leq 0$.

Let conditions H4.16, H4.18–H4.20 hold and let $\{s_m'\}$ be an arbitrary sequence of real numbers. Then there exists a subsequence $\{s_l\}$, $s_l = s_{m_l}'$ such that the system (4.2.20) moves to system

$$
\begin{cases}
\dot{u}_i(t) = u_i(t)\left[r_i^s(t) - a_i^s(t)u_i(t) - \sum_{j=1, j\neq i}^{n} a_{ij}^s(t)u_j(t)\right] + \\
\qquad + \sum_{j=1}^{n} b_{ij}^s(t)(u_j(t) - u_i(t)),\ t \neq t_k^s, \\
\Delta u_i(t_k^s) = d_k^s u_i(t_k^s),\ k = \pm 1, \pm 2, \dots.
\end{cases}
\tag{4.47}
$$

In the proof of the main results we shall use the following lemmas for the system (4.46).

Lemma 4.10. *Let the following conditions hold:*

1. Conditions H4.16, H4.18–H4.20 are met.

2. There exist functions P_i, $Q_i \in PC^1[\mathbb{R}, \mathbb{R}]$ such that

$$P_i(t_0^+) \leq u_i(t_0^+) \leq Q_i(t_0^+),$$

where $t_0 \in \mathbb{R}$, $i = 1, 2, \ldots, n$.

Then

$$P_i(t) \leq u_i(t) \leq Q_i(t) \tag{4.48}$$

for all $t \geq t_0$ *and* $i = 1, 2, \ldots, n$.

Proof. First we shall proof that

$$u_i(t) \leq Q_i(t) \tag{4.49}$$

for all $t \geq t_0$ and $i = 1, 2, \ldots, n$, where $Q_i(t)$ is the maximal solution of the logistic system

$$\begin{cases} \dot{q}_i(t) = q_i(t)\big[r_i(t) - a_i(t)q_i(t)\big], \ t \neq t_k, \\ q_i(t_0^+) = q_{i0} > 0, \\ \Delta q_i(t_k) = d^M q_i(t_k), \ k = \pm 1, \pm 2, \ldots, \end{cases} \tag{4.50}$$

and $d^M = \max\limits_{k=\pm 1, \pm 2, \ldots} d_k$.

The maximal solution $Q_i(t) = Q_i(t; t_0, q_0)$, $q_0 = col(q_{10}, q_{20}, \ldots, q_{n0})$ of (4.50) is defined by the equality

$$Q_i(t; t_0, q_0) = \begin{cases} Q_i^0(t; t_0, Q_i^{0+}), \ t_0 < t \leq t_1, \\ Q_i^1(t; t_1, Q_i^{1+}), \ t_1 < t \leq l_2, \\ \cdots\cdots\cdots\cdots\cdots\cdots\cdots\cdots \\ Q_i^k(t; t_k, Q_i^{k+}), \ t_k < t \leq t_{k+1}, \\ \cdots\cdots\cdots\cdots\cdots\cdots\cdots\cdots, \end{cases}$$

where $Q_i^k(t; t_k, Q_i^{k+})$ is the maximal solution of the equation without impulses

$$\dot{q}_i(t) = q_i(t)\big[r_i(t) - a_i(t)q_i(t)\big],$$

in the interval $(t_k, t_{k+1}]$, $k = \pm 1, \pm 2, \ldots$, for which

$$Q_i^{k+} = (1 + d^M)Q_i^k(t_k; t_{k-1}, Q_i^{k-1+}), \ k = 1, 2, \ldots, \ 1 \leq i \leq n$$

and $Q_i^{0+} = q_{i0}$.

By [189], it follows for (4.46) that

$$\dot{u}_i(t) \leq u_i(t)\big[r_i(t) - a_i(t)u_i(t))\big], \ t \neq t_k. \tag{4.51}$$

Now, let $t \in (t_0, t_1]$. If $0 < u_{i0} \le Q_i(t_0^+)$, then elementary differential inequality [94] yields that

$$u_i(t) \le Q_i(t),$$

for all $t \in (t_0, t_1]$, i.e. the inequality (4.49) is valid for $t \in (t_0, t_1]$.

Suppose that (4.49) is satisfied for $t \in (t_{k-1}, t_k]$.

Then, from H4.18 and the fact that (4.49) is satisfied for $t = t_k$, we obtain

$$u_i(t_k^+) = u_i(t_k) + d_k u_i(t_k) \le u_i(t_k) + d^M u_i(t_k)$$

$$\le Q_i(t_k) + d^M Q_i(t_k) = Q_i(t_k^+).$$

We apply again the comparison result (4.51) in the interval $(t_k, t_{k+1}]$, and obtain

$$u_i(t; t_0, u_0) \le Q_i^k(t; t_k, Q_i^{k+}) = Q_i(t; t_0, q_0)$$

i.e. the inequality (4.49) is valid for $(t_k, t_{k+1}]$.

The proof of (4.49) is completed by induction.

Further, by analogous arguments, using [177], we obtain from (4.46) and (4.51) that

$$
\begin{cases}
\dot{u}_i(t) \ge u_i(t)\left[r_i(t) - a_i(t)u_i(t) - \displaystyle\sum_{j=1, j\ne i}^{n} a_{ij}(t) \sup_{t\in\mathbb{R}} Q_i(t) \right] \\
\quad - \displaystyle\sum_{j=1}^{n} b_{ij}(t) \sup_{t\in\mathbb{R}} Q_i(t), \; t \ne t_k, \\
\Delta u_i(t_k) \ge d_L u_i(t_k), \; k = \pm 1, \pm 2, \dots,
\end{cases}
$$

$i = 1, \dots, n$, $n \ge 2$, and hence $u_{i0} \ge P_i(t_0^+)$ implies that

$$u_i(t) \ge P_i(t) \tag{4.52}$$

for all $t \in \mathbb{R}$ and $i = 1, 2, \dots, n$, where $P_i(t)$ is the minimal solution of the logistic system

$$
\begin{cases}
\dot{p}_i(t) = p_i(t)\left[r_{iL} - a_i(t)p_i(t) - \displaystyle\sum_{\substack{j=1 \\ j\ne i}}^{n} a_{ij}(t) \sup_{t\in\mathbb{R}} Q_i(t) \right] \\
\quad - \displaystyle\sum_{j=1}^{n} b_{ij}(t) \sup_{t\in\mathbb{R}} Q_i(t), \; t \ne t_k, \\
p_i(t_0^+) = p_{i0} > 0, \\
\Delta p_i(t_k) = d_L p_i(t_k), \; k = \pm 1, \pm 2, \dots,
\end{cases}
\tag{4.53}
$$

$i = 1, \dots, n$, and $d_L = \displaystyle\min_{k=\pm 1, \pm 2, \dots} d_k$ for $1 \le i \le n$. Thus, the proof follows from the last system and (4.49). $\square$

Lemma 4.11. *Let the following conditions hold:*

1. *Conditions H4.16, H4.18–H4.20 are met.*
2. $u(t) = col(u_1(t), u_2(t), \ldots, u_n(t))$ *be a solution of (4.46) such that* $u_i(t_0^+) > 0, \ 1 \le i \le n.$

Then:

1. $u_i(t) > 0, \ 1 \le i \le n, \ t \in \mathbb{R}.$
2. *For* $t \in \mathbb{R}$ *and* $1 \le i \le n$ *there exist constants* $A > 0, \ B > 0,$ *such that*

$$A \le u_i(t) \le B.$$

Proof. Under hypotheses H4.16, H4.18–H4.20, we consider the non impulsive Lotka–Volterra system

$$
\begin{cases}
\dot{y}_i(t) = y_i(t)\left[r_i(t) - A_i(t)y_i(t) - \displaystyle\sum_{j=1, j\neq i}^{n} A_{ij}(t)y_j(t)\right] + \\
\quad + \displaystyle\sum_{j=1}^{n} B_{ij}[y_j(t) - y_i(t)], \ t \neq t_k, \ t > t_0,
\end{cases}
\tag{4.54}
$$

where

$$A_i(t) = a_i(t) \prod_{0 < t_k < t} (1 + d_k), \quad A_{ij}(t) = a_{ij}(t) \prod_{0 < t_k < t} (1 + d_k),$$

$$B_{ij}(t) = b_{ij}(t) \prod_{0 < t_k < t} (1 + d_k).$$

We shall prove that if $y_i(t)$ is a solution of (4.54), then $u_i = y_i \displaystyle\prod_{0 < t_k < t} (1 + d_k)$ is a solution of (4.46), $1 \le i \le n$.

In fact, for $t \neq t_k$ it follows

$$\dot{u}_i(t) - u_i(t)\left[r_i(t) - a_i(t)u_i(t) - \sum_{j=1, j\neq i}^{n} a_{ij}(t)u_j(t)\right] - \sum_{j=1}^{n} b_{ij}[u_j(t) - u_i(t)]$$

$$= \dot{y}_i(t) \prod_{0 < t_k < t} (1 + d_k) - y_i(t) \prod_{0 < t_k < t} (1 + d_k)\Big[r_i(t)$$

$$- a_i(t)y_i(t) \prod_{0 < t_k < t} (1 + d_k) - \sum_{j=1, j\neq i}^{n} a_{ij}(t)y_j(t) \prod_{0 < t_k < t} (1 + d_k)\Big]$$

$$- \sum_{j=1}^{n} b_{ij}\Big[y_j(t) \prod_{0 < t_k < t} (1 + d_k) - y_i(t) \prod_{0 < t_k < t} (1 + d_k)\Big]$$

$$= \prod_{0<t_k<t} (1+d_k)\Big[\dot{y}_i(t) - y_i(t)\Big[r_i(t) - A_i(t)y_i(t) - \sum_{j=1,j\neq i}^{n} A_{ij}(t)y_j(t)\Big]$$

$$- \sum_{j=1}^{n} B_{ij}(t)[y_j(t) - y_i(t)]\Big] \equiv 0. \tag{4.55}$$

For $t = t_k$, we have

$$u_i(t_k^+) = \lim_{t\to t_k^+} \prod_{0<t_k<t} (1+d_k)y_i(t) = \prod_{0<t_k<t} (1+d_k)y_i(t_k),$$

and

$$u_i(t_k) = \prod_{0<t_k<t} (1+d_k)y_i(t_k).$$

Thus, for every $k = \pm 1, \pm 2, \ldots$

$$u_i(t_k^+) = \prod_{0<t_k<t} (1+d_k)y_i(t_k). \tag{4.56}$$

From (4.55) and (4.56) it follows that $u_i(t)$ is the solution of (4.46).

The proof that if $u_i = y_i \prod_{0<t_k<t} (1 + d_k)$ is a solution of (4.46), then $y_i(t)$, $1 \leq i \leq n$ is a solution of (4.54) is analogous.

From [189] it follows that for the system without impulses (4.54) exists a positive solution on $t \in \mathbb{R}$.

Then, from (4.55) and (4.56) it follows that $u_i(t) > 0, 1 \leq i \leq n, t \in \mathbb{R}$.

Again from [189] under the conditions of the lemma for the solutions of (4.50) and (4.53) it is valid that

$$\alpha_i \leq P_i(t), \ Q_i(t) \leq \beta_i,$$

where $\alpha_i > 0$, $0 < \beta_i < \infty$ for all $t \neq t_k$, $1 \leq i \leq n$, and then

$$\alpha_i \leq u_i(t) \leq \beta_i.$$

Also, since the solution $u_i(t)$ is left continuous, and at $t = t_k$, we have for $t = t_1$ that

$$\alpha_i \leq u_i(t_1) \leq \beta_i.$$

On the other hand,

$$(1 + d_1)a_i < u_i(t_1^+) \leq (1 + d_1)\beta_i \leq \beta_i.$$

By analogous arguments for the $t \in (t_{k-1}, t_k]$ it follows

$$\prod_{l=1}^{k}(1+d_l)\alpha_i \le u_i(t_k^+) \le \beta_i.$$

Then for all $t \in \mathbb{R}$, we have

$$A \le u_i(t) \le B,$$

where

$$A = \min_i\left\{\alpha_i \prod_{k=\pm 1,\pm 2,\dots}(1+d_k)\right\}, \quad B = \min_i \beta_i. \qquad \square$$

Theorem 4.5. *Let the following conditions hold:*

1. *Conditions H4.16, H4.18–H4.20 are met.*
2. *There exist nonnegative almost periodic functions $\delta_\nu(t)$, $1 \le \nu \le n$ such that*

$$a_\nu(t) - \sum_{i=1,i\neq\nu}^{n} a_{i\nu}(t) - \frac{1}{A}\sum_{i=1}^{n} b_{i\nu}(t) \ge \delta_\nu(t), \ t \neq t_k,$$

for $t \in \mathbb{R}$, $A > 0$, $k = \pm 1, \pm 2, \dots$.

Then:

1. *For the system (4.46) there exists a unique strictly positive almost periodic solution.*
2. *If there exists a constant $c \ge 0$ such that*

$$\int_{t_0}^{t}\delta(t)ds = c(t-t_0),$$

where $\delta(t) = min(\delta_1(t),\delta_2(t),\dots,\delta_N(t))$, then the almost periodic solution is globally exponentially stable.

Proof. The proof follows from Lemmas 4.10 and 4.11 the same way like the proof of Theorem 4.4. $\qquad \square$

4.2.3 Impulsive Lotka–Volterra Models with Delays

Gopalsamy [59] has studied the existence of periodic solution of the following Lotka–Volterra system

$$\dot{u}_i(t) = u_i(t)\Big[b_i(t) - a_{ii}(t)u_i(t)$$

$$- \sum_{j=1,j\neq i}^{n}\int_{-\infty}^{t} k_i(t,s)h_{ij}(t)u_j(s)\,ds\Big], \qquad (4.57)$$

where $i = 1, 2, \ldots, n$ and the delay kernel $k_i(t, s) = k_i(t - s)$ is of convolution type.

Ahmad and Rao [1] have investigated the existence of periodic asymptotically stable solution of the next system of integro-differential equations

$$\dot{u}_i(t) = u_i(t)\Big[b_i(t) - f_i(t, u_i(t))$$

$$-\sum_{j=1}^{n} \int_{-\infty}^{t} k_i(t, s)h_{ij}(t, u_j(s))\,ds\Big], \tag{4.58}$$

$i = 1, 2, \ldots, n$. The paper [1] improves the results of Gopalsamy and some of earlier results on this topic of interest.

If at certain moments of time the evolution of the process from (4.57) or (4.58) is subject to sudden changes, then the population number vary by jumps. So, in this part we shall consider the impulsive nonautonomous competitive Lotka–Volterra system of integro-differential equations with infinite delays and fixed moments of impulsive perturbations in the form

$$\begin{cases} \dot{u}_i(t) = u_i(t)\Big[r_i(t) - f_i(t, u_i(t)) \\ \quad -\displaystyle\sum_{j=1}^{n} \int_{-\infty}^{t} k_i(t, \sigma)h_{ij}(t, u_j(s))\,ds\Big], \ t \ne t_k, \\ \Delta u_i(t_k) = d_{ik}u_i(t_k) + c_i, \ k = \pm 1, \pm 2, \ldots, \end{cases} \tag{4.59}$$

where $1 \le i \le n$, $n \ge 2$, $t \in \mathbb{R}$ and $\{t_k\} \in \mathcal{B}$.

We shall assume that the functions $r_i \in C[\mathbb{R}, \mathbb{R}^+]$, $k_i \in C[\mathbb{R} \times \mathbb{R}, \mathbb{R}^+]$, and $f_i, h_{ij} \in C[\mathbb{R} \times \mathbb{R}^+, \mathbb{R}^+]$, $1 \le i, j \le n$, the constants $d_{ik} \in \mathbb{R}$, $c_i \in \mathbb{R}^+$, $1 \le i \le n$, $k = \pm 1, \pm 2, \ldots$.

Let $u^{t_0} : (-\infty, t_0] \to \mathbb{R}^n$, $u^{t_0} = col(u_1^{t_0}, u_2^{t_0}, \ldots, u_n^{t_0})$ is a continuous function. We denote by

$$u(t) = u(t; t_0, u^{t_0}) = col(u_1(t; t_0, u^{t_0}), u_2(t; t_0, u^{t_0}), \ldots, u_n(t; t_0, u^{t_0}))$$

the solution of the system (4.59), satisfying the initial conditions

$$\begin{cases} u(s; t_0, u^{t_0}) = u^{t_0}(s), \ s \in (-\infty, t_0], \\ u(t_0^+; t_0, u^{t_0}) = u(t_0). \end{cases} \tag{4.60}$$

Note that the solution $u(t) = u(t; t_0, u^{t_0})$ of the problem (4.59), (4.60) is a piecewise continuous function with points of discontinuity of the first kind at the momento t_k, $k = +1, +2, \ldots$, at which it is left continuous, i.e. the following relations are satisfied:

$$u_i(t_k^-) = u_i(t_k),$$

$$u_i(t_k^+) = u_i(t_k) + d_{ik}u_i(t_k) + c_i, \quad k = \pm 1, \pm 2, \ldots, \ 1 \le i \le n.$$

In our subsequent analysis, we shall consider only initial functions that belong to a class of bounded continuous function.

Let $BC = BC[(-\infty, t_0], \mathbb{R}^n]$ be the set of all bounded continuous functions from $(-\infty, t_0]$ into $\mathbb{R}^n$, and let $u^{t_0}(.) \in BC$. If $u(t)$ is an $\mathbb{R}^n$-valued function on $(-\infty, \beta)$, $\beta \le \infty$, we define for each $t \in (-\infty, \beta)$, $u^t(.)$ to be the restriction of $u(s)$ given by $u^t(s) = u(t+s)$, $-\infty < s \le t$, and the norm is defined by

$$||u^t(.)|| = \sup_{-\infty < s \le t} ||u(s)||.$$

It is clear that $||u(t)|| \le ||u^t(.)||$.

Introduce the following conditions:

H4.21.   The functions $r_i(t)$, $1 \le i \le n$ are nonnegative, almost periodic and $r_{iL} > 0$, $r_{im} < \infty$.

H4.22.   The functions $k_i(t, \sigma) > 0$ are uniformly continuous, almost periodic with respect to t, integrable with respect to σ on $(-\infty, t_0]$ and there exist positive numbers μ_i such that

$$\int_{-\infty}^t k_i(t, \sigma)\, d\sigma \le \mu_i < \infty,$$

for all $t \in \mathbb{R}$, $t \ne t_k$, $k = \pm 1, \pm 2, \ldots$, and $1 \le i \le n$.

H4.23.   The functions $f_i(t, u_i)$ are almost periodic on t uniformly with respect to $u_i \in \mathbb{R}^+$, $f_i(t, u_i) > 0$ for $u_i > 0$, $f_i(t, 0) = 0$, and there exist positive almost periodic continuous functions $L_i(t)$ such that

$$|f_i(t, u_i) - f_i(t, \nu_i)| \ge L_i(t)|u_i - \nu_i|,$$

for all (t, u_i), $(t, \nu_i) \in \mathbb{R} \times \mathbb{R}^+$, and $(u_i - \nu_i)|f_i(t, u_i) - f_i(t, \nu_i)| > 0$ where $u_i \ne \nu_i$, $1 \le i \le n$.

H4.24.   The functions $h_{ij}(t, u_i)$ are almost periodic on t uniformly with respect to $u_i \in \mathbb{R}^+$, $1 \le i, j \le n$, $h_{ij}(t, u_i) > 0$ for $u_i > 0$, $h_{ij}(t, 0) = 0$, and there exist positive almost periodic continuous functions $L_{ij}(t)$ such that

$$|h_{ij}(t, u_i) - h_{ij}(t, \nu_i)| \le L_{ij}(t)|u_i - \nu_i|,$$

for all (t, u_i), $(t, \nu_i) \in \mathbb{R} \times \mathbb{R}^+$, and $L_{ij}(t)$ are nonincreasing for $t \in \mathbb{R}$ and $1 \le i, j \le n$, $i \ne j$.

H4.25. $c^M < \infty$, $c_L > 0$, where $c^M = max\{c_i\}$ and $c_L = min\{c_i\}$ for $1 \le i \le n$.

Let conditions H4.18, H4.21–H4.25 hold and let $\{s'_m\}$ be an arbitrary sequence of real numbers. Then there exists a subsequence $\{s_l\}$, $s_l = s'_{m_l}$ such that the system (4.59) moves to the system

$$
\begin{cases}
\dot{u}_i(t) = u_i(t)\Big[r_i^s(t) - f_i^s(t, u_i(t)) \\
\quad - \displaystyle\sum_{j=1}^{n} \int_{-\infty}^{t} k_i^s(t,s)h_{ij}^s(t, u_j(s))\,ds\Big],\ t \neq t_k^s, \\
\Delta u_i(t_k^s) = d_{ik}^s u_i(t_k^s) + c_i,\ k = \pm 1, \pm 2, \ldots.
\end{cases}
\tag{4.61}
$$

In the proof of the main results we shall use the next lemmas.

Lemma 4.12. *Let the conditions H4.17, H4.18, H4.21–H4.25 hold, and*

$$
\int_{-\infty}^{t} k_i(t,s)h_{ij}(t, u_j(s))ds
$$

be continuous for $t \in [t_0, \infty)$, $i, j = 1, 2, \ldots, n$.

$$
J^+(t_0, u^{t_0}) = [t_0, \infty).
$$

Proof. If conditions H4.17, H4.18, H4.21–H4.25 hold and $\int_{-\infty}^{t} k_i(t,s)h_{ij}(t, u_j(s))ds$ is continuous for $t \in [t_0, \infty)$, $i, j = 1, 2, \ldots, n$, then for the initial problem (4.59), (4.60) there exists a unique solution [1, 37] which is defined on $[t_0, t_1] \cup (t_k, t_{k+1}]$, $k = \pm 1, \pm 2, \ldots$. From H4.25 and H4.18 it follows that $J^+(t_0, u^{t_0}) = [t_0, \infty)$.   □

The proof of the next lemma is similar to the proof of Lemma 4.11.

Lemma 4.13. *Let the following conditions hold:*

1. *Conditions H4.17, H4.18 and H4.21–H4.25 are met.*
2. *There exist functions P_i, $Q_i \in PC^1[[t_0, \infty), \mathbb{R}]$ such that*

$$
P_i(t_0^+) \leq u_i^{t_0}(s) \leq Q_i(t_0^+),
$$

where $s \leq t_0$, $t_0 \in \mathbb{R}$, $1 \leq i \leq n$.

Then

$$
P_i(t) \leq u_i(t) \leq Q_i(t)
\tag{4.62}
$$

for all $t \geq t_0$ and $1 \leq i \leq n$.

Lemma 4.14 ([177]). *Let the following conditions hold:*

1. *Conditions of Lemma 4.11 are met.*
2. *$u_i(t) = u_i(t; t_0, u_i^{t_0})$ be a solution of (4.59), (4.60) such that*

$$
u_i(s) = u_i^{t_0}(s) \geq 0,\ \sup u_i^{t_0}(s) < \infty,\ u_i^{t_0} > 0,\ 1 \leq i \leq n.
\tag{4.63}
$$

3. *For $1 \leq i \leq n$ and $k = \pm 1, \pm 2, \dots$*

$$1 + d_{ik} > 0.$$

Then:

1. *$u_i(t) > 0$, $1 \leq i \leq n$, $t > t_0$.*
2. *There exist positive constants α_i and β_i such that*

$$\alpha_i \leq u_i(t) \leq \beta_i,$$

for all $t > t_0$, $1 \leq i \leq n$, and if in addition

$$-d_{ik}\alpha_i < c_i < -d_{ik}\beta_i$$

then

$$\alpha_i \leq u_i(t) \leq \beta_i,$$

for $t \geq t_0$ and $1 \leq i \leq n$.

The proof of the next lemma is similar of the proof of Lemma 4.9.

Lemma 4.15. *Let the following conditions hold:*

1. *Conditions H4.17, H4.18 and H4.21–H4.25 are met.*
2. *$\{s_n\}$ be an arbitrary sequence of real numbers.*
3. *For the system (4.61) there exists a strictly positive solution.*

Then the system (4.59) has a unique strictly positive almost periodic solution.

Let $u_i(t; t_0, u^{t_0})$ and $v_i(t, t_0, v^{t_0})$, $1 \leq i \leq n$, (t_0, u^{t_0}), $(t_0, v^{t_0}) \in \mathbb{R} \times BC$ be any two solutions of (4.59) such that

$$u_i(\sigma) = u_i^{t_0}(\sigma) \geq 0, \quad \sup_{\sigma \in (-\infty, t_0]} u_i^{t_0}(\sigma) < \infty, \ u_i^{t_0}(t_0) > 0.$$

$$v_i(\sigma) = v_i^{t_0}(\sigma) \geq 0, \quad \sup_{\sigma \in (-\infty, t_0]} v_i^{t_0}(\sigma) < \infty, \ v_i^{t_0}(t_0) > 0.$$

We shall use the Lyapunov function

$$V(u(t), v(t)) = \sum_{i=1}^{n} \left| \ln \frac{u_i(t)}{v_i(t)} \right|. \tag{4.64}$$

By Mean Value Theorem it follows that for any closed interval contained in $(t_{k-1}, t_k]$, $k = \pm 1, \pm 2, \dots$ there exist positive numbers r and R such that for $1 \leq i \leq n$, it follows that $r \leq u_i(t)$, $v_i(t) \leq R$ and

$$\frac{1}{R}|u_i(t) - v_i(t)| \leq \left| \ln u_i(t) - \ln v_i(t) \right| \leq \frac{1}{r}|u_i(t) - v_i(t)|. \tag{4.65}$$

Theorem 4.6. *Let the following conditions hold:*

1. *Conditions H4.17, H4.18 and H4.21–H4.25 are met.*
2. *There exist nonnegative almost periodic continuous functions $\delta_\nu(t)$, $1 \leq \nu \leq n$ such that*

$$rL_\nu(t) - R\sum_{j=1}^{n} \mu_\nu L_{\nu j}(t) > \delta_\nu(t), \ t \neq t_k, \ k = \pm 1, \pm 2, \ldots. \tag{4.66}$$

Then:

1. *For the system (4.59) there exists a unique strictly positive almost periodic solution.*
2. *If there exists a constant $c \geq 0$ such that*

$$\int_{t_0}^{t} \delta(s)ds = c(t - t_0),$$

where $\delta(t) = min(\delta_1(t), \delta_2(t), \ldots, \delta_n(t))$, then the almost periodic solution is globally exponentially stable.

Proof. Let the conditions H4.17, H4.18 and H4.21–H4.25 hold, and let $\{s_l\}$ be an arbitrary sequence of real numbers. If $u^s(t)$ is a solution of (4.61), from Lemma 4.14, we get

$$0 < \inf_{t \geq t_0} u_i^s(t) \leq \sup_{t \geq t_0} u_i^s(t) < \infty, \ 1 \leq i \leq n. \tag{4.67}$$

Suppose that the system (4.61) has two arbitrary strictly positive solutions

$$u^s = col(u_1^s(t), u_2^s(t), \ldots, u_n^s(t)), \ v^s = col(v_1^s(t), v_2^s(t), \ldots, v_n^s(t)).$$

Consider the Lyapunov function

$$V^s(u^s(t), v^s(t)) = \sum_{i=1}^{n} \left| \ln \frac{u_i^s(t)}{v_i^s(t)} \right|.$$

Then for $t \in \mathbb{R}$, $t \neq t_k^s$, $k = \pm 1, \pm 2, \ldots$, and hypotheses H4.17, H4.18 and H4.21–H4.25, we have

$$D^+V^s(u^s(t), v^s(t)) = \sum_{i=1}^{n} \left(\frac{\dot{u}_i^s(t)}{u_i^s(t)} - \frac{\dot{v}_i^s(t)}{v_i^s(t)} \right) sgn\left(u_i^s(t) - v_i^s(t) \right)$$

$$\leq \sum_{i=1}^{n} \Big[-L_i^s(t)|u_i^s(t) - v_i^s(t)|$$

$$+ \sum_{j=1}^{n} \int_{-\infty}^{t} k_i^s(t,\sigma)L_{ij}^s(t)|u_j^s(\sigma) - v_j^s(\sigma)|d\sigma \Big].$$

Thus in view of hypothesis (4.66) we obtain

$$D^+ V^s(u^s(t), v^s(t)) \leq -\delta^s(t)m^s(t), \; t \in \mathbb{R}, \; t \neq t_k^s, \qquad (4.68)$$

where $\delta(t + s_l) \to \delta^s(t), \; l \to \infty.$

$$\delta^s(t) = min(\delta_1^s(t), \delta_2^s(t), \ldots, \delta_n^s(t)), \; m^s(t) = \sum_{i=1}^{n} |u_i^s - v_i^s|.$$

On the other hand, for $t = t_k^s$, we have

$$V^s(u^s(t_k^{s+}), v^s(t_k^{s+})) = \sum_{i=1}^{n} \left| \ln \frac{u_i^s(t_k^{s+})}{v_i^s(t_k^{s+})} \right| = \sum_{i=1}^{n} \left| \ln \frac{(1+d_{ik}^s)u_i^s(t_k^s) + c_i}{(1+d_{ik}^s)v_i^s(t_k^s) + c_i} \right|$$

$$\leq \sum_{i=1}^{n} \left| \ln \frac{(1+d_{ik}^s)R - d_{ik}^s R}{(1+d_{ik}^s)r - d_{ik}^s r} \right| = \sum_{i=1}^{n} \left| \ln \frac{R}{r} \right|$$

$$= \sum_{i=1}^{n} \left| -\ln \frac{R}{r} \right| = \sum_{i=1}^{n} \left| \ln \frac{r}{R} \right| \leq \sum_{i=1}^{n} \left| \ln \frac{u_i^s(t_k^s)}{v_i^s(t_k^s)} \right|$$

$$= V^s(u^s(t_k^s), v^s(t_k^s)). \qquad (4.69)$$

From (4.68) and (4.69) it follows that for $t < t_0$,

$$\int_{t}^{t_0} \delta^s(t)m^s(t)dt \leq V^s(u^s(t), v^s(t)) - V^s(u^{st_0}, v^{st_0}).$$

Then from the almost periodicity of the right hand of system (4.59) and definition of $V^s(u^s(t), v^s(t))$, for the last inequality it follows that

$$\int_{-\infty}^{t_0} |u_i^s(\sigma) - v_i^s(\sigma)|d\sigma < \infty, \; 1 \leq i \leq n,$$

and then $|u_i^s(t) - v_i^s(t)| \to 0$ as $t \to -\infty.$

Hence, from (4.69), we obtain $V^s(u^s(t), v^s(t)) \to 0$ for $t \to -\infty.$
Analogously, from (4.68), we find

$$\int_{t_0}^{t} |u^s(\sigma) - v^s(\sigma)|d\sigma \leq V^s(u^{st_0}, v^{st_0}) - V^s(u^s(t), v^s(t)), \; t \geq t_0.$$

Therefore,

$$\int_{t_0}^{\infty} |u^s(\sigma) - v^s(\sigma)| d\sigma < \infty, \ 1 \le i \le n.$$

Let

$$\nu^s = \inf_{t \in \mathbb{R}} \{u_i^s(t), v_i^s(t), \ 1 \le i \le n\}.$$

From definition of $V^s(u^s(t), v^s(t))$, we have

$$V^s(u^s(t), v^s(t)) = \sum_{i=1}^{n} \left| \ln \frac{u_i^s(t)}{v_i^s(t)} \right| \le \sum_{i=1}^{n} \frac{1}{\nu^s} |u_i^s(t) - v_i^s(t)|.$$

Hence $V^s(u^s(t), v^s(t)) \to 0, \ t \to \infty$.

We have that $V^s(u^s(t), v^s(t))$ is non increasing nonnegative function on $\mathbb{R}$ and consequently

$$V^s(u^s(t), v^s(t)) \equiv 0, \tag{4.70}$$

for $t \ne t_k^s, \ t \in \mathbb{R}$. From (4.68), (4.69) and (4.70) it follows that $u_i^s(t) \equiv v_i^s(t)$ for all $t \in \mathbb{R}$ and $1 \le i \le n$. Then, every system from (4.60) has at least one strictly positive solution.

From Lemma 4.15, analogously it follows that system (4.59) has a unique strictly positive almost periodic solution.

Let for the system (4.59) there exists another bounded strictly positive solution $v_i(t; t_0, v^{t_0}), \ 1 \le i \le n, \ (t_0, v^{t_0}) \in \mathbb{R} \times BC$.

Now we consider again the Lyapunov function $V(u(t), v(t))$ and obtain

$$V(u^{t_0}, v^{t_0}) = \sum_{i=1}^{n} \left| \ln \frac{u_i(t_0)}{v_i(t_0)} \right| \le \frac{1}{r} ||u^{t_0} - v^{t_0})||. \tag{4.71}$$

On the other hand, for $t \in \mathbb{R}, \ t \ne t_k$,

$$D^+V(u(t), v(t)) \le -\delta(t)m(t) \le -\delta(t)rV(u(t), v(t)). \tag{4.72}$$

For $t \in \mathbb{R}, \ t = t_k$, it follows

$$V(u(t_k^+), v(t_k^+)) = \sum_{i=1}^{n} \left| \ln \frac{u_i(t_k^+)}{v_i(t_k^+)} \right| = \sum_{i=1}^{n} \left| \ln \frac{(1+d_{ik})u_i(t_k) + c_i}{(1+d_{ik})v_i(t_k) + c_i} \right|$$

$$\le \sum_{i=1}^{n} \left| \ln \frac{(1+d_{ik})R - d_{ik}R}{(1+d_{ik})r - d_{ik}r} \right| = \sum_{i=1}^{n} \left| \ln \frac{R}{r} \right|$$

$$= \sum_{i=1}^{n} \left| -\ln \frac{R}{r} \right| = \sum_{i=1}^{n} \left| \ln \frac{r}{R} \right| \le \sum_{i=1}^{n} \left| \ln \frac{u_i(t_k)}{v_i(t_k)} \right|$$

$$= V^s(u(t_k), v(t_k)). \tag{4.73}$$

From (4.71), (4.72) and (4.73), it follows

$$V(u(t), v(t)) \leq V(u^{t_0}, v^{t_0}) \exp\left\{ -r \int_{t_0}^{t} \delta(\sigma)d\sigma \right\}. \tag{4.74}$$

Therefore, from (4.71), (4.72) and (4.73) we deduce the inequality

$$\sum_{i=1}^{N} |u_i(t) - v_i(t)| \leq \frac{R}{r} ||u^{t_0} - v^{t_0})|| e^{-rc(t-t_0)}, \ t \geq t_0.$$

This shows that the unique almost periodic solution $u(t)$ of the system (4.59) is globally exponentially stable. ☐

Example 4.3. We shall consider the impulsive nonautonomous competitive Lotka–Volterra system

$$
\begin{cases}
\dot{u}_1(t) = u_1(t)\left[5\sqrt{2} - 9u_1(t) - \int_{-\infty}^{t} k_1(t,\sigma)u_2(\sigma)\,d\sigma \right], \\
\dot{u}_2(t) = u_2(t)\left[\dfrac{5\sqrt{2}}{9} - \dfrac{2}{3}\int_{-\infty}^{t} k_2(t,\sigma)u_1(\sigma)\,d\sigma - \dfrac{16}{3}u_2(t) \right], \ t \neq t_k, \\
u_1(t_k^+) = \dfrac{3\sqrt{2} - 2u_1(t_k)}{4}, \\
u_2(t_k^+) = \dfrac{\sqrt{2} - 2u_2(t_k)}{3}, \ k = \pm 1, \pm 2, \dots .
\end{cases}
\tag{4.75}
$$

Let for the sequence $\{t_k\} \in \mathcal{B}$ condition H4.18 holds.
From

$$\int_{-\infty}^{t} k_1(t,\sigma)d\sigma = \mu_1 = 2, \quad \int_{-\infty}^{t} k_2(t,\sigma)d\sigma = \mu_2 = 1,$$

we have

$$
\begin{cases}
rL_1(t) - \nu_1 RL_{12}(t) = \dfrac{1}{4}.9 - 2.1.1 = \dfrac{1}{4} = \delta_1, \\
rL_2(t) - \nu_2 RL_{21}(t) = \dfrac{1}{4}.\dfrac{16}{3} - 1.1.\dfrac{2}{3} = \dfrac{2}{3} = \delta_2, \\
\delta = min(\delta_1, \delta_2) = \dfrac{1}{4}, \ \int_{t_0}^{t} \delta ds = \int_{t_0}^{t} \dfrac{1}{4}ds = \dfrac{1}{4}(t - t_0), \\
-1 < d_{1k} = -\dfrac{1}{2} < 0, \ -1 < d_{2k} = -\dfrac{1}{3} < 0, \ c_1 = \dfrac{3\sqrt{2}}{4}, \ c_2 = \dfrac{\sqrt{2}}{3}.
\end{cases}
$$

Then for the system (4.75) all conditions of Theorem 4.6 hold, and consequently, there exists a unique positive almost periodic globally exponentially
stable solution of (4.75).

4.3 Neural Networks

In the present Section, the problems of existence and uniqueness of almost
periodic solutions for impulsive neural networks are considered.

Neural networks have been successfully employed in various areas such
as pattern recognition, associative memory and combinatorial optimization
[30–34, 38, 44, 48, 60, 70, 72–74, 81, 120, 121, 133, 169, 175].

While an artificial neural network has been known insofar for its transient
processing behavior, its circuit design has never been disentangled from
destabilizing factors such as impulses.

Impulses can make unstable systems stable so they have been widely used
in many fields such as physics, chemistry, biology, population dynamics, and
industrial robotics. The abrupt changes in the voltages produced by faulty
circuit elements are exemplary of impulse phenomena that can affect the
transient behavior of the network. Some results for impulsive neural networks
have been given, for example, see [6, 169, 175–177, 202] and references therein.

4.3.1 Impulsive Neural Networks

In this part we shall investigate the problem of existence of almost periodic
solutions of the system of impulsive Hopfield neural networks

$$
\begin{cases}
\dot{x}_i(t) = \sum_{j=1}^{n} a_{ij}(t)x_j(t) + \sum_{j=1}^{n} \alpha_{ij}(t)f_j(x_j(t)) + \gamma_i(t), \ t \neq t_k, \\
\Delta x(t_k) = A_k x(t_k) + I_k(x(t_k)) + p_k, \ k = \pm 1, \pm 2, \ldots,
\end{cases}
\tag{4.76}
$$

where $t \in \mathbb{R}$, $\{t_k\} \in \mathcal{B}$, $a_{ij}, \alpha_{ij}, \ f_j, \ \gamma_i \in C[\mathbb{R}, \mathbb{R}]$, $i = 1, 2, \ldots, n$, $j =$
$1, 2, \ldots, n$, $x(t) = col(x_1(t), x_2(t), \ldots, x_n(t))$, $A_k \in \mathbb{R}^{n \times n}$, $I_k \in C[\Omega, \mathbb{R}^n]$, $p_k \in$
$\mathbb{R}^n$, $k = \pm 1, \pm 2, \ldots$.

The solution $x(t) = x(t; t_0, x_0)$ of (4.76) with the initial condition $x(t_0^+) =$
x_0 is a piecewise continuous function with points of discontinuity at the
moments t_k, $k = \pm 1, \pm 2, \ldots$ at which it is continuous from the left.

Together with system (4.76) we shall consider the linear system

$$
\begin{cases}
\dot{x}(t) = A(t)x(t), \ t \neq t_k, \\
\Delta x(t_k) = A_k x(t_k,), \ k = \pm 1, \pm 2, \ldots,
\end{cases}
\tag{4.77}
$$

where $A(t) = (a_{ij}(t))$, $i = 1, 2, \ldots n$, $j = 1, 2, \ldots, n$.

Introduce the following conditions:

H4.26.   The matrix function $A \in C[\mathbb{R}, \mathbb{R}^{n \times n}]$ is almost periodic in the sense of Bohr.

H4.27. $det(E + A_k) \neq 0$ and the sequence $\{A_k\}$, $k = \pm 1, \pm 2, \ldots$ is almost periodic, E is the identity matrix in $\mathbb{R}^{n \times n}$.

H4.28.   The set of sequences $\{t_k^j\}$, $t_k^j = t_{k+j} - t_k$, $k, j = \pm 1, \pm 2, \ldots$ is uniformly almost periodic, and there exists $\theta > 0$ such that $\inf\limits_{k} t_k^1 = \theta > 0$.

From Chap. 1 it follows that if $U_k(t, s,)$ is the Cauchy matrix for system

$$\dot{x}(t) = A(t)x(t), \ t_{k-1} < t \leq t_k,$$

then the Cauchy matrix of the system (4.77) is in the form

$$W(t, s) = \begin{cases} U_k(t, s) \ \text{as} \ t, s \in (t_{k-1}, t_k], \\ U_{k+1}(t, t_k^+)(E + A_k)U_k(t_k, s) \ \text{as} \ t_{k-1} < s \leq t_k < t \leq t_{k+1}, \\ U_k(t, t_k)(E + A_k)^{-1}U_{k+1}(t_k^+, s) \ \text{as} \ t_{k-1} < t \leq t_k < s \leq t_{k+1}, \\ U_{k+1}(t, t_k^+) \prod\limits_{j=k}^{i+1} (E + A_j)U_j(t_j, t_{j-1}^+)(E + A_i)U_i(t_i, s) \\ \quad \text{as} \ t_{i-1} < s \leq t_i < t_k < t \leq t_{k+1}, \\ U_i(t, t_i) \prod\limits_{j=i}^{k-1} (E + A_j)^{-1}U_{j+1}(t_j^+, t_{j+1})(E + A_k)^{-1}U_{k+1}(t_k^+, s) \\ \quad \text{as} \ t_{i-1} < t \leq t_i < t_k < s \leq t_{k+1}, \end{cases}$$

and the solutions of (4.77) can to write in the form

$$x(t; t_0, x_0) = W(t, t_0)x_0.$$

Introduce the next conditions:

H4.29.   The functions $f_j(t)$ are almost periodic in the sense of Bohr, and

$$0 < \sup_{t \in \mathbb{R}} |f_j(t)| < \infty, \ f_j(0) = 0,$$

and there exists $L_1 > 0$ such that for $t, s \in \mathbb{R}$

$$\max_{j=1,2,\ldots,n} |f_j(t) - f_j(s)| < L_1 |t - s|.$$

H4.30.   The functions $\alpha_{ij}(t)$ are almost periodic in the sense of Bohr, and

$$0 < \sup_{t \in \mathbb{R}} |\alpha_{ij}(t)| = \overline{\alpha}_{ij} < \infty.$$

H4.31.   The functions $\gamma_i(t)$, $i = 1, 2, \ldots, n$ are almost periodic in the sense of Bohr, the sequence $\{p_k\}$, $k = \pm 1, \pm 2, \ldots$ is almost periodic and there exists $C_0 > 0$ such that

$$max\{\max_{i=1,2,\ldots,n} |\gamma_i(t)|, \max_{k=\pm 1,\pm 2,\ldots,n} ||p_k||\} \leq C_0.$$

H4.32.   The sequence of functions $\{I_k(x)\}$, $k = \pm 1, \pm 2, \ldots$ is almost periodic uniformly with respect to $x \in \Omega$, and there exists $L_2 > 0$ such that

$$||I_k(x) - I_k(y)|| \leq L_2 ||x - y||,$$

for $k = \pm 1, \pm 2, \ldots$, $x, y \in \Omega$.

Now we need the following lemmas.

Lemma 4.16. *Let the following conditions hold:*

1. *The conditions H4.26–H4.28 are met.*
2. *For the Cauchy matrix $W(t, s)$ of the system (4.77) there exist positive constants K and λ such that*

$$||W(t, s)|| \leq K e^{-\lambda(t-s)}, t \geq s, \quad t, s \in \mathbb{R}.$$

Then for any $\varepsilon > 0$ there exists a relatively dense set $\overline{T}$ of ε-almost periods of the matrix $A(t)$ and a positive constant Γ such that for $\tau \in \overline{T}$ it follows

$$||W(t + \tau, s + \tau) - W(t, s)|| \leq \varepsilon \Gamma e^{-\frac{\lambda}{2}(t-s)}.$$

Proof. The proof is analogous to the proof of Lemma 2 in [138]. $\square$

Now from Lemma 4.16 we have the following lemmas.

Lemma 4.17. *Let the following conditions hold:*

1. *For the matrix $A(t) = diag[-a_1(t), -a_2(t), \ldots, -a_n(t)]$ it follows that $a_i(t)$, $a_i \in PC[\mathbb{R}, \mathbb{R}]$, $i = 1, 2, \ldots, n$ is an almost periodic function in the sense of Bohr, and*

$$\lim_{A \to \infty} \frac{1}{A} \int_t^{t+A} a_i(t) dt > 0, \quad i = 1, 2, \ldots, n.$$

2. *The conditions H4.27 and H4.28 are met.*

Then:

1. *For the Cauchy's matrix $W(t, s)$ it follows*

$$||W(t, s)|| \leq K e^{-\lambda(t-s)},$$

where $t \in \mathbb{R}$, $s \in \mathbb{R}$, $t \geq s$, and K, λ are positive constants.

2. *For each $\varepsilon > 0$ there exists a relatively dense set $\overline{T}$ from ε-almost periods of matrix $A(t)$, and a positive constant Γ such that for $\tau \in \overline{T}$ it follows*

$$\|W(t + \tau, s + \tau) - W(t, s)\| \leq \varepsilon \Gamma e^{-\frac{\lambda}{2}(t-s)}.$$

Lemma 4.18. *Let the following conditions hold:*

1. *Conditions H4.26–H4.29 are met.*
2. *There exists a constant $\lambda > 0$ such that for $t \in \mathbb{R}$ the eigenvalues λ_i, $i = 1, 2, \ldots, n$ of matrix $A(t)$ satisfy the conditions*

$$Re\lambda_i(t) < -\lambda.$$

Then:

1. *For the Cauchy's matrix $W(t, s)$ it follows*

$$\|W(t, s)\| \leq Ke^{-\lambda(t-s)},$$

where $t \in \mathbb{R}$, $s \in \mathbb{R}$, $t \geq s$, K is a positive constant.
2. *For each $\varepsilon > 0$ there exists a relatively dense set $\overline{T}$ from ε-almost periods of matrix $A(t)$, and a positive constant Γ such that for $\tau \in \overline{T}$ it follows*

$$\|W(t + \tau, s + \tau) - W(t, s)\| \leq \varepsilon \Gamma e^{-\frac{\lambda}{2}(t-s)}.$$

The proof of the next lemma is similar to the proof of Lemma 1.7.

Lemma 4.19. *Let conditions H4.26–H4.31 hold. Then for each $\varepsilon > 0$ there exist ε_1, $0 < \varepsilon_1 < \varepsilon$, a relatively dense set $\overline{T}$ of real numbers, and a set P of integer numbers, such that the following relations hold:*

(a) $\|A(t + \tau) - A(t)\| < \varepsilon$, $t \in \mathbb{R}$, $\tau \in \overline{T}$.
(b) $|\alpha_{ij}(t + \tau) - \alpha_{ij}(t)| < \varepsilon$, $\tau \in \overline{T}$, $t \in \mathbb{R}$, $i, j = 1, 2, \ldots, n$.
(c) $|f_j(t + \tau) - f_j(t)| < \varepsilon$, $t \in \mathbb{R}$, $\tau \in \overline{T}$, $j = 1, 2, \ldots, n$.
(d) $|\gamma_j(t + \tau) - \gamma_j(t)| < \varepsilon$, $\tau \in \overline{T}$, $t \in \mathbb{R}$, $j = 1, 2, \ldots, n$.
(e) $\|A_{k+q} - A_k\| < \varepsilon$, $q \in P$, $k = \pm 1, \pm 2, \ldots$.
(f) $|p_{k+q} - p_k| < \varepsilon$, $q \in P$, $k = \pm 1, \pm 2, \ldots$.
(g) $|\tau_{k+q} - \tau| < \varepsilon_1$, $q \in P$, $\tau \in \overline{T}$, $k = \pm 1, \pm 2, \ldots$.

Now, we are at the position to proof the main theorem.

Theorem 4.7. *Let the following conditions hold:*

1. *Conditions H4.26–H4.32 are met.*
2. *For the Cauchy matrix $W(t, s)$ of system (4.77) there exist positive constants K and λ such that*

$$||W(t,s)|| \le Ke^{-\lambda(t-s)}, \ t \ge s, \ t, s \in \mathbb{R}.$$

3. The number

$$r = K\left\{ \max_{i=1,2,\dots,n} \lambda^{-1}L_1 \sum_{j=1}^{n} \overline{\alpha}_{ij} + \frac{L_2}{1-e^{-\lambda}} \right\} < 1.$$

Then:

(1)   There exists a unique almost periodic solution $x(t)$ of (4.76).
(2)   If the following inequalities hold

$$1 + KL_2 < e, \ \lambda - KL_1 \max_{i=1,2,\dots,n} \sum_{j=1}^{n} \overline{\alpha}_{ij} - N\ln(1+KL_2) > 0,$$

then the solution $x(t)$ is exponentially stable.

Proof. We denote by AP, $AP \subset PC[\mathbb{R}, \mathbb{R}^n]$ the set of all almost periodic functions $\varphi(t)$, satisfying the inequality $|\varphi|_\infty < \overline{K}$, where

$$|\varphi|_\infty = \sup_{t \in \mathbb{R}} ||\varphi(t)||, \ \overline{K} = KC_0\left(\frac{1}{\lambda} + \frac{1}{1-e^{-\lambda}}\right).$$

Let

$$\varphi_0 = \int_{-\infty}^{t} W(t,s)\gamma(s)ds + \sum_{t_k < t} W(t,t_k)p_k,$$

where $\gamma(t) = (\gamma_1(t), \gamma_2(t), \dots, \gamma_n(t))$.
  Then

$$|\varphi_0|_\infty = \sup_{t \in \mathbb{R}} \left\{ \max_{i=1,2,\dots,n} \int_{-\infty}^{t} ||W(t,s)|| |\gamma_i(s)| ds \right.$$

$$\left. + \sum_{t_k < t} ||W(t,t_k)|| ||p_k|| \right\}$$

$$\le \sup_{t \in \mathbb{R}} \left\{ \max_{i=1,2,\dots,n} \int_{-\infty}^{t} Ke^{-\lambda(t-s)} |\gamma_i(s)| ds + \sum_{t_k < t} Ke^{-\lambda(t-t_k)} ||p_k|| \right\}$$

$$\le K\left(\frac{C_0}{\lambda} + \frac{C_0}{1-e^{-\lambda}}\right) = \overline{K}. \tag{4.78}$$

Set

$$F(t,x) = col\{F_1(t,x), F_2(t,x), \dots, F_n(t,x)\},$$

where

$$F_i(t,x) = \sum_{j=1}^{n} \alpha_{ij}(t)f_j(x_j), \ i = 1, 2, \dots, n.$$

Now, we define in AP an operator S,

$$S\varphi = \int_{-\infty}^{t} W(t,s)\big[F(s,\varphi(s)) + \gamma(s))\big]\,ds$$

$$+ \sum_{t_k < t} W(t,t_k)\big[I_k(\varphi(t_k)) + p_k\big] \qquad (4.79)$$

and consider a subset AP^*, $AP^* \subset AP$, where

$$AP^* = \Big\{\varphi \in AP : |\varphi - \varphi_0|_\infty \le \frac{r\overline{K}}{1-r}\Big\}.$$

Consequently, for an arbitrary $\varphi \in AP^*$ from (4.78) and (4.79) it follows

$$|\varphi|_\infty \le |\varphi - \varphi_0|_\infty + |\varphi_0|_\infty \le \frac{r\overline{K}}{1-r} + \overline{K} = \frac{\overline{K}}{1-r}.$$

Now, we are proving that S is self-mapping from AP^* to AP^*. For $\varphi \in AP^*$ it follows

$$|S\varphi - \varphi_0|_\infty = \sup_{t\in\mathbb{R}} \Big\{ \max_{i=1,2,\ldots,n} \int_{-\infty}^{t} \|W(t,s)\| \sum_{j=1}^{n} |\alpha_{ij}(s)| |f_j(\varphi_j(s))|\,ds$$

$$+ \sum_{t_k < t} \|W(t,t_k)\| \|I_k(\varphi(t_k))\| \Big\}$$

$$\le \Big\{ \max_{i=1,2,\ldots,n} \int_{-\infty}^{t} K e^{-\lambda(t-s)} \sum_{j=1}^{n} \overline{\alpha}_{ij} L_1\,ds + \sum_{t_k<t} K e^{-\lambda(t-t_k)} L_2 \Big\} |\varphi|_\infty$$

$$\le K\Big\{ \max_{i=1,2,\ldots,n} \lambda^{-1} L_1 \sum_{j=1}^{n} \overline{\alpha}_{ij} + \frac{L_2}{1-e^{-\lambda}} \Big\} |\varphi|_\infty = r|\varphi|_\infty \le \frac{r\overline{K}}{1-r}.$$

$$(4.80)$$

Let $\tau \in \overline{T}$, $q \in P$, where the sets $\overline{T}$ and P are determined in Lemma 4.19. Then

$$|S\varphi(t+\tau) - S\varphi(t)|_\infty \le \sup_{t\in\mathbb{R}} \Big\{ \max_{i=1,2,\ldots,n} \Big(\int_{-\infty}^{t} \|W(t+\tau,s+\tau)$$

$$- W(t,s)\| \Big| \sum_{j=1}^{n} \alpha_{ij}(s+\tau) f_j(\varphi_j(s+\tau)) \Big|\,ds$$

$$+ \int_{-\infty}^{t} \|W(t,s)\| \Big| \sum_{j=1}^{n} \alpha_{ij}(s+\tau) f_j(\varphi_j(s+\tau)) - \sum_{j=1}^{n} \alpha_{ij}(s) f_j(\varphi_j(s)) \Big|\,ds \Big)$$

$$+ \sum_{t_k < t} ||W(t+\tau, t_{k+q}) - W(t, t_k)|| ||I_{k+q}(\varphi(t_{k+q}))||$$

$$+ \sum_{t_k < t} ||W(t, t_k)|| ||I_{k+q}(\varphi(t_{k+q})) - I_k(\varphi(t_k))|| \Big\} \le \varepsilon C_1, \qquad (4.81)$$

where

$$C_1 = \frac{L_1}{\lambda} \Big(\max_{i=1,2,\dots,n} \sum_{j=1}^{n} (2\Gamma + K)\overline{\alpha}_{ij} + K \Big) + \frac{L_2 \Gamma N}{1 - e^{-\lambda}}.$$

and the number N is from Lemma 1.2.

Consequently, after (4.80) and (4.81), we obtain that $S\varphi \in AP^*$.

Let $\varphi \in AP^*$, $\psi \in AP^*$.

Then

$$|S\varphi - S\psi|_\infty$$

$$\le \sup_{t \in \mathbb{R}} \Big\{ \max_{i=1,2,\dots,n} \int_{-\infty}^{t} ||W(t,s)|| \sum_{j=1}^{n} |\alpha_{ij}(s)| |f_j(\varphi_j(s)) - f_j(\psi_j(s))| ds$$

$$+ \sum_{t_k < t} ||W(t, t_k)|| ||I_k(\varphi(t_k)) - I_k(\psi(t_k))|| \Big\}$$

$$\le K \Big(\max_{i=1,2,\dots,n} \lambda^{-1} L_1 \sum_{j=1}^{n} \overline{\alpha}_{ij} + \frac{L_2}{1 - e^{-\lambda}} \Big) |\varphi - \psi|_\infty = r|\varphi - \psi|_\infty.$$

$$(4.82)$$

Then from (4.82) it follows that S is a contracting operator in AP^*, and there exists a unique almost periodic solution of (4.76).

Let now $y(t)$ be an arbitrary solution of (4.76). Then from (4.78), we obtain

$$y(t) - x(t) = W(t, t_0)\big(y(t_0) - x(t_0)\big) + \int_{t_0}^{t} W(t, s)\big[F(s, y(s)) - F(s, x(s))\big] ds$$

$$+ \sum_{t_0 < t_k < t} W(t, t_k)[I_k(y(t_k)) - I_k(x(t_k))].$$

Hence,

$$||y(t) - x(t)|| \le K e^{-\lambda(t-t_0)} ||y(t_0) - x(t_0)||$$

$$+ \max_{i=1,2,\dots,n} \Big(\int_{t_0}^{t} K e^{-\lambda(t-s)} L_1 \sum_{j=1}^{n} \overline{\alpha}_{ij} |y_i(s) - x_i(s)| ds \Big)$$

$$+ \sum_{t_0 < t_k < t} K e^{-\lambda(t-t_k)} L_2 ||y(t_k) - x(t_k)||.$$

Set $u(t) = ||y(t) - x(t)||e^{\lambda t}$ and from Gronwall–Bellman's inequality and Theorem 1.9, for the last inequality we have

$$||y(t) - x(t)||$$

$$\leq K||y(t_0) - x(t_0)||(1+KL_2)^{i(t_0,t)} \exp\left\{ -\lambda + KL_1 \max_{i=1,2,\ldots,n} \sum_{j=1}^{n} \overline{\alpha}_{ij} \right\}(t-t_0). \;\; \square$$

Example 4.4. Now, we shall consider the classical model of impulsive Hopfield neural networks

$$\begin{cases} \dot{x}_i(t) = -\dfrac{1}{R_i}x_i(t) + \displaystyle\sum_{j=1}^{n} \alpha_{ij} f_j(x_j(t)) + \gamma_i(t), \; t \neq t_k, \; i = 1, 2, \ldots, n, \\ \Delta x(t_k) = Gx(t_k) + I_k(x(t_k)) + p_k, \; k = \pm 1, \pm 2, \ldots, \end{cases}$$

(4.83)

where $t \in \mathbb{R}$, $\{t_k\} \in \mathcal{B}$, $R_i > 0$, $\alpha_{ij} \in \mathbb{R}$, $i = 1, 2, \ldots, n$, $j = 1, 2, \ldots, n$, $\gamma_i, f_i \in C[\mathbb{R}, \mathbb{R}]$, $i = 1, 2, \ldots, n$, $x(t) = col(x_1(t), x_2(t), \ldots, x_n(t))$, $I_k \in C[\Omega, \mathbb{R}^n]$, $G = diag[g_i]$, $g_i \in \mathbb{R}$, $i = 1, 2, \ldots, n$, $p_k \in \mathbb{R}^n$.

Theorem 4.8. *Let the following conditions hold:*

1. *Conditions H4.28, H4.29, H4.31 and H4.32 are met.*
2. *The following inequalities hold*

$$\lambda = \min_{i=1,2,\ldots,n} \frac{1}{R_i} - N \max_{i=1,2,\ldots,n} \ln(1 + |g_i|) > 0,$$

$$r = \exp\left\{ N \max_{i=1,2,\ldots,n} \ln(1 + |g_i|) \right\}$$

$$\times \left\{ \max_{i=1,2,\ldots,n} \lambda^{-1} L_1 \sum_{j=1}^{n} \alpha_{ij} + \frac{L_2}{1 - e^{-\lambda}} \right\} < 1.$$

Then:

1. *There exists a unique almost periodic solution $x(t)$ of (4.83).*
2. *If the following inequalities hold*

$$1 + \exp\left\{ N \max_{i=1,2,\ldots,n} \ln(1 + |g_i|) \right\} L_2 < e,$$

$$\lambda - \exp\left(N \max_{i=1,2,\ldots,n} \ln(1 + |g_i|) \right) L_1 \sum_{j=1}^{n} \overline{\alpha}_{ij}$$

$$- N \ln\left(1 + \{ N \max_{i=1,2,\ldots,n} \ln(1 + |g_i|) \} L_2 \right) > 0,$$

then the solution $x(t)$ is exponentially stable.

Proof. Let

$$
\begin{cases}
\dot{x}_i(t) = -\dfrac{1}{R_i} x_i(t), \ t \neq t_k, \\
\Delta x(t_k) = G x(t_k), \ k = \pm 1, \pm 2, \ldots
\end{cases}
\tag{4.84}
$$

is the linear part of (4.83).
Recall [138] the matrix $W(t,s)$ of (4.84) is in the form

$$
W(t,s) = e^{A(t-s)}(E+G)^{i(s,t)}, \ A = diag\left[-\frac{1}{R_1}, -\frac{1}{R_2}, \ldots, -\frac{1}{R_n}\right].
$$

Then

$$
\|W(t,s)\| \leq e^{\displaystyle N \max_{i=1,2,\ldots,n} \ln(1+|g_i|)} e^{-\lambda(t-s)},
$$

$t > s, \ t, s \in \mathbb{R}$, and the proof follows from Theorem 4.7. $\qquad \square$

4.3.2 *Impulsive Neural Networks with Delays*

In this part, we shall investigate the existence and attractivity of almost periodic solutions for impulsive cellular neural networks with delay. The results obtained are a generalization of the results for the dynamics behavior of Hopfield neural networks with delay [34, 38].

We shall investigate the system of impulsive Hopfield neural networks with delay

$$
\begin{cases}
\dot{x}_i(t) = \displaystyle\sum_{j=1}^{n} a_{ij}(t) x_j(t) + \sum_{j=1}^{n} \alpha_{ij}(t) f_j(x_j(t-h)) + \gamma_i(t), \ t \neq t_k, \\
\Delta x(t_k) = A_k x(t_k) + I_k(x(t_k)) + p_k, \ k = \pm 1, \pm 2, \ldots,
\end{cases}
\tag{4.85}
$$

where $t \in \mathbb{R}$, $\{t_k\} \in \mathcal{B}$, $a_{ij}, \alpha_{ij}, f_j, \gamma_i \in C[\mathbb{R}, \mathbb{R}]$, $i = 1, 2, \ldots, n$, $j = 1, 2, \ldots, n$, $h > 0$, $x(t) = col(x_1(t), x_2(t), \ldots, x_n(t))$, $A_k \in \mathbb{R}^{n \times n}$, $I_k \in C[\Omega, \mathbb{R}^n]$, $p_k \in \mathbb{R}^n$, $k = \pm 1, \pm 2, \ldots$.

Let $t_0 \in \mathbb{R}$. Introduce the following notations:

$PC(t_0)$ is the space of all functions $\phi : [t_0 - h, t_0] \to \Omega$ having points of discontinuity at $\xi_1, \xi_2, \ldots, \xi_s \in (t_0 - h, t_0)$ of the first kind and are left continuous at these points.

Let $\phi_0 \in PC(t_0)$. Denote by $x(t) = x(t; t_0, \phi_0)$ the solution of system (4.85), satisfying initial conditions:

$$
\begin{cases}
x(t; t_0, \phi_0) = \phi_0(t), \ t_0 - h \leq t \leq t_0, \\
x(t_0^+; t_0, \phi_0) = \phi_0(t_0).
\end{cases}
\tag{4.86}
$$

The solution $x(t) = x(t; t_0, \phi_0)$ of the initial value problem (4.85), (4.86) is characterized by the following:

(a) For $t_0 - h \leq t \leq t_0$ the solution $x(t)$ satisfied the initial conditions (4.86).

(b) For $t > t_0$ the solution $x(t; t_0, \varphi_0)$ of problem (4.85), (4.86) is a piecewise continuous function with points of discontinuity of the first kind at the moments $t = t_k$, $k = \pm 1, \pm 2, \ldots$ at which it is continuous from the left, i.e., the following relations hold

$$x(t_k^-) = x(t_k), \ x(t_k^+) = x(t_k) + \Delta x(t_k) = x(t_k) + A_k x(t_k) + I_k(x(t_k)) + p_k.$$

(c) If for some integer j we have $t_k < t_j + h < t_{k+1}$, $k = \pm 1, \pm 2, \ldots$, then in the interval $[t_j + h, t_{k+1}]$ the solution $x(t)$ of problem (4.85), (4.86) coincides with the solution of the problem

$$
\begin{cases}
\dot{y}_i(t) = \displaystyle\sum_{j=1}^{n} a_{ij}(t) y_j(t) + \sum_{j=1}^{n} \alpha_{ij}(t) f_j(x_j(t - h^+)) + \gamma_j(t), \\
y(t_j + h) = x(t_j + h),
\end{cases}
$$

and if $t_j + h \equiv t_k$ for $j = 0, 1, 2, \ldots$, $k = \pm 1, \pm 2, \ldots$, then in the interval $[t_j + h, t_{k+1}]$ the solution $x(t)$ coincides with the solution of the problem

$$
\begin{cases}
\dot{y}_i(t) = \displaystyle\sum_{j=1}^{n} a_{ij}(t) y_j(t) + \sum_{j=1}^{n} \alpha_{ij}(t) f_j(x_j(t - h^+)) + \gamma_j(t), \\
y(t_j + h) = x(t_j + h) + A_k x(t_j + h) + I_k(x(t_j + h)) + p_k.
\end{cases}
$$

The proof of the next theorem is similar to the proof of Theorem 4.7.

Theorem 4.9. *Let the following conditions hold:*

1. *Conditions H4.26–H4.32 are met.*
2. *For the Cauchy matrix $W(t, s)$ of the system (4.85) there exist positive constants K and λ such that*

$$\|W(t, s)\| \leq K e^{-\lambda(t-s)}, \ t \geq s, \ t, s \in \mathbb{R}.$$

3. *The number*

$$r = K \left\{ \max_{i=1,2,\ldots,n} \lambda^{-1} L_1 \sum_{j=1}^{n} \overline{\alpha}_{ij} + \frac{L_2}{1 - e^{-\lambda}} \right\} < 1.$$

Then:

1. *There exists a unique almost periodic solution $x(t)$ of (4.85).*
2. *If the following inequalities hold*

$$1 + KL_2 < e, \quad \lambda - N\ln(1 + KL_2) - L_1 \max_{i=1,2,\dots,n} \sum_{j=1}^{n} \overline{\alpha}_{ij} e^{\lambda h} > 0,$$

   then the solution $x(t)$ is exponentially stable.

We note that the main inequalities which are used in Theorem 4.9 are connected with the properties of matrix $W(t,s)$ for system (4.77). Now we shall consider a special case in which these properties are accomplished.

Example 4.5. Consider an impulsive generalization of the classical model of impulsive Hopfield neural networks with delay [70],

$$\begin{cases} C_i \dot{x}_i(t) = -\frac{1}{R_i} x_i(t) + \sum_{j=1}^{n} T_{ij} f_j(x_j(t-h)) + \gamma_i(t), \ t \neq t_k, \\ \Delta x(t_k) = G x(t_k) + I_k(x(t_k)) + p_k, \ k = \pm 1, \pm 2, \dots, \end{cases} \quad (4.87)$$

where $t \in \mathbb{R}$, $\{t_k\} \in \mathcal{B}$, $C_i > 0$, $R_i > 0$, $T_{ij} \in \mathbb{R}$, $f_j, \gamma_i \in C[\mathbb{R}, \mathbb{R}]$, $h > 0$, $i = 1,2,\dots,n$, $j = 1,2,\dots,n$, $I_k \in C[\Omega, \mathbb{R}^n]$, $G = diag[g_1, g_2, \dots, g_n]$, $g_i \in \mathbb{R}$, $i = 1,2,\dots,n$, $p_k \in \mathbb{R}^n$, $k = \pm 1, \pm 2, \dots$.

Theorem 4.10. *Let the following conditions hold:*

1. *Conditions H4.28, H4.29, H4.31 and H4.32 are met.*
2. *The following inequalities hold*

$$\lambda = \min_{i=1,2,\dots,n} \frac{1}{C_i R_i} - N \max_{i=1,2,\dots,n} \ln\left(1 + |g_i|\right) > 0,$$

$$r = \exp\left\{ N \max_{i=1,2,\dots,n} \ln(1 + |g_i|) \right\}\left\{ \max_{i=1,2,\dots,n} \lambda^{-1} L_1 \sum_{j=1}^{n} \frac{T_{ij}}{C_i} + \frac{L_2}{1 - e^{-\lambda}} \right\} < 1.$$

Then:

1. *There exists a unique almost periodic solution $x(t)$ of (4.87).*
2. *If*

$$1 + \exp\left\{ N \max_{i=1,2,\dots,n} \ln\left(1 + |g_i|\right) \right\} L_2 < e,$$

$$\lambda - \exp\left(N \max_{i=1,2,\dots,n} \ln\left(1 + |g_i|\right) \right) L_1 \sum_{j=1}^{n} \frac{T_{ij}}{C_i} e^{\lambda h}$$

$$- N\ln\left(1 + \left\{ N \max_{i=1,2,\dots,n} \ln\left(1 + |g_i|\right) \right\} L_2 \right) > 0,$$

   then the solution $x(t)$ is exponentially stable.

Proof. Let the system

$$\begin{cases} \dot{x}_i(t) = -\frac{1}{C_i R_i} x_i(t), \ t \neq t_k, \\ \Delta x(t_k) = Gx(t_k), \ k = \pm 1, \pm 2, \ldots, \end{cases} \tag{4.88}$$

is the linear part of (4.87). Recall [138] the matrix $W(t, s)$ for the linear system (4.88) is in the form

$$W(t, s) = e^{A(t-s)} (E + G)^{i(s,t)},$$

where

$$A = diag\left[-\frac{1}{C_1 R_1}, -\frac{1}{C_2 R_2}, \ldots, -\frac{1}{C_n R_n} \right].$$

Then

$$||W(t, s)|| \leq e^{N \max\limits_{i=1,2,\ldots,n} \ln\left(1 + |g_i|\right)} e^{-\lambda(t-s)}, \ t > s, \ t, s \in \mathbb{R},$$

and the proof follows from Theorem 4.9.   □

4.3.3 Impulsive Neural Networks of a General Type

We shall investigate the existence of almost periodic solutions of the system of impulsive cellular neural networks with finite and infinite delays

$$\begin{cases} \dot{x}_i(t) = \sum_{j=1}^{n} a_{ij}(t)x_j(t) + \sum_{j=1}^{n} \alpha_{ij}(t)f_j(x_j(t - h)) + \\ \quad + \sum_{j=1}^{n} \beta_{ij}(t)f_j\left(\mu_j \int_0^\infty k_{ij}(u)x_j(t - u)du\right) + \gamma_i(t), \ t \neq t_k, \\ \Delta x(t_k) = A_k x(t_k) + I_k(x(t_k)) + p_k, \ k = \pm 1, \pm 2, \ldots, \end{cases} \tag{4.89}$$

where $t \in \mathbb{R}$, $\{t_k\} \in \mathcal{B}$, $a_{ij}, \alpha_{ij}, \ f_j, \ \beta_{ij}, \ \gamma_i \in C[\mathbb{R}, \mathbb{R}]$, $\mu_j > 0$, $i = 1, 2, \ldots, n$, $j = 1, 2, \ldots, n$, $h > 0$, $k_{ij} \in C[\mathbb{R}^+, \mathbb{R}^+]$, $x(t) = col(x_1(t), x_2(t), \ldots, x_n(t))$, $A_k \in \mathbb{R}^{n \times n}$, $I_k \in C[\Omega, \mathbb{R}^n]$, $p_k \in \mathbb{R}^n$, $k = \pm 1, \pm 2, \ldots$.

For $t_0 \in \mathbb{R}$, the initial conditions associated with (4.89) are in the form

$$\begin{cases} x(t; t_0, \phi_0) = \phi_0(t), \ -\infty < t \leq t_0, \\ x(t_0^+; t_0, \phi_0) = \phi_0(t_0). \end{cases} \tag{4.90}$$

where $\phi_0(t) \in PC[(-\infty, t_0], \mathbb{R}^n]$ is a piecewise continuous function with points of discontinuity of first kind at the moments t_k, $k = \pm 1, \pm 2, \ldots$.

Introduce the following conditions:

H4.33.   The functions $\beta_{ij}(t)$, $i = 1, 2, \ldots, n$, $j = 1, 2, \ldots, n$ are almost periodic in the sense of Bohr, and

$$0 < \sup_{t \in \mathbb{R}} |\beta_{ij}(t)| = \overline{\beta}_{ij} < \infty.$$

H4.34.   The functions $k_{ij}(t)$ satisfy

$$\int_0^\infty k_{ij}(s)ds = 1, \quad \int_0^\infty s k_{ij}(s)ds < \infty, \quad i, j = 1, 2, \ldots, n.$$

H4.35.   The function $\phi_0(t)$ is almost periodic.

The proof of the next lemma is similar to the proof of Lemma 1.7.

Lemma 4.20. *Let the following conditions hold:*

1. *Conditions of Lemma 4.18 are met.*
2. *Conditions H4.33–H4.35 are met.*

Then for each $\varepsilon > 0$ there exist ε_1, $0 < \varepsilon_1 < \varepsilon$ and relatively dense sets $\overline{T}$ of real numbers and Q of integer numbers, such that the following relation holds:

(a) $|\beta_{ij}(t + \tau) - \beta_{ij}(t)| < \varepsilon$, $t \in \mathbb{R}$, $\tau \in \overline{T}$, $i, j = 1, 2, \ldots, n$;
(b) $|\phi_0(t + \tau) - \phi_0(t)| < \varepsilon$, $t \in \mathbb{R}$, $\tau \in \overline{T}$, $|t - t_k| > \varepsilon$, $k = \pm 1, \pm 2, \ldots$.

The proof of the next theorem follows from Lemma 4.20 to the same way like Theorem 4.7.

Theorem 4.11. *Let the following conditions hold:*

1. *Conditions H4.26–H4.32 are met.*
2. *For the Cauchy matrix $W(t, s)$ of the system (4.89) there exist positive constants K and λ such that*

$$||W(t, s)|| \leq K e^{-\lambda(t-s)}, \quad t \geq s, \quad t, s \in \mathbb{R}.$$

3. *The number*

$$r = K\left\{ \max_{i=1,2,\ldots,n} \lambda^{-1} L_1 \sum_{j=1}^n (\overline{\alpha}_{ij} + \overline{\beta}_{ij}\mu_j) + \frac{L_2}{1 - e^{-\lambda}} \right\} < 1.$$

Then:

1. *There exists a unique almost periodic solution $x(t)$ of (4.89).*
2. *If the following inequalities hold*

$$1 + KL_2 < e,$$

$$\lambda - KL_1 \max_{i=1,2,\ldots,n} \sum_{j=1}^{n} (\overline{\alpha}_{ij} + \overline{\beta}_{ij}\mu_j) - N\ln(1 + KL_2) > 0,$$

then the solution $x(t)$ is exponentially stable.

Example 4.6. Consider the next model of impulsive neural networks

$$
\begin{cases}
\dot{x}_i(t) = -a_i(t)x_i(t) + \displaystyle\sum_{j=1}^{n} \alpha_{ij} f_j(x_j(t - h)) \\
\qquad + \displaystyle\sum_{j=1}^{n} \beta_{ij} f_j\left(\mu_j \int_0^{\infty} k_{ij}(u)x_j(t - u)du\right) + \gamma_i(t),\ t \neq t_k, \\
\Delta x(t_k) = A_k x(t_k) + I_k(x(t_k)) + p_k,\ k = \pm 1, \pm 2, \ldots,
\end{cases}
\tag{4.91}
$$

where $t \in \mathbb{R}$, $\{t_k\} \in \mathcal{B}$, $a_i, f_j, \in C[\mathbb{R}, \mathbb{R}]$, $\alpha_{ij}, \beta_{ij} \in \mathbb{R}$, $\mu_j \in \mathbb{R}^+$, $k_{ij} \in C[\mathbb{R}^+, \mathbb{R}^+]$, $\gamma_i \in C[\mathbb{R}, \mathbb{R}]$, $i = 1, 2, \ldots, n$, $j = 1, 2, \ldots, n$, $A_k \in \mathbb{R}^{n \times n}$, $I_k \in C[\Omega, \mathbb{R}^n]$, $p_k \in \mathbb{R}^n$, $k = \pm 1, \pm 2, \ldots$.

Theorem 4.12. *Let the following conditions hold:*

1. *Conditions of Lemma 4.16 are met.*
2. *Conditions H4.28–H4.32 hold*
3. *The number*

$$r = K\left\{ \max_{i=1,2,\ldots,n} \lambda^{-1} L_1 \sum_{j=1}^{n} (\alpha_{ij} + \beta_{ij}\mu_j) + \frac{L_2}{1 - e^{-\lambda}} \right\} < 1.$$

Then:

1. *There exists a unique almost periodic solution $x(t)$ of (4.91).*
2. *If the following inequalities hold*

$$1 + KL_2 < e,\ \lambda - KL_1 \sum_{j=1}^{n} (\alpha_{ij} + \beta_{ij}\mu_j) - N\ln(1 + KL_2) > 0,$$

then the solution $x(t)$ is exponentially stable.

References

1. Ahmad, S., Rao, M.R.M.: Asymptotically periodic solutions of N-competing species problem with time delays. J. Math. Anal. Appl. **186**, 559–571 (1994)
2. Ahmad, S., Stamov, G.Tr.: Almost periodic solutions of N-dimensional impulsive competitive systems. Nonlinear Anal. Real World Appl. **10**, 1846–1853 (2009)
3. Ahmad, S., Stamov, G.Tr.: On almost periodic processes in impulsive competitive systems with delay and impulsive perturbations. Nonlinear Anal. Real World Appl. **10**, 2857–2863 (2009)
4. Ahmad, S., Stamova, I.M.: Asymptotic stability of an N-dimensional impulsive competitive system. Nonlinear Anal. Real World Appl. **8**, 654–663 (2007)
5. Ahmad, S., Stamova, I.M.: Global exponential stability for impulsive cellular neural networks with time-varying delays. Nonlinear Anal. **69**, 786–795 (2008)
6. Akca, H., Alassar, R., Covachev, V., Covacheva, Z., Al-Zahrani, E.: Continuous-time additive Hopfield-type neural networks with impulses. J. Math. Anal. Appl. **290**, 436 451 (2004)
7. Akhmet, M.U., Beklioglu, M., Ergenc, T., Tkachenko, V.I.: An impulsive ratio-dependent predator prey system with diffusion. Nonlinear Anal. Real World Appl. **7**, 1255–1267 (2006)
8. Akhmetov, M.U.: Recurrent and almost-periodic solutions of nonautonomous systems with impulse. Izv. Akad. Nauk Kaz. SSR. **3**, 8–10 (1988)
9. Akhmetov, M.U., Perestyuk, N.A.: Almost periodic solutions of nonlinear impulse systems. Ukrainian Math. J. **41**, 291–296 (1989)
10. Alzabut, J.O., Nieto, J.J., Stamov, G.Tr.: Existence and exponential stability of positive almost periodic solutions for a model of hematopoiesis. Bound. Value Probl. 2009, 1–10 (2009)
11. Alzabut, J.O., Stamov, G.Tr., Sermutlu, E.: On almost periodic solutions for an impulsive delay logarithmic population model. Math. Comput. Model. **51**, 625–631 (2010)
12. Amerio, L.: Soluzioni quasi-periodiche, o limitate, di sistemi differenziali non lineari quasi-periodici, o limitati. Ann. Mat. Pura. Appl. **39**, 97–119 (1955)
13. Andronov, A.A., Vitt, A.A., Haykin, S.E.: Oscillation Theory. Nauka, Moscow (1981); (in Russian)
14. Bachar, M., Arino, O.: Stability of a general linear delay-differential equation with impulses. Dyn. Contin. Discrete Impuls. Syst. Ser. A Math. Anal. **10**, 973–990 (2003)
15. Bainov, D.D., Simeonov, P.S.: Impulsive Differential Equations: Periodic Solutions and Applications. Longman, Harlow (1993)

G.T. Stamov, *Almost Periodic Solutions of Impulsive Differential Equations*,
Lecture Notes in Mathematics 2047, DOI 10.1007/978-3-642-27546-3,
© Springer-Verlag Berlin Heidelberg 2012

16. Bainov, D.D., Kostadinov, S.I., Myshkis, A.D.: Bounded periodic solutions of differential equations with impulsive effect in a Banach space. Differ. Integr. Equat. **1**, 223–230 (1988)

17. Bainov, D.D., Myshkis, A.D., Stamov, G.T.: Dichotomies and almost periodicity of the solutions of systems of impulsive differential equations. Dynam. Syst. Appl. **5**, 145–152 (1996)

18. Bainov D.D., Dishliev, A.B., Stamov, G.T.: Almost periodic solutions of hyperbolic systems of impulsive differential equations. Kumamoto J. Math. **10**, 1–10 (1997)

19. Bellman, R., Cooke, K.L.: Differential-Difference Equations. Academic Press, New York (1963)

20. Benchohra, M., Henderson, J., Ntouyas, S.: Impulsive Differential Equations and Inclusions. Hindawi, New York (2006)

21. Besicovitch, A.S.: Almost Periodic Functions. Dover, New York (1954)

22. Bochner, S.: Beitrage zur theorie der fastperiodischen funktionen, I: funktionen einer variaben. Math. Ann. **96**, 119–147 (1927); (in German)

23. Bochner, S.: Homogeneous systems of differential equations with almost periodic coefficients. J. London Math. Soc. **8**, 283–288 (1933)

24. Bochner, S., von Neumann, J.: Almost periodic functions of groups. II. Trans. Amer. Math. Soc. **37**, 21–50 (1935)

25. Bogolyubov, N.N., Mitropolskii, Y.A.: Asimptotic Methods in the Theory of Nonlinear Variations. Nauka, Moscow (1974); (in Russian)

26. Bohr, H.: Zur theorie der fastperiodischen funktionen. II: Zusammenhang der fastperiodischen funktionen mit funktionen von unendlich vielen variabeln; gleichmssige approximation durch trigonometrische summen. Acta Math. **46**, 101–214 (1925); (in German)

27. Bohr, H., Neugebauer, O.: Uber lineare differentialgleichungen mit konstanten koeffizienten und fastperiodischer rechter seite. Nachr. Ges. Wiss. Geottingen. Math.-Phys. Klasse. 8–22 (1926); (in German)

28. Burton, T.A. , Zhang, B.: Uniform ultimate boundedness and periodicity in functional differential equations. Tohoku Math. J. **42**, 93–100 (1990)

29. Butler, G., Freedman, H.I., Waltman, P.: Uniformly persistent systems. Proc. Amer. Math. Soc. **96**, 425–430 (1986)

30. Cao, J.: On stability of delayed cellular neural networks. Phys. Lett. A **261**, 303–308 (1999)

31. Cao, J.: Global exponential stability of Hopfield neural networks. Internat. J. Syst. Sci. **32**, 233–236 (2001)

32. Cao, J., Wang, J.: Global exponential stability and periodicity of recurrent neural networks with times delays. IEEE Trans. Cir. Syst. I Regul. Pap. **52**, 920–931 (2005)

33. Cao, J., Chen, A., Huang, X.: Almost periodic attractor of delayed neural networks with variable coefficients. Phys. Lett. A **340**, 104–120 (2005)

34. Chen, A., Cao, J.: Existence and attractivity of almost periodic solutions for cellular neural networks with distributed delays and variable coefficients. Appl. Math. Comput. **134**, 125–140 (2003)

35. Chen, G.: Control and stabilization for the wave equation in a bounded domain. I. SIAM J. Contr. Optim. **17**, 66–81 (1979)

36. Chen, G., Shen, J.: Boundedness and periodicity for impulsive functional differential equations with applications to impulsive delayed Hopfield neuron networks. Dyn. Contin. Discrete Impuls. Syst. Ser. A Math. Anal. **14**, 177–188 (2007)

37. Chen, M.P., Yu, J.S., Shen, J.H.: The persistence of nonoscillatory solutions of delay differential equations under impulsive perturbations. Comput Math Appl **27** 1–6 (1994)

38. Chen, T.: Global exponential stability of delayed Hopfield neural networks. Neutral Netw. **14**, 977–980 (2001)

39. Chetayev, N.G.: The Stability of Motion. Pergamon Press, Oxford (1961)

40. Chua, L.O.: CNN: A Paradigm for Complexity. World Scientific, Singapore (1998)
41. Chua, L.O., Roska, T.: Stability of a class of nonreciprocal cellular neural networks. IEEE Trans. Circ. Syst. I **37**, 1520–1527 (1990)
42. Chua, L.O., Yang, L.: Cellular neural networks: theory. IEEE Trans. Circ. Syst. **35**, 1257–1272 (1988)
43. Chua, L.O., Yang, L.: Cellular neural networks: applications: IEEE Trans. Circ. Syst. **35**, 1273–1290 (1988)
44. Civalleri, P.P., Gilli, M.: A set of stability criteria for delayed cellular neural networks. IEEE Trans. Circ. Syst. I **48**, 494–498 (2001)
45. Coddington, E.A., Levinson, N.: Theory of Ordinary Differential Equations. McGraw-Hill, New York (1955)
46. Coppel, W.: Dichotomies and reducibility. J. Differ. Equat. **3**, 500–521 (1967)
47. Corduneanu, C.: Almost Periodic Functions. Interscience Publication, New York (1968)
48. Cui, W.: Global stability of a class of neural networks model under dynamical thresholds with delay. J. Biomath. **15**, 420–424 (2000)
49. Dafermos, C.M.: Almost periodic processes and almost periodic solutions of evolution equations. In: Dynamical Systems (Proceedings of International Symposium, University of Florida, Gainesville, Florida, 1976), pp. 43–57. Academic Press, New York (1977)
50. Dalec'kii, Ju.L., Krein, M.G.: Stability of Solutions of Differential Equations in Banach Space. American Mathematical Society, Providence (1974)
51. Dannan, F., Elaydi, S.: Lipschitz stability of nonlinear systems of differential equations. J. Math. Anal. Appl., **113**, 562–577 (1986)
52. Demidovich, B.P.: Lectures on the Mathematical Theory of Stability. Nauka, Moscow (1967); (In Russian)
53. Driver, R.: Ordinary and Delay Differential Equations. Springer, New York (1977)
54. Fan, M., Wang, K., Jiang, D.: Existence and global attractivity of positive periodic solutions of periodic species Lotka–Volterra competition systems with several deviating arguments. Math. Biosci. **160**, 47–61 (1999)
55. Fink, A.M.: Almost Periodic Differential Equations. Lecture Notes in Mathematics. **377**, Springer, Berlin (1974)
56. Fink, A.M.: Almost periodic solutions to forced Lienard equations. In: Nonlinear Vibration Problems, No. 15 (Proceedings of Sixth International Conference on Nonlinear Oscillations, Pozna, 1972, Part II), pp. 95–105. PWN-Polish Sci. Publ., Warsaw (1974)
57. Fink, A.M., Seifert, G.: Lyapunov functions and almost periodic solutions for almost periodic systems. J. Differ. Equat. **5**, 307–313 (1969)
58. Friedman, A.: Partial Differential Equations. Holt, Rinehart and Winston, New York (1969)
59. Gopalsamy, K.: Stability and Oscillation in Delay Differential Equations of Population Dynamics. Kluwer, Dodrecht (1992)
60. Gopalsamy, K., Leung, I.K.C.: Convergence under dynamical thresholds with delays. IEEE Trans. Neural Netw. **8**, 341–348 (1997)
61. Gopalsamy, K., Zhang, B.: On delay differential equations with impulses. J. Math. Anal. Appl. **139**, 110–122 (1989)
62. Gurgulla, S.I., Perestyuk, N.A.: On Lyapunov's second method in systems with impulse action. Dokl. Akad. Nauk Ukrain. SSR Ser. A **10**, 11–14 (1982); (in Russian)
63. Halanay, A., Wexler, D.: Qualitative Theory of Impulse Systems. Mir, Moscow (1971); (in Russian)
64. Hale, J.K.: Theory of Functional Differential Equations. Springer, New York (1977)
65. Hartman, P.: Ordinary Differential Equations. Wiley, New York (1964)

66. He, M., Chen, F., Li, Z.: Almost periodic solution of an impulsive differential equation model of plankton allelopathy. Nonlinear Anal. Real World Appl. **11**, 2296–2301 (2010)

67. Hekimova, M.A., Bainov, D.D.: Almost periodic solutions of singularly perturbed systems of differential equations with impulse effect. Forum Math. **1**, 323–329 (1989)

68. Henry, D.: Geometric Theory of Semilinear Parabolic Equations. Springer, Berlin (1981)

69. Hino, Y.: Stability and existence of almost periodic solutions of some functional differential equations. Tohoku Math. J. **28**, 389–409 (1976)

70. Hopfield, J.J.: Neurons with graded response have collective computational properties like those of two-stage neurons. Proc. Natl. Acad. Sci. USA **81**, 3088–3092 (1984)

71. Hristova, S.G., Bainov, D.D.: Integral surfaces for hyperbolic ordinary differential equations with impulses effect. COMPEL **4**, 1–18 (1995)

72. Hu, D., Zhao, H., Zhu, H.: Global dynamics of Hopfield neural networks involving variable delays. Comput. Math. Applicat. **42**, 39–45 (2001)

73. Huang, H., Cao, J.: On global asymptotic stability of recurrent neural networks with time-varying delays. Appl. Math. Comput. **142**, 143–154 (2003)

74. Huang, X., Cao, J.: Almost periodic solutions of shunting inhibitory cellular neural networks with time-varying delays. Phys. Lett. A **314**, 222–231 (2003)

75. Jiang, G., Lu, Q.: Impulsive state feedback control of a predator–prey model. J. Comput. Appl. Math. **200**, 193–207 (2007)

76. Jin, Z., Maoan, H., Guihua, L.: The persistence in a Lotka–Volterra competition systems with impulsive perturbations. Chaos Solut. Fractals **24**, 1105–1117 (2005)

77. Jost, C., Ariono, O., Arditi, R.: About deterministic extinction in ratio-dependent predator–prey models. Bull. Math. Biol. **61**, 19–32 (1999)

78. Kapur, J.N.: Mathematical Modelling. Wiley, New York (1988)

79. Khadra, A., Liu, X., Shen, X.: Application of impulsive synchronization to communication security. IEEE Trans. Circ. Syst. I Fund. Theor. Appl. **50**, 341–351 (2003)

80. Khadra, A., Liu, X., Shen, X.: Robuts impulsive synchronization and application to communication security. Dyn. Contin. Discrete Impuls. Syst. **10**, 403–416 (2003)

81. Kim, S., Campbell, S., Liu, X.: Stability of a class of linear switching systems with time delay. IEEE Trans. Circ. Syst. I **53**, 384–393 (2006)

82. Kirlinger, G.: Permanence in Lotka-Voltera equations: Linked prey- predator systems. Math. Biosci. **82**, 165–191 (1986)

83. Kolmanovskii, V.B., Nosov, V.R.: Stability of Functional-Differential Equations. Academic Press, London (1986)

84. Krasnosel'skii, M.A., Burd, V.Sh., Kolesov, Yu.S.: Nonlinear Almost Periodic Oscillations. Wiley, New York (1973)

85. Krasovskii, N.N.: Certain Problems in the Theory of Stability of Motion. Fiz.-Mat. Lit., Moscow (1959); (in Russian)

86. Krasovskii, N.N.: Stability of Motion. Stanford University Press, Stanford (1963)

87. Krishna, S., Vasundhara, J., Satyavani, K.: Boundedness and Dichotomies for Impulsive Equations. J. Math. Anal. Appl. **158**, 352–375 (1991)

88. Kuang, Y.: Delay Differential Equations with Applications in Population Dynamics. Academic Press, Boston (1993)

89. Kulenovic, M.R.S., Ladas, G.: Linearized oscillations in population dynamics. Bull. Math. Biol. **49**, 615–627 (1987)

90. Kulev, G.K., Bainov, D.D.: Strong stability of impulsive systems. Internat. J. Theoret. Phys. **27**, 745–755 (1988)

91. Lakshmikantham, V., Leela, S.: Differential and Integral Inequalities: Theory and Applications. Academic Press, New York (1969)

92. Lakshmikantham, V., Liu, X.: Stability Analysis in Terms of Two Measures. World Scientific, River Edge (1993)

93. Lakshmikantham, V., Rao, M.R.M.: Theory of Integro-Differential Equations. Gordon and Breach, Lausanne (1995)
94. Lakshmikantham, V., Bainov, D.D., Simeonov, P.S.: Theory of Impulsive Differential Equations. World Scientific, Teaneck (1989)
95. Lakshmikantham, V., Leela, S., Martynyuk, A.A.: Stability Analysis of Nonlinear Systems. Marcel Dekker, New York (1989)
96. Lakshmikantham, V., Leela, S., Martynyuk, A.A.: Practical Stability Analysis of Nonlinear Systems. World Scientific, Singapore (1990)
97. Levitan, B.M.: Almost Periodic Functions. Gostekhizdat, Moscow (1953); (in Russian)
98. Levitan, B.M., Zhikov, V.V.: Almost Periodic Functions and Differential Equations. Cambridge University Press, Cambridge (1983)
99. Li, M., Duan, Y., Zhang, W., Wang, M.: The existence of positive periodic solutions of a class of Lotka–Volterra type impulsive systems with infinitely distributed delay. Comput. Math. Appl. **49**, 1037–1044 (2005)
100. Liao, X., Ouyang, Z., Zhou, S.: Permanence of species in nonautonomous discrete Lotka–Volterra competitive system with delays and feedback controls. J. Comput. Appl. Math. **211**, 1–10 (2008)
101. Lisena, B.: Extinction in three species competitive systems with periodic coefficients. Dynam. Syst. Appl. **14**, 393–406 (2005)
102. Liu, J.: Bounded and periodic solutions of finite delay evolution equations. Nonlinear Anal. **34**, 101–111 (1998)
103. Liu, X.: Stability results for impulsive differential systems with applications to population growth models. Dynam. Stabil. Syst. **9**, 163–174 (1994)
104. Liu, X.: Stability of impulsive control systems with time delay. Math. Comput. Model. **39**, 511–519 (2004)
105. Liu, X., Ballinger, G.: Existence and continuability of solutions for differential equations with delays and state-dependent impulses. Nonlinear Anal. **51**, 633–647 (2002)
106. Liu, Y., Ge, W.: Global attractivity in delay "food-limited" models with exponential impulses. J. Math. Anal. Appl. **287**, 200–216 (2003)
107. Liu, Z.J.: Positive periodic solutions for delay multispecies Logarithmic population model. J. Engrg. Math., **19**, 11–16 (2002); (in Chinese).
108. Liu, B., Liu, X., Liao X.: Robust stability of uncertain dynamical systems. J. Math. Anal. Appl. **290**, 519–533 (2004)
109. Lotka, A.: Elements of Physical Biology. Williams and Wilkins, Baltimore (1925); [Reprinted as: Elements of Mathematical Biology. Dover, New York (1956)]
110. Luo, Z., Shen, J.: Stability and boundedness for impulsive functional differential equations with infinite delays. Nonlinear Anal. **46**, 475–493 (2001)
111. Lyapunov, A.M.: General Problem on Stability of Motion. Grostechizdat, Moscow (1950); (in Russian)
112. Mackey, M.C., Glass, L.: Oscillation and chaos in physiological control system. Science **197**, 287–289 (1977)
113. Malkin, I.G.: Theory of Stability of Motion. Nauka, Moscow (1966); (in Russian)
114. Markoff, A.: Stabilitt im Liapounoffschen Sinne und Fastperiodizitt. Math. Z. **36**, 708–738 (1933); (in German)
115. Martin, R.H.: Nonlinear Operators and Differential Equations in Banach Spaces. Wiley, New York (1976)
116. Massera, J.L.: Contributions to stability theory. Ann. of Math. **64**, 182–206 (1956)
117. Maynard-Smith, J.: Models in Ecology. Cambridge University Press, Cambridge (1974)
118. McRae, F.: Practical stability of impulsive control systems. J. Math. Anal. Appl. **181**, 656–672 (1994)
119. Mil'man, V.D., Myshkis, A.D.: On the stability of motion in the presence of impulses. Siberian Math. J. **1**, 233–237 (1960); (in Russian)

120. Mohamad, S.: Global exponential stability of continuous-time and discrete-time delayed bidirectional neural networks. Phys. Nonlinear Phenom. **159**, 233–251 (2001)

121. Mohamad, S., Gopalsamy, K.: A unified treatment for stability preservation in computer simulation of impulsive BAM networks. Comput. Math. Appl. **55**, 2043–2063 (2008)

122. Neugebauer, O.: The Exact Sciences in Antiquity. Braun University Press, Providence (1957)

123. Nicholson, A.J.: The balance of animal population. J. Anim. Ecol. **2**, 132–178 (1933)

124. Nieto, J.: Periodic boundary value problems for first-order impulsive ordinary differential equations. Nonlinear Anal. **51**, 1223–1232 (2002)

125. Nindjin, A.F., Aziz-Alaoui, M.A., Cadivel, M.: Analysis of predator–prey model with modified Leslie-Gower and Holling-type II schemes with time delay. Nonlinear Anal. Real World Appl. **7**, 1104–1118 (2006)

126. Pazy, A.: Semigroups of Linear Operators and Applications to Partial Differential Equations. Springer, New York (1983)

127. Perestyuk, N.A., Ahmetov, M.U.: On almost periodic solutions of a class of systems with periodic impulsive action. Ukrainian Math. J. **36**, 486–490 (1984)

128. Perestyuk, N.A., Chernikova, O.S.: On the stability of integral sets of impulsive differential systems. Math. Notes (Miskolc) **2**, 49–60 (2001)

129. Randelovic, B.M., Stefanovic, L.V., Dankovic, B.M.: Numerical solution of impulsive differential equations. Facta Univ. Ser. Math. Inform. **15**, 101–111 (2000)

130. Rao M.R.M., Rao, V.S.H.: Stability of impulsively perturbed systems. Bull. Austral. Math. Soc. **16**, 99–110 (1977)

131. Rao, M.R.M., Sathanantham, S. and Sivasundaram, S.: Asymptotic behavior of solutions of impulsive integro-differential systems. Appl. Math. Comput. **34**, 195–211 (1989)

132. Razumikhin, B.S.: Stability of Hereditary Systems. Nauka, Moscow (1988); (in Russian)

133. Roska, T., Wu, C.W., Balsi, M., Chua, L.O.: Stability and dynamics of delay-type general cellular neural networks. IEEE Trans. Circuits Syst. I **39**, 487–490 (1992)

134. Rouche, H., Habets, P., Laloy, M.: Stability Theory by Lyapunov's Direct Method. Springer, New York (1977)

135. Saaty, T.L., Joyce, M.: Thinking with Models: Mathematical Models in the Physical, Biological, and Social Sciences. Pergamon Press, Oxford (1981)

136. Samoilenko, A.M., Perestyuk, N.A.: Stability of the solutions of differential equations with impulse effect. Diff. Eqns. **11**, 1981–1992 (1977); (in Russian)

137. Samoilenko, A.M., Perestyuk, N.A.: Periodic and almost periodic solutions of differential equations with impulses. Ukrainian Math. J. **34**, 66–73 (1982)

138. Samoilenko, A.M., Perestyuk, N.A.: Differential Equations with Impulse Effect. World Scientific, Singapore (1995)

139. Samoilenko, A.M., Trofimchuk, S.: Spaces of piecewise-continuous almost-periodic functions and of almost-periodic sets on the line I. Ukrainian Math. J. **43**, 1613–1619 (1991); (in Russian)

140. Samoilenko, A.M., Trofimchuk, S.: Spaces of piecewise-continuous almost-periodic functions and of almost-periodic sets on the line II. Ukrainian Math. J., **44**, 389–400 (1992); (in Russian)

141. Samoilenko, A.M., Perestyuk, N.A., Akhmetov, M. U.: Almost Periodic Solutions of Differential Equations with Impulse Action. Akad. Nauk Ukrain. SSR Inst. Mat., Kiev (1983); (in Russian)

142. Seifert, G.: A condition for almost periodicity with some applications to functional differential equations. J. Differ. Equat. **1**, 393–408 (1965)

143. Seifert, G.: Almost periodic solutions for almost periodic systems of ordinary differential equations. J. Differ. Equat. **2**, 305–319 (1966)

144. Seifert, G.: Nonlinear evolution equation with almost periodic time depence. SIAM J. Math.l Anal. **18**, 387–392 (1987)
145. Shen, J.: Razumikhin techniques in impulsive functional differential equations. Nonlinear Anal. **36**, 119–130 (1999)
146. Shen, J., Li, J.: Impulsive control for stability of Volterra functional differential equations. Z. Anal. Anwendungen **24**, 721–734 (2005)
147. Siljak, D.D., Ikeda, M., Ohta, Y.: Parametric stability. In: Proceedings of the Universita di Genova-Ohio State University Joint Conference, pp. 1–20. Birkhauser, Boston (1991)
148. Simeonov, P.S., Bainov, D.D.: Estimates for the Cauchy matrix of perturbed linear impulsive equation. Internat. J. Math. Math. Sci. **17**, 753–758 (1994)
149. Stamov, G.T.: Almost periodic solutions for systems of impulsive integro-differential equations. Appl. Anal. **64**, 319–327 (1997)
150. Stamov, G.T.: Almost periodic solutions and perturbations of the linear part of singularly impulsive differential equations. Panamer. Math. J. **9**, 91–101 (1999)
151. Stamov, G.T.: Semi-separated conditions for almost periodic solutions of impulsive differential equations. J. Tech. Univ. Plovdiv Fundam. Sci. Appl. Ser. A Pure Appl. Math. **7**, 89–98 (1999)
152. Stamov, G.T.: Almost periodic solutions for forced perturbed impulsive differential equations. Appl. Anal. **74**, 45–56 (2000)
153. Stamov, G.T.: On the existence of almost periodic Lyapunov functions for impulsive differential equations. Z. Anal. Anwendungen **19**, 561–573 (2000)
154. Stamov, G.T.: Separated conditions for almost periodic solutions of impulsive differential equations with variable impulsive perturbations. Comm. Appl. Nonlinear Anal. **7**, 73–82 (2000)
155. Stamov, G.T.: Existence of almost periodic solutions for strong stable impulsive differential equations. IMA J. Math. Contr. Inform. **18**, 153–160 (2001)
156. Stamov, G.T.: Separated and almost periodic solutions for impulsive differential equations. Note Mat. **20**, 105–113 (2001)
157. Stamov, G.T.: Asymptotic stability of almost periodic systems of impulsive differential-difference equations. Asymptot. Anal. **27**, 1–8 (2001)
158. Stamov, G.Tr.: Existence of almost periodic solutions for impulsive differential equations with perturbations of the linear part. Nonlinear Stud. **9**, 263–273 (2002)
159. Stamov, G.Tr.: Second method of Lyapunov for existence of almost periodic solutions for impulsive integro-differential equations. Kyungpook Math. J. **43**, 221–231 (2003)
160. Stamov, G.T.: Families of Lyapunov's functions for existence of almost periodic solutions of (h_0, h)- stable impulsive differential equations. Nonlinear Stud. **10**, 135–150 (2003)
161. Stamov, G.Tr.: Lyapunov's functions for existence of almost periodic solutions of impulsive differential equations. Adv. Stud. Contemp. Math. (Kyungshang) **8** (2004), 35–46 (2004)
162. Stamov, G.Tr.: Impulsive cellular neural networks and almost periodicity. Proc. Japan Acad. Ser. A Math. Sci. **80**, 198–203 (2004)
163. Stamov, G.Tr.: Asymptotic stability in the large of the solutions of almost periodic impulsive differential equations. Note Mat. **24**, 75–83 (2005)
164. Stamov, G.Tr.: Almost periodic solutions of impulsive differential equations with time-varying delay on the PC-space. Nonlinear Stud. **14**, 269–279 (2007)
165. Stamov, G.Tr.: Almost periodic impulsive equations in a Banach space. J. Tech. Uni. Sliven **2**, 3–11 (2007)
166. Stamov, G.T.: Almost periodic models in impulsive ecological systems with variable diffusion. J. Appl. Math. Comput. **27**, 243–255 (2008)
167. Stamov, G.Tr.: Existence of almost periodic solutions for impulsive cellular neural networks. Rocky Mt. J. Math. **38**, 1271–1285 (2008)

168. Stamov, G.Tr.: On the existence of almost periodic solutions for impulsive Lasota-Wazewska model. Appl. Math. Lett. **22**, 516–520 (2009)

169. Stamov, G.Tr.: Almost periodic models of impulsive Hopfield neural networks. J. Math. Kyoto Univ. **49**, 57–67 (2009)

170. Stamov, G.Tr.: Almost periodic processes in ecological systems with impulsive perturbations. Kyungpook Math. J. **49**, 299–312 (2009)

171. Stamov, G.Tr.: Almost periodic solutions in impulsive competitive systems with infinite delays. Publ. Math. Debrecen **76**, 89–100 (2010)

172. Stamov, G.Tr.: Almost periodicity and Lyapunov's functions for impulsive functional differential equations with infinite delays. Canad. Math. Bull. **53**, 367–377 (2010)

173. Stamov, G.Tr., Alzabut, J.O.: Almost periodic solutions for abstract impulsive differential equations. Nonlinear Anal. **72**, 2457–2464 (2010)

174. Stamov, G.Tr., Petrov, N.: Lyapunov-Razumikhin method for existence of almost periodic solutions of impulsive differential-difference equations. Nonlinear Stud. **15**, 151–161 (2008)

175. Stamov, G.Tr., Stamova, I.M.: Almost periodic solutions for impulsive neural networks with delay. Appl. Math. Model. **31**, 1263–1270 (2007)

176. Stamova, I.M.: Global asymptotic stability of impulse delayed cellular neural networks with dynamical threshold. Nonlinear Stud. **13**, 113–122 (2006)

177. Stamova, I.M.: Stability Analysis of Impulsive Functional Differential Equations. Walter de Gruyter, Berlin (2009)

178. Stamova, I.M., Stamov, G.T.: Lyapunov-Razumikhin method for impulsive functional differential equations and applications to the population dynamics. J. Comput. Appl. Math. **130**, 163–171 (2001)

179. Sternberg, S.: Celestial Mechanics. Part I. W. A. Benjamin, New York (1969)

180. Taam, C.T.: Asymptotically Periodic and Almost Periodic Polutions of Nonlinear Differential Equations in Banach Spaces. Technical Reports, Georgetown University, Washington (1966)

181. Takeuchi, Y.: Global Dynamical Properties of Lotka–Volterra Systems. World Scientific, Singapore (1996)

182. Tineo, A.: Necessary and sufficient conditions for extinction of one species. Adv. Nonlinear Stud. **5**, 57–71 (2005)

183. Veech, W.A.: Almost automorphic functions on groups. Amer. J. Math. **87**, 719–751 (1965)

184. Volterra, V.: Fluctuations in the abundance of a species considered mathematically. Nature **118**, 558–560 (1926)

185. Wang, L., Chen, L., Nieto, J.J.: The dynamics of an epidemic model for pest control with impulsive effect. Nonlinear Anal. Real World Appl. **11**, 1374–1386 (2010)

186. Wazewska-Czyzewska, M., Lasota, A.: Mathematical problems of the dynamics of a system of red blood cells. Mat. Stos. **6**, 23–40 (1976)

187. Wei, F., Wang, K.: Asymptotically periodic solution of n-species cooperation system with time delay. Nonlinear Anal. Real World Appl. **7**, 591–596 (2006)

188. Xia, Y.: Positive periodic solutions for a neutral impulsive delayed Lotka–Volterra competition system with the effect of toxic substance. Nonlinear Anal. Real World Appl. **8**, 204–221 (2007)

189. Xinzhu, M.: Almost periodic solution for a class of Lotka–Volterra type N-species evological systems with time delay. J. Syst. Sci. Complex. **18**, 488–497 (2005)

190. Xu, W., Li, J.: Global attractivity of the model for the survival of red blood cells with several delays. Ann. Differ. Equat. **14**, 357–363 (1998)

191. Xue, Y., Wang, J., Jin, Z.: The persistent threshold of single population under pulse input of environmental toxin. WSEAS Trans. Math. **6**, 22–29 (2007)

192. Ye, D., Fan, M.: Periodicity in impulsive predator–prey system with Holling III functional response. Kodai Math. J., **27**, 189–200 (2004)

193. Yoshizawa, T.: Stability Theory by Lyapunov's Second Method. The Mathematical Society of Japan, Japan (1966)
194. Yoshizawa, T.: Asymptotically almost periodic solutions of an almost periodic system. Funkcial. Ekvac. **12**, 23–40 (1969)
195. Yoshizawa, T.: Some remarks on the existence and the stability of almost periodic solutions. SIAM Studies in Apl. Math. **5**, 166–172 (1969)
196. Zanolin, F.: Permanence and positive periodic solutions for Kolmogorov competing species systems. Results Math. **21**, 224–250 (1992)
197. Zhang, B.: Boundedness in functional differential equations. Nonlinear Anal. **22**, 1511–1527 (1994)
198. Zhang, B.G., Gopalsamy, K.: Global attractivity in the delay logistic equation with variable parameters. Math. Proc. Cambridge Philos. Soc. **170**, 579–590 (1990)
199. Zhang, Y., Sun, J.: Stability of impulsive delay differential equations with impulses at variable times. Dyn. Syst. **20** (2005), 323–331 (2005)
200. Zhao, C., Wang, K.: Positive periodic solutions of a delay model in population. Appl. Math. Lett. **16**, 561–565 (2003)
201. Zhikov, V.V.: The problem of almost periodicity for differential and operator equations. Matematika **8**, 94–188 (1969)
202. Zhong, W., Lin, W., Jiong, R.: The stability in neural networks with delay and dynamical threshold effects. Ann. Differ. Equat. **17**, 93–101 (2001)

Index

G.T. Stamov, *Almost Periodic Solutions of Impulsive Differential Equations*,
Lecture Notes in Mathematics 2047, DOI 10.1007/978-3-642-27546-3,
© Springer-Verlag Berlin Heidelberg 2012

LECTURE NOTES IN MATHEMATICS  Springer

Edited by J.-M. Morel, B. Teissier; P.K. Maini

Editorial Policy (for the publication of monographs)

1. Lecture Notes aim to report new developments in all areas of mathematics and their applications - quickly, informally and at a high level. Mathematical texts analysing new developments in modelling and numerical simulation are welcome.

   Monograph manuscripts should be reasonably self-contained and rounded off. Thus they may, and often will, present not only results of the author but also related work by other people. They may be based on specialised lecture courses. Furthermore, the manuscripts should provide sufficient motivation, examples and applications. This clearly distinguishes Lecture Notes from journal articles or technical reports which normally are very concise. Articles intended for a journal but too long to be accepted by most journals, usually do not have this "lecture notes" character. For similar reasons it is unusual for doctoral theses to be accepted for the Lecture Notes series, though habilitation theses may be appropriate.

2. Manuscripts should be submitted either online at www.editorialmanager.com/lnm to Springer's mathematics editorial in Heidelberg, or to one of the series editors. In general, manuscripts will be sent out to 2 external referees for evaluation. If a decision cannot yet be reached on the basis of the first 2 reports, further referees may be contacted: The author will be informed of this. A final decision to publish can be made only on the basis of the complete manuscript, however a refereeing process leading to a preliminary decision can be based on a pre-final or incomplete manuscript. The strict minimum amount of material that will be considered should include a detailed outline describing the planned contents of each chapter, a bibliography and several sample chapters.

   Authors should be aware that incomplete or insufficiently close to final manuscripts almost always result in longer refereeing times and nevertheless unclear referees' recommendations, making further refereeing of a final draft necessary.

   Authors should also be aware that parallel submission of their manuscript to another publisher while under consideration for LNM will in general lead to immediate rejection.

3. Manuscripts should in general be submitted in English. Final manuscripts should contain at least 100 pages of mathematical text and should always include

   – a table of contents;
   – an informative introduction, with adequate motivation and perhaps some historical remarks: it should be accessible to a reader not intimately familiar with the topic treated;
   – a subject index: as a rule this is genuinely helpful for the reader.

   For evaluation purposes, manuscripts may be submitted in print or electronic form (print form is still preferred by most referees), in the latter case preferably as pdf- or zipped psfiles. Lecture Notes volumes are, as a rule, printed digitally from the authors' files. To ensure best results, authors are asked to use the LaTeX2e style files available from Springer's web-server at:

   ftp://ftp.springer.de/pub/tex/latex/svmonot1/ (for monographs) and
   ftp://ftp.springer.de/pub/tex/latex/svmultt1/ (for summer schools/tutorials).

Additional technical instructions, if necessary, are available on request from lnm@springer.com.

4. Careful preparation of the manuscripts will help keep production time short besides ensuring satisfactory appearance of the finished book in print and online. After acceptance of the manuscript authors will be asked to prepare the final LaTeX source files and also the corresponding dvi-, pdf- or zipped ps-file. The LaTeX source files are essential for producing the full-text online version of the book (see http://www.springerlink.com/openurl.asp?genre=journal&issn=0075-8434 for the existing online volumes of LNM). The actual production of a Lecture Notes volume takes approximately 12 weeks.

5. Authors receive a total of 50 free copies of their volume, but no royalties. They are entitled to a discount of 33.3 % on the price of Springer books purchased for their personal use, if ordering directly from Springer.

6. Commitment to publish is made by letter of intent rather than by signing a formal contract. Springer-Verlag secures the copyright for each volume. Authors are free to reuse material contained in their LNM volumes in later publications: a brief written (or e-mail) request for formal permission is sufficient.

Addresses:
Professor J.-M. Morel, CMLA,
École Normale Supérieure de Cachan,
61 Avenue du Président Wilson, 94235 Cachan Cedex, France
E-mail: morel@cmla.ens-cachan.fr

Professor B. Teissier, Institut Mathématique de Jussieu,
UMR 7586 du CNRS, Équipe "Géométrie et Dynamique",
175 rue du Chevaleret
75013 Paris, France
E-mail: teissier@math.jussieu.fr

For the "Mathematical Biosciences Subseries" of LNM:

Professor P. K. Maini, Center for Mathematical Biology,
Mathematical Institute, 24-29 St Giles,
Oxford OX1 3LP, UK
E-mail : maini@maths.ox.ac.uk

Springer, Mathematics Editorial, Tiergartenstr. 17,
69121 Heidelberg, Germany,
Tel.: +49 (6221) 4876-8259

Fax: +49 (6221) 4876-8259
E-mail: lnm@springer.com